- ***To the Reader***

This book, The Book of Peace, is a gift to humanity. It was written with the sincere intention of being freely accessible to all who seek truth, awareness, and the path to inner and collective peace.

To honor this intention, it is available through multiple channels:

• ***Free access*** *to the full PDF is offered* ***through Lulu, Internet Archive, and Smashwords****, allowing anyone to read and share it without cost.*

• *For those who prefer to purchase a copy as a gesture of support or convenience, a* ***symbolic price of $0.99*** *has been set for the* ***eBook edition on Lulu.***

• *The* ***printed edition*** *is available* ***at cost****, with no profit added—only covering the actual expenses of printing and production.*

No matter how you access this book—freely or by purchase—what matters most is that its message touches the hearts of those who feel called to it.

Peace be upon all beings—without exception.

-Bahram Siavash

The Book of Peace

A Gateway to the NUSAWVO Worldview

Bahram Siavash

Title:
The Book of Peace: A Gateway to the NUSAWVO Worldview

Author: Bahram Siavash

Publisher: Bahram Siavash
Printed and Distributed via Lulu.com
Location: Toronto, Ontario, Canada
Date: May 2025
ISBN (eBook): 978-1-0694652-0-7
ISBN (Paperback): 978-1-0694652-2-1

Contents

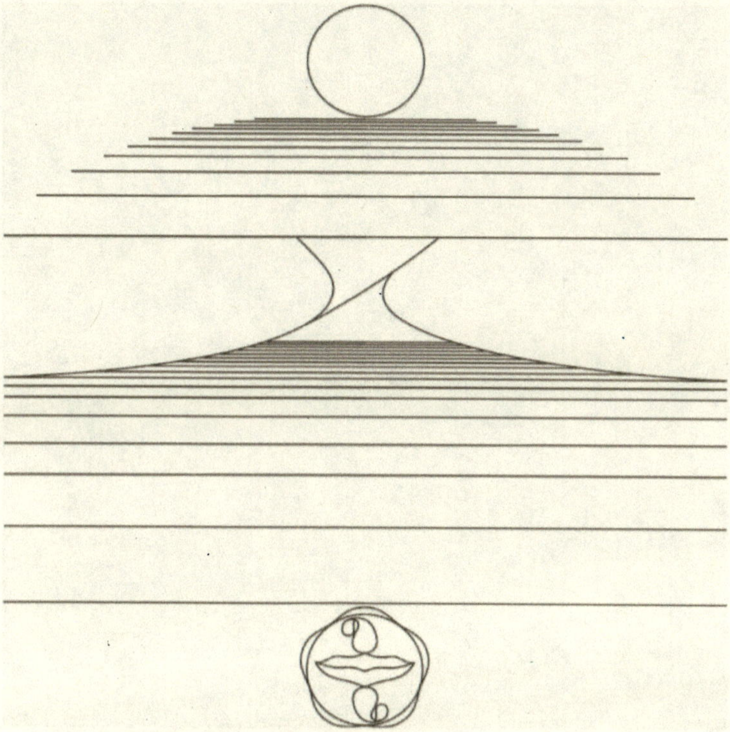

Introduction

Humanity is in an ongoing process of conscious evolution.
At this time, we are taking a giant step in consciousness-a great leap in our revolutionary process. This transformation is a pivotal event in our reality, making an exciting phase in human evolution.

-Shakti Gawain

Peace begins with a smile.
-Mother Teresa

Introduction

Mapping is a fascinating field of knowledge. It allows us to create two-dimensional representations of the real world, which can be helpful for various tasks. A map can display features such as ponds, forests, valleys, mountains, paths, and streams while remaining compact enough to fit into your pocket.

Typically, the mapping process begins by creating a virtual polygon around the area of interest. This polygon is formed by positioning the survey camera at specific points around the area and recording the coordinates of each point. Permanent markers are placed at these points to indicate the polygon's vertices.

A crucial aspect of forming this polygon is ensuring that, as you navigate the area and return the camera to the starting point, the coordinates of the first and last points match. This step is essential for accuracy and completeness. In this case, the polygon is closed, confirming that your work is correct. Once this is established, the next task is straightforward: defining the features of the area—such as elevations, roads, rivers, and ravines—based on your pre-made polygon and the points you established on the ground. Finally, you present the results on paper according to your chosen scale. If you succeed in this task, you have symbolically captured that area on your map, allowing anyone to explore it safely and engage in various activities.

The process of creating the *NUSAWVO* school of thought or worldview, initially known in Persian as *TANJOUHPANDA*, is similar to mapping. It is like preparing a map of the entire world (universe), one that enables us to understand its secrets and complexities, allowing us to explore without fear of getting lost. Interestingly, the contents of this map encompass not only the universe at large but also our own very existence, as an essential part of it.

This map, which we refer to as the *NUSAWVO* school of thought or worldview, has been created by integrating various scientific hypotheses, philosophical perspectives, and mystical viewpoints. The uniqueness of these ideas lies in their ability to explain the nature of the world (universe) and our existence clearly and logically, while also remaining compatible with one another. Our goal is to unify diverse efforts to understand both the universe and us, ultimately leading to a comprehensive map of the most significant concepts in human knowledge.

Like any undertaking, this work is not without its mistakes. Some of the assumptions and viewpoints—or perhaps even all of them—may be incorrect. At this moment, we simply do not know. The truth will be revealed in time.

Before we delve into the details of the work done, we need a more accurate and comprehensive definition of the term "worldview." Generally, worldviews can be divided into two types: the "religious worldview" and the "non-religious worldview."

The religious worldview is rooted in the teachings of prophets and religious leaders, conveyed through sermons, revelations, and other forms of communication. This perspective often employs metaphors and enlightening stories to provide an allegorical representation of the universe, humanity, and concepts such as salvation, the afterlife, and divinity. Additionally, it offers practical advice for addressing various life challenges, helping individuals distinguish between good and evil. The religious worldview is often straightforward and accessible, making it understandable for the public. It also touches upon philosophical and mystical elements.

A non-religious worldview has emerged in response to our ongoing quest for a logical understanding of the universe and humanity's place within it. This perspective intentionally distances itself from myths and metaphors, focusing instead on the collection of scientific and empirical data. It aims to create theoretical frameworks grounded in scientific, philosophical, and sometimes mystical viewpoints.

This worldview seeks to uncover truths about the universe and our origins, while also assessing the potential and limitations of the human mind in grasping the essence of truth. It addresses fundamental human questions, particularly concerning the pursuit of happiness. Furthermore, a non-religious worldview proposes ideals and goals for improving both individual and social well-being, along with actionable steps for addressing the famous question, "What Is To Be Done?"

When we examine these two worldviews together, we can conclude that the religious worldview does not fundamentally conflict with scientific and empirical issues, even though its primary focus is on theology. Conversely, the non-religious worldview, which encompasses science and philosophy, is not a competitor to the religious perspective. In fact, these two worldviews are not contradictory; rather, they are complementary. Their differences enrich, balance, and beautify human culture, and together, they help prevent misinterpretations and misunderstandings on both sides. Additionally, they share common ground in many areas, including philosophy and mysticism. Ultimately, it is fair to say that the public, whether consciously or not, is influenced by both worldviews.

After briefly defining worldview, we will discuss the "NUSAWVO" worldview, which represents a non-religious perspective. Like other non-religious worldviews, this school seeks answers to some of humanity's most important questions.

Questions such as: What is the essence of the world (universe)? Where does its past point to? Is there any truth in it, or is it merely a mirage? What is the history of life and the living beings that inhabit our world? If it is true that living beings have changed over time and that some, like the dinosaurs, have gone extinct, is there a force guiding these changes? What is the origin and history of human beings? What is our nature and temperament? What is our place in the wonders of the universe? Where are we heading, and what does the future hold for us? Is happiness a genuine concept? Is there a purpose to people's lives?

Is there life after death? Is there something implied in the universe—or beyond it—that our minds cannot perceive? Could something unperceived or unanalyzable influence the universe and us specifically? Are good and evil relative concepts that change under different conditions? Does conscience indeed exist? What will happen to us and our environment? In the context of human civilization today, which economic and political systems provide the most efficiency and satisfaction?

These questions, along with many others, are fundamental concerns for every non-religious worldview. One of the most essential inquiries we must make is about the truthfulness and reliability of the answers provided by a specific worldview. It's vital to recognize that no answer can be considered universally correct; instead, we uncover the shortcomings and limitations of various perspectives through investigation and testing. This point emphasizes the nature of a worldview in terms of its capacity for skepticism or dogmatic limitations, and it also evaluates its potential for dynamism, durability, and long-term impact.

Until now, numerous non-religious worldviews have emerged throughout human history, each contributing to the advancement of culture and civilization. However, three key issues plague these worldviews, rendering them vulnerable, ineffective, and even harmful:

- We are right.
- The rest are wrong in every way.
- We are better than everyone.

Worldviews must address these three points; otherwise, they risk deviating from their primary goals and providing misleading guidance to people.

Now, if you want to know in summary what the NUSAWVO worldview is, what it is not, where it works, and how far it is from the three evils we have mentioned, we can say:

NUSAWVO does not align with any specific religious viewpoint, nor can it be classified as a philosophy. It is neither a well-established nor thoroughly tested school of thought, and it likely contains many mistakes and structural flaws that can only be assessed through practical application. Are there valuable texts within it? Yes, particularly those passed down by humanity's cultural pioneers. However, one can wonder whether those pioneers would agree with how their ideas are referenced within this school!

Is this worldview new or exclusive? By no means! It does not present any new concepts. In fact, the teachings of this school are a blend of ideas with ancient roots, although they have been somewhat scattered in the past.

Is this book worth reading? Possibly! The school claims to have gathered significant insights regarding human knowledge and related topics, and these insights may warrant at least a cursory examination.

Can the teachings of NUSAWVO be viewed as temporary guidelines—like a plumb bob or compass—for those seeking the truth? At the beginning of your journey, perhaps! NUSAWVO serves primarily as an introduction to your exploration and makes no claims beyond that. Harper Lee once said, "The book to read is not the one that thinks for you, but the one that makes you think." Ideally, NUSAWVO aims to fulfill a similar purpose. It presents certain concepts that inspire you to develop your understanding and worldview. Afterward, it can serve as a preliminary map for your journey. You may refer to it, explore any aspect of the universe, and decide on your next steps accordingly. - (see Note 1)

Note 1- *"To walk the wilderness is better than sitting in vain,*
For if I do not reach the goal, I'll strive with all my gain." – Saadi
"The beloved delights in this sweet unrest,
For futile striving outshines idle rest." – Rumi

From this point on, you will continuously refine the map of NUSAWVO, and in the end, you will have something meaningful that reflects your own discoveries. Leibniz remarked that Descartes's philosophy is the entrance to the corridor of truth; one must enter it but not stop there. Similarly, the same may be true for the school of NUSAWVO: perhaps one day, it will be interpreted as a corridor. They say: Enter it, but don't stop there.

Can NUSAWVO alleviate the world's pain, such as poverty, injustice, and war, ultimately ending suffering? In the long run, perhaps, to some extent! The solution to these problems lies in the growth of human consciousness. Every school fostering balanced and comprehensive human awareness contributes to this goal. However, it is ultimately up to individuals to take action, transforming suffering into opportunities and valuable lessons. And what would happen if there were no NUSAWVO worldview at all? The world would not be missing anything; rather, you would be lacking. How can you express your thoughts if you don't have a blank piece of paper in your hands?

Additionally, consider the symbol of this worldview—the smile. Wouldn't it be a shame for such a smile to not exist in the world?

It is important to emphasize that the NUSAWVO worldview makes no claims in any field and is open to declaring its non-validity. Furthermore, this school of thought is not intended for business purposes; therefore, this book is available to everyone at just the cost of printing. You can also access its free electronic version online. What is the goal of this work? The goal is to promote human knowledge and peace in the world.

Who has sponsored it? No one has funded this project; it was created as a hobby, or in the best case, as a personal duty.

What is the benefit for the writer? A better world for everyone, including their descendants! What benefit could be higher than this?

The following is an Extended Abstract from this book or worldview that, while it may seem lengthy, is perhaps worth reading.

Extended Abstract

General concepts lack meaning without supporting details.
Details can be unclear when they are not connected to a general idea.

After providing the necessary context in the introduction, we will examine a summary of this worldview, which consists of excerpts from this book. It is important to note that this summary may not be concise enough and could be perceived as somewhat broad in scope. However, the need to clarify the interconnectedness of the various elements of this school of thought has left the author with little choice.

Here is the summary:

The NUSAWVO worldview is formed around the following eight pillars:

1. Belief in "meaning" (God).
2. Belief in "consciousness" as the essence of the world (universe).
3. Belief in "truth" under three concepts:
   - Consistency in seeking the truth in science, art, philosophy, and mysticism.
   - Adherence to truth, correctness, honesty, and being truth oriented.
   - Conditional persistence in belief, rights, and truth.
4. Belief in the noble and peaceful (transcendent) human nature.
5. Belief in innate human happiness.
6. The effort to achieve "The human's full-scale nature" as the foundation of human authority over their happiness (within the framework of six natural human characteristics known as the "NUSAWVO core").
7. Belief in the power of "love and compassion" (and the law of attraction) as a factor in achieving the sixth pillar (i.e., the full-scale nature of humanity and the NUSAWVO core).
8. Inviting and accompanying human happiness with self-awareness and vigilance.

Subsequently, these pillars have expanded into 13 fundamental insights or beliefs, as presented below which ultimately shape the NUSAWVO worldview. This book is also organized according to the same order.

First Insight: "Meaning"

Second Insight: "The Universe, a Truth Grounded in Consciousness"

Third Insight: "Happy Human"

Fourth Insight: "The Full-Scale Nature of Human" (Transcendent Nature of Human)

Fifth Insight: "Natural" (N)

Sixth Insight: "Unaltered" (U)

Seventh Insight: "Seeker of Truth" (S)

Eighth Insight: "Abidance in Truth" (A)

Ninth Insight: "Wishing Happiness for Everyone" (W)

Tenth Insight: "Volunteer to Serve Everyone" (VO)

Eleventh Insight: "The Power of Love and Compassion" (And the Law of Attraction)

Twelfth Insight: "Self-Aware Happiness"

Thirteenth Insight: "Meaning"

Some of the insights presented in this book are derived from the core of other insights. However, due to their significance, they are introduced as independent ideas. The book also includes several appendices, primarily aimed at elaborating on the second insight. This insight, which explores the nature and essence of the universe, requires additional details for a fuller explanation. In the following section, we will summarize the 13 insights of this school.

The First Insight: "Meaning"

The first insight explains the school's perspective on "meaning"-God.

This insight posits that there is a "meaning" beyond the universe and its everyday concepts and themes. This meaning exists far outside the realm of our everyday thoughts and perceptions; nonetheless, people discover and experience it, cherish it, seek help from it, and find refuge in it. This relationship requires no preconditions other than the individual's desire.

However, a challenge arises in this relationship: individuals may attribute specific characteristics and qualities to the meaning. These can include positive attributes, such as pure goodness, absolute love, and truth, or even negative formulations, such as not being unkind, not being oppressive, or not being limited. Such anthropomorphism and imagery leads to a more transactional and expectation-based relationship, rather than a genuine experience of meaning.

The NUSAWVO school's perspective on meaning encompasses five key principles:

1. **Belief in meaning** as a loving, mutual, and unconditional relationship.
2. **Avoiding the attribution** of any specific action, trait, or quality to meaning.
3. **Naturally enriching the relationship** with meaning by activating the full-scale nature of humans, primarily through acts of kindness and voluntary service to others (as elaborated in the school's 9th and 10th insights).
4. **Complete and unconditional trust in meaning,** which involves suspending self-centeredness and selfishness (essentially, setting aside personal disruptive volition) and allowing it to guide all aspects of life.

5. **Optional and arbitrary rituals of NUSAWVO,** which include prayer and communion with "meaning." These rituals feature a simple dance and the repetition of the healing hymn, "Oh God."- (see Note 2)

The symbol of the first insight is a circle, which resembles the ring in the hand of the wise older man in the Zoroastrian symbol of Farvahar and also mirrors the wedding ring. This circle represents communication, love, and friendship between man and "meaning" (God). - (see Note 3)

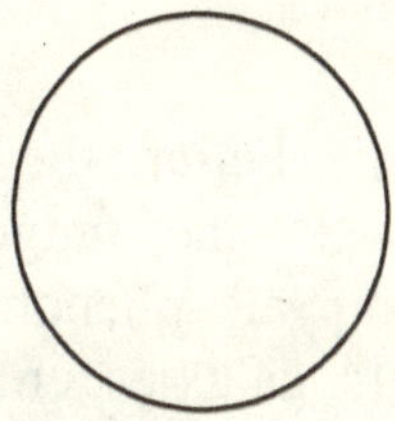

The notes of the hymn "O God" are shown below (Lyrics in Persian):

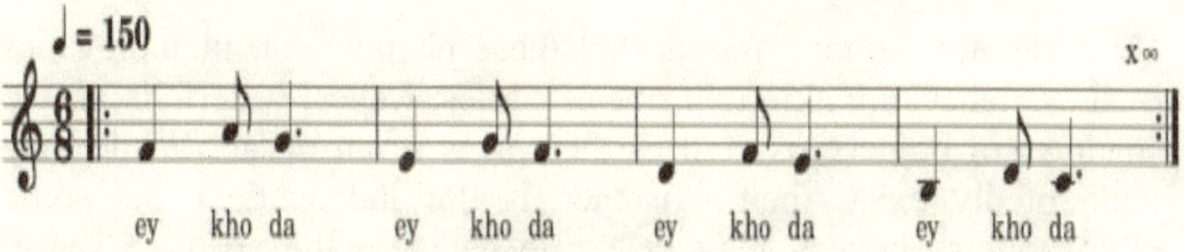

***Note 2**-This dance and hymn will likely bring you happiness and joy, which is wonderful, as this is the essence of this insight. After all, the most beautiful and practical relationship is the one that makes you cheerful and lighthearted, lifting a burden from your shoulders! However, these rituals can be performed in any other way that feels right for you.*

***Note 3**-If the concepts presented here conflict with your existing religious beliefs, please feel free to disregard them. Your prior beliefs should always take precedence. Since the NUSAWVO is a non-religious worldview, it does not seek to engage with religious matters. Therefore, the ideas discussed regarding "Meaning" (God) in this context can be seen as a reflection of the story of "Moses and the Shepherd," which ultimately expresses the shepherd's deep emotional connection and love for God.*

The Second Insight: "The Universe, a truth grounded in consciousness."

The second insight aims to explain the essence of the universe. (Apologies for the extensive discussion on this topic!)

According to the perspective of the NUSAWVO school, the origin, structure, and essence of the world are ultimately grounded in consciousness, which, in our daily lives, manifests as a persistent illusion. - (see Note 4)

According to this school's belief, the universe consists of two realms. The first realm is the tangible world of our daily experiences, encompassing all phenomena, manifestations, and contents—time, space, stars, galaxies, creatures, thoughts, emotions, and more. These elements possess a persistent reality that does not rely on our individual or human consciousness; their existence is undeniable and self-evident.

However, this school asserts that these phenomena, including our own existence, are akin to a river's eddies, waves, waterfalls, or the shadows of objects, as well as the waves of an ocean. While they undoubtedly exist, their existence is not independent. As some philosophers suggest, they lack true being in the absolute sense. Instead, their existence depends on an underlying truth, identified as the basis and ground of all phenomena: a foundational consciousness.

In other words, just as the eddies and waves of a river rely on the river's flow, or the shadow of an object depends on the object itself, the waves of an ocean exist only with the ocean. Similarly, the objective existence of our world is not autonomous; it is sustained by a more profound truth, referred to as "consciousness."

Note 4- *"Reality is merely an illusion, albeit a very persistent one."*
– Albert Einstein

This consciousness is integral, singular, all-encompassing, and continuous, transcending any linear or causal logic. It envelops everything, including the first realm, and since it constitutes the ultimate essence of the universe, there is nothing beyond it. - (see Note 5) By its very nature, consciousness is indescribable. - (see Note 6)

In this framework, the term "consciousness" represents the second realm of the universe, which serves as the ultimate foundation underlying the first realm. This second realm could also be referred to as the "Basis Realm."

A significant portion of this book explores the strengths and limitations of the claim regarding the first and second realms of the universe. The results of these investigations are briefly summarized below.

The First Realm of the Universe

The investigations into the first realm aimed to review the history of the world, including our own, through scientific and experimental findings. This rational and logical approach seeks to uncover the "objective" and "real" foundations of the complex that constitutes our familiar world—and ourselves—thus providing an introduction to understanding the nature of reality. To begin, we turned to cosmology.

Note 5- *As stated in the first insight, we discover "meaning" (or God) beyond the realm of existence—whether the first or second realm—and outside the confines of our mental constructs.*

Note 6-*To describe any phenomenon, something beyond it is necessary. To perceive anything, a background is required. (Imagine painting with black paint on a black canvas!)*

In this context, we examined various hypotheses, - (see Note 7) including the Big Bang theory. The Big Bang theory forms the foundation of a cosmology centered on the "expansion of the universe," a concept developed over the past century. Originating in the early 20th century, it models the origin and evolution of the universe, stars, and beyond. Today, it is considered the most widely accepted hypothesis in this field.

We also delved into the fascinating subject of life, drawing on findings from geology, biology, paleontology, and anthropology. This allowed us to investigate the history of the Earth, the processes that led to the origin of life, and the evolution of living beings—particularly the natural evolution of the human species.

In summary, the story of the origin and evolution of ourselves and our universe in the first realm unfolds as follows:

13.8 billion years ago, our universe emerged in an instant through a phenomenon known as the Big Bang, originating from a "quantum vacuum" (or, in other words, from "nothing"- (see Note 8). It was born in an infinite expanse and has been expanding ever since at a relatively constant rate. This hypothesis is based on three key cosmological facts:

Note 7*- Even the most definitive scientific findings are always humbly accepted as hypotheses, subject to ongoing refinement and revision as new discoveries emerge.*

Note 8*- It is possible that the "quantum vacuum" or "nothing" corresponds to the characteristics of the hypothesis proposed by this school, i.e., the "second realm." This will be clarified in the future.*

1. Measurements related to the universe's expansion (specifically, the expansion of space).
2. The detection of waves caused by the universe's birth event— the Big Bang—uniformly distributed throughout the cosmos.
3. The specific ratio of the universe's fundamental building blocks—chemical elements—which can only be explained by the "Big Bang"- (see Note 9) cosmological model.

Now, if we wish to obtain information about the state of the universe before its birth, it must be acknowledged that our knowledge is effectively zero. From a scientific perspective, we currently lack the tools to investigate this state because it is inherently inaccessible, tied to the concept of the quantum vacuum (or, for instance, "nothing").

Note 9- *From the very beginning, the term "Big Bang" may mislead the mind by suggesting a purely spatial event: "the explosion of a single point expanding into surrounding space." Instead, a more intuitive term, such as "Big Birth," could better guide thinking. The term "Birth" is not necessarily tied to a physical point but can be understood as a broader phenomenon of emergence and transformation. Much like the concept in the movie* "A Star Is Born", *where Lady Gaga and Bradley Cooper's characters experience profound rebirths and transformations, "birth" evokes an origin and emergence, rather than a localized explosion. It can be said, among other things, that the universe's birth emerged from the "womb" of the quantum vacuum, appearing in an infinite expanse. Alternatively, according to the "Inflationary Cosmological Model," the universe expanded instantly to infinity. The critical point is that, regardless of which model we consider, we must remember that, according to Einstein's theory of relativity, space and time are intertwined. Therefore, in the absence of our universe (i.e., before its birth), space itself did not exist. This means it was impossible to define a point or position. Consequently, by any account, the "Big Bang"—or "Big Birth"—occurred in an infinite expanse, not a finite spatial location.*

However, scientists, through the hypothesis of the "multiverse," attempt to explore the conditions before the "Big Bang." Within this framework, they propose a model of infinite universes existing independently, without interference in one another's processes. From eternity to eternity, these universes emerge from one another's "wombs." Some of these universes may possess the necessary conditions and parameters to foster organic or inorganic life, while others, lacking such conditions, remain barren and incapable of generating life.

Returning to the moment of birth, it is believed that at the very first instant of creation (approximately 13.8 billion years ago), energy and its various manifestations—including forces and carriers such as photons, gluons, bosons, and gravitons—were born, alongside fields, electrons, protons, and neutrons. In just a few minutes, protons and neutrons combined to form hydrogen nuclei and, subsequently, helium nuclei. At this stage, the universe was opaque and existed in the form of plasma, with temperatures so high that electrons were free and not bound to nuclei.

After about 380,000 years, as the universe expanded and cooled, the temperature dropped to around 3,000 degrees Kelvin, allowing electrons to bind with the nuclei of hydrogen and helium, making the universe transparent. As the universe continued to expand and cool, hydrogen and helium gases began to clump together under gravitational forces. These gas clouds compressed, and around 100 million years after the universe's birth, the first generation of stars formed. - (see Note 10)

Note 10- *In fact, the oldest stars discovered so far are approximately 13.5 billion years old. Some of these ancient stars, including those exceeding 13 billion years in age, are scattered throughout the universe—even within our own Milky Way galaxy.*

Some of these stars, several times the mass of our Sun, gradually produced heavier elements up to iron through nuclear fusion processes. Stars with much higher masses accelerated their fusion processes and, after a few tens of millions of years, when their fuel was depleted, they collapsed inward. During this collapse, many elements heavier than iron were produced. These stars then exploded in a massive event known as a supernova (evidence of which is abundant throughout the universe), dispersing the elements they had created into the surrounding space.

Our Sun and the planets that orbit it, which formed around 4.6 billion years ago, originated from the same materials produced in the stars of previous generations.

It is widely believed that the Sun formed first, followed by the creation of the planets in our solar system, including Earth, from the surrounding dust. Initially, Earth was a molten mass, but over time, it cooled, forming a solid crust, and the oceans eventually appeared.

It took millions—possibly even hundreds of millions—of years before the first living organisms emerged around 4.1 billion years ago in the form of single-celled life. Fossils of these ancient organisms have been discovered in Earth's deep layers. This extraordinary phenomenon resulted from the natural trial and error of elements with high affinities, such as carbon, interacting within Earth's vast natural laboratory—its oceans, lagoons, and other dynamic environments, driven by various energy sources.

For approximately 2.5 billion years, life in the oceans remained at the level of these primitive single-celled organisms. Over time, under the fundamental pressures of survival, self-preservation, self-replication, and evolutionary mechanisms—such as random genetic mutations and natural selection—the first multicellular organisms appeared around 1.5 billion years ago. However, it took another billion years—until about 550 million years ago—for life to venture onto dry land.

During this period, life evolved into diverse forms, eventually giving rise to five distinct biological kingdoms: Protista, Monera,

Fungi, Plants, and Animals. Within the animal kingdom, countless species emerged, broadly categorized into two main groups: invertebrates and vertebrates. Invertebrates include sponges, corals, worms, mollusks, and arthropods, while vertebrates comprise fish, amphibians, reptiles, birds, and mammals—the latter being most closely related to us. Humans belong to the class of mammals, which first appeared around 200 million years ago. However, it was only after the extinction of the dinosaurs 66 million years ago that mammals diversified and evolved into many different forms. It is widely believed that around 5 million years ago, we shared a common ancestor with bonobos and chimpanzees. Due to geological changes, our ancestors eventually branched off, setting them on a distinct evolutionary path.

Paleontological evidence suggests that human evolution over the past 5 million years has been extraordinarily complex, with multiple human species emerging throughout this period. The modern human species, *Homo sapiens*, appeared around 300,000 years ago. Through interbreeding with other human species, such as Neanderthals, *Homo sapiens* gradually became dominant. Today, humans have spread across the entire Earth.

With this, we conclude our historical investigation into the first realm. Next, we turn our attention to the second realm, beginning with an exploration of the physical foundations of our world. This will serve as a foundation for familiarizing ourselves with the second realm from a similar perspective.

The Second Realm of the Universe (Basis Realm)

Investigations into the second realm have focused on uncovering the true essence of our **"consciousness-based"** world. In this pursuit, we have explored various fields, including physics—particularly quantum physics and relativity—along with philosophy and mysticism. We have drawn on the groundbreaking work of pioneering scientists, the profound insights of philosophers, and the collective wisdom accumulated over thousands of years by ancient mystics. Our approach integrates rational understanding with intuitive knowledge to explore this second aspect of the universe. We hypothesize that the universe consists of two realms, and we aim to investigate this claim.

To examine it from a physical perspective, our research must begin with the first realm. For this purpose, it is essential to focus on the universe's most fundamental components, including- (see Note 11) **electrons, protons, neutrons,** and energy carriers such as **photons, gluons, bosons,** and **gravitons.**

Fundamental quantum- (see Note 12) physics experiments—such as the **double- slit experiment**—suggest that these components can be conceptualized as **"wave-particles."** The term **"wave-particle"** reflects their ability to manifest in various forms—waves, particles, and field interactions—depending on the conditions

Note 11- *While there are even more fundamental elements, the essence of all these components ultimately reduces to* ***"wave-particles"*** *or, more fundamentally,* ***"energy."*** *Therefore, we have focused on the more well-known components.*

Note 12- Quantum *refers to the smallest possible unit of a physical quantity. For example, a* ***photon*** *is a quantum of light—meaning the smallest indivisible unit of light.*

These **wave- particles** exhibit remarkable properties, possessing a form of **intrinsic intelligence** – (see Note 13) and **awareness-** (see Note 14) of their nature and the fields in which they exist.

When a **wave-particle** interacts with another **wave-particle** or with a field possessing sufficient energy (i.e., an energy exchange occurs), it may instantaneously collapse to a single point and manifest as a particle. This process transcends the constraints of time and the speed of light, as demonstrated in experiments regarding the instant collapse of wave-particles.

These particles remain aware of their nature and respond accordingly to their surroundings. From the perspective of the NUSAWVO worldview, awareness and consciousness are not mere attributes but intrinsic aspects of existence at all levels, including the subatomic.

***Note 13- Intelligence**: Intelligence is the capacity to acquire, process, and apply knowledge and skills in various contexts. It involves the ability to learn from experience, adapt to new situations, solve problems, and think critically.*

***Note 14- Awareness**: Awareness, in its everyday sense, contrasts with ignorance, unconsciousness, and irrationality. Technically, awareness refers to the cognitive ability of a system to perceive, recognize, and understand both internal states and external stimuli, while enabling measured and appropriate responses to its environment.*

At the quantum level, a single wave-particle exists in a state of superposition—simultaneously occupying multiple potential states— until it interacts with its environment. The moment this interaction occurs, the entire wave-particle collapses instantaneously, regardless of how vast its spread may be—even if it extends across billions of light-years-a phenomenon called non-locality.

This collapse defies the limitations of light-speed transmission, occurring as a unified whole without delay. Furthermore, when such particles become entangled and then move apart—no matter how great the distance between them—they remain connected in a way that transcends space and time, without requiring the transfer of information.

This suggests a fundamental interrelation between them, and their seemingly random properties (such as electric charge, polarization, etc.) manifest in relation to one another, as confirmed by experiments on quantum entanglement. Additionally, we can highlight two remarkable features of quantum mechanics. The first is the "uncertainty principle," which states that at the subatomic level, many interrelated properties of wave-particles, such as speed and position, cannot be measured simultaneously due to their inherent indeterminacy.

The second is the phenomenon known as the quantum leap, particularly concerning electrons. This refers to the discrete transition of an electron from one atomic orbital to another without traversing the space in between, as dictated by quantum mechanics.

To describe the physical nature of the universe in its most fundamental terms, we can refer to two key physical relations. These relations allow us to express the wave and energy characteristics of all components of the universe.

(see Note 15) The first is the experimentally validated relationship proposed by Louis de Broglie, represented by $\boldsymbol{\lambda = h/p}$, - (see Note 16) which expresses the wave nature and wavelength of every particle, whether microscopic or macroscopic (including humans). The second is Einstein's famous equation, $\mathbf{E = mc^2}$, - (see Note 17) which establishes the equivalence of mass and energy for all particles and components of the universe, illustrating the potential for mass to convert into energy and vice versa. - (see Note 18)

Note 15- *This statement is often attributed to Nikola Tesla: "If you want to understand the secrets of the universe, think in terms of energy, vibration, and frequency."*

Note 16- *In this equation, λ represents the wavelength of each wave-particle (or object), **p** denotes its momentum (the product of mass and velocity or its equivalent energy), and **h** stands for Planck's constant.*

Note 17- *In this equation, **E** signifies the energy of each wave-particle (or object), **m** represents its mass, and **c** is the constant speed of light.*

Note 18-*Unfortunately, one of the most well-known examples of mass-to-energy conversion is the atomic bomb, a subject widely recognized. This discovery has played a major role in the global arms race, and even the most skeptical observers of science do not dispute its validity.*

Now, referring to the cases mentioned and considering the subject of consciousness, the nature of the world can be summarized as follows: The universe, encompassing everything—including you and me—in our everyday experience, can be described as a dynamic interplay of "waves" or "wave-particles." These waves, fundamental bundles of energy, exhibit behaviors suggesting a form of "energy-consciousness," characterized by continuous fluctuations between states of compression and expansion.

The interactions associated with these wave-particles can be understood in two distinct ways. The first involves the transfer of information or energy through measurable processes, governed by the familiar laws of physics. The second represents a deeper, more enigmatic form of instantaneous connection observed in quantum phenomena, such as wave-particle instantaneous collapse and entanglement. This second type does not rely on the transfer of information, force, or energy, and exists in a domain beyond the limitations of time and space, transcending even the speed of light. These correlations ensure coherence and interconnectedness both within a single wave-particle and between entangled wave-particles, enabling the universe to function as a unified whole.

Current cosmological theory tells us that at the moment of the universe's inception—the Big Bang—all its components were inherently integrated and entangled. Consequently, we can conclude that, in principle, this foundational interconnectedness—whether viewed as correlation or consciousness—permeates every part of the universe, linking all elements in ways that transcend our conventional understanding of reality.

If we aim to examine the relationship between energy and consciousness—the two most fundamental aspects of the universe—and lay the groundwork for a "logical-rational" understanding of the nature of reality, we must consider the following: There appears to be a causal (or perhaps meta-causal) relationship between consciousness and energy. This relationship underpins the emergence and continuity of existence in our observable world (the first realm).

The critical question arises: Which of these is more fundamental or real—"consciousness" or "energy"? It becomes evident that energy cannot be the source of consciousness, as energy without consciousness lacks meaning. Even if energy could exist independently, it would remain inherently ignorant.

How, then, could unconscious and unintelligent phenomena produce any action or effect—let alone give rise to consciousness, the complexity of the universe, or even a single "wave-particle" endowed with profound awareness?

Therefore, energy cannot be the origin of consciousness. On the other hand, "consciousness" can be regarded as the source of everything, including energy. - (see Note 19)

"Consciousness" is a profound and fundamental reality, inherently aware of its completeness. As noted in the principles of quantum physics, consciousness governs the most essential feature of the universe: the awareness and knowledge possessed by "wave-particles" (or, in other terms, the building blocks of our world, such as photons, electrons, protons, atoms, and molecules). It endows these entities with the capacity to recognize their own characteristics and completeness, as well as their surroundings. Moreover, this awareness enables them to respond in measurable and deliberate ways to their environment.

Note 19- *"I regard consciousness as fundamental. I regard matter as derivative from consciousness. We cannot get behind consciousness. Everything that we talk about, everything that we regard as existing, postulates consciousness."*

-Max Planck

This same consciousness permeates all realms, including our world, as an indivisible totality and an integral, indeterminate truth. Quantum phenomena, such as the "instantaneous collapse of the wave function" (observed in the wave behavior of the universe's components) and "quantum entanglement," persist continuously—without the need for energy, force, or the transfer of information.

As mentioned at the beginning of this insight, cosmologists propose that the origin of the universe lies in the "quantum vacuum," or, more simply, "nothing." At the same time, they calculate that the current balance of the universe's "matter-energy" content equals zero. Why zero? This conclusion can be understood through the following points:

1. The sum of the universe's "mass and energy" is calculated to be zero because gravitational energy is negative.
2. The total electric charge of the universe is zero.
3. The universe as a whole does not rotate, and the rotational dynamics of galaxies also balance to zero.

These findings suggest that a universe created from "nothing," with a net matter-energy content of zero, aligns with the foundational hypothesis of our second insight.

In other words, energy and its manifestations (waves, particles, Fields, matter, etc.) do not possess true existence or, at the very least, do not have a self-sustaining existence. Their origin and continuity depend on a deeper truth or principle—what we might term "pure consciousness," or, as cosmologists describe, the "quantum vacuum" or "nothing."

However, this so-called "quantum vacuum," often referred to as "nothing," is not truly empty. It is not a void in the literal sense but a dynamic, flowing essence. This essence, which defies precise scientific definition, can be more accurately understood as "pure consciousness." Symbolically, we may refer to this "pure consciousness" as the "second realm" (Basis realm) of the universe— a fundamental truth that underpins all existence.

In the second insight and the three appendices of this book, we will explore the claim that the essence of the world is "consciousness." This investigation will incorporate perspectives from physics, philosophy, and mysticism to offer a thorough understanding of this profound concept. To provide a brief overview of the potential implications and spiritual outcomes of this view, we can outline the following:

- **If this claim holds true**—that nothing in the world has an independent existence and that everything is ultimately grounded in consciousness, despite its diverse forms and manifestations—then the secret of our unity with others and all of existence is unveiled. From this realization, our inner awareness, spirituality, and conscience can emerge in their truest sense. Within this perspective, spirituality reflects the impermanence and invalidity of the self-centered ego, allowing humility, empathy, loyalty, love, and compassion to naturally arise and flourish.
- **With the blossoming of genuine spirituality** in this concept, we come to realize that what we perceive as "others" are, in truth, reflections of ourselves. Their pain is our pain, and the path to healing may lie within us. Moreover, any irritation we feel toward these so-called "others" begins to dissolve, for the deeper truth is that those we cherish and those we feel indifferent toward are simply ourselves in another form.
- **With the blossoming of true spirituality** within this concept, the allure of material things fades, and life begins to resemble a sweet and delightful game. It is a game that requires neither winners nor losers, and as a result, competition and the urge to dominate—along with their companions of failure, envy, anger, anxiety, and despair—lose their grip. In their place, a harmonious sense of cooperation, acceptance, diligence, contentment, humility, and hope emerges, filling life with deeper meaning and joy.
- **And at the end**, we can consider that by understanding the unity of the world through an essence known as "pure consciousness," we come to realize that we are, in essence, "Immortal."

In this perspective, "death" is not an end but a profound transformation—a shift in certain aspects of our consciousness, such as its expansion or the dissolution of the focused awareness we experience in daily life. Rather than perceiving "death" as our disappearance—like a drop vanishing into the river of existence or a wave dissolving into the ocean—it is more accurate to see it as a reconnection: the wave merging with the ocean, the waterfall flowing back to the river, and ultimately, a return to eternity and our true self, which is "pure consciousness," the "soul of the universe." Undoubtedly, even with this understanding, we as humans will grieve the loss of our loved ones. We will shed tears and feel sorrow. Yet, with the mystery of death unveiled, this noble grief need not lead to spiritual collapse or depression. Instead, it can awaken latent abilities within us, enabling a purer connection to spirituality and perhaps even igniting a profound desire to serve others. Thus, the seemingly bitter inevitability of death may transform into a sacred platform for the exchange of knowledge, faith, and love.

Ultimately, it is important to remember that the term "consciousness" does not refer to "meaning" (God) as described in the first insight. Of course, it is evident that spirituality, beyond the concepts mentioned earlier, also encompasses the human connection with God—a profound experience that transcends fear, hope, and the ordinary aspects of life.

The symbol for the second insight is a Möbius surface, represented as follows:

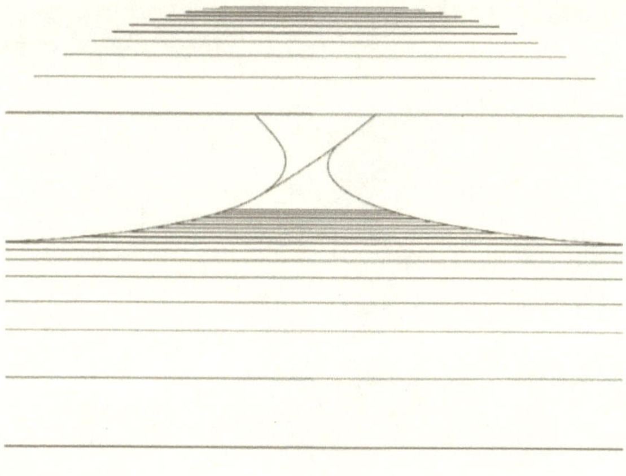

Which, when combined with the first insight, will be as follows:

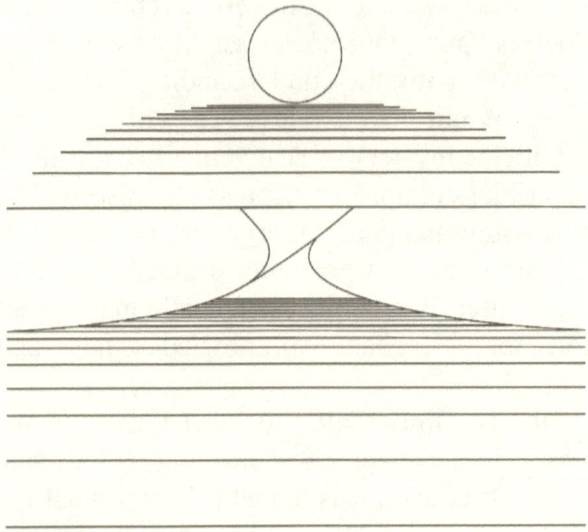

The Third Insight: "The Happy Human"

The third insight of this book delves into the origin, history, and nature of humanity. We will explore the biological, cultural, social, and spiritual evolution of human beings while also investigating the foundations of happiness—bliss. We acknowledge the length of this discussion, but because the fourth through twelfth insights are rooted in this exploration, a thorough explanation is essential.

Through this insight, we understand that our distant ancestors, approximately 5 million years ago, were isolated in the savannas of eastern Africa due to significant geological changes. This separation marked their divergence from the common ancestors of chimpanzees and bonobos. From that moment onward, their biological and cultural evolution followed a remarkably distinct trajectory compared to these close relatives. According to paleontological evidence, the story of our biological and cultural evolution from 5 million years ago until now has unfolded as follows:

Around 5 million years ago, our ancestors began adapting to the challenging conditions of the African savanna. The gradual evolution of bipedalism, first evident in species like *Australopithecus afarensis*, enabled them to walk on two legs with increasing efficiency. This shift freed their hands for manipulating objects such as bones, wood, and stones, used for scavenging, small hunts, and self-defense against predators. Over hundreds of thousands of years, genetic mutations refined the structure of thumbs and toes, enhancing dexterity and making early humans more adept at running, hunting, and crafting tools.

In this harsh environment, hunting emerged as a key survival strategy. Physical traits that improved endurance, agility, and strength, combined with social behaviors like cooperation and deeper social bonds, provided critical evolutionary advantages.

The intricate interplay between the brain, hands, feet, and tools fostered the development of early technologies, driving the success of those who excelled in these areas.

Women played a pivotal role in this evolutionary story—not only as mothers and gatherers and, later, as the first domesticators of animals but also as innovators in food preparation and social organization. Their contributions to fostering cooperation, establishing complex social relationships, and nurturing offspring were fundamental in shaping early human communities. These efforts laid the groundwork for the emergence of evolved human culture and society, including the nuclear family, which became a cornerstone of human civilization.

In the prehistoric state of our ancestors, any vocal system that enhanced hunting techniques and group coordination likely conferred a significant evolutionary advantage. These systems, though simple at first, gradually became embedded in the human genetic makeup through the slow and persistent process of natural selection. Over millions of years, the evolution of brain regions associated with language and speech, alongside the development of the larynx, unfolded in tandem. This intricate co-evolution of the brain, vocal apparatus, and cultural behaviors ultimately culminated in the emergence of language—a profound treasure of human culture.

Language did not emerge in isolation; therefore, it acted as a catalyst, propelling the growth of human culture in unprecedented ways. With the expansion of communication came the flourishing of art, a deeper consideration of the future, and an acute awareness of mortality. From this awareness, the fear of death and the quest for meaning gave rise to spirituality, marking a pivotal shift in the human experience. Through these interconnected developments, language became not merely a tool for survival but the foundation of a rich and complex cultural legacy that continues to define humanity.

Now we understand that over the past hundreds of thousands of years, biological evolution has played a diminishing role in shaping the trajectory of humanity. Instead, cultural evolution—fueled by the transmission of knowledge, technological advancements, and social organization—has emerged as the dominant force in determining our destiny. In the modern era, this profound shift has enabled unprecedented levels of interaction and coexistence among billions of people. Moreover, it has equipped humanity, particularly through the rise of our relatively recent civilization, with astonishing possibilities and capabilities, from exploring the cosmos to harnessing artificial intelligence. These advancements continue to redefine the boundaries of what we can achieve while presenting new challenges and opportunities for our collective future.

Anyhow, within this perspective, in addition to exploring the origin and nature of humanity, attention is also given to the concept of human bliss. The NUSAWVO school, like other humanistic schools, holds that humanity, in its full potential, is a noble, social, kind, and peaceful entity with a rich and beautiful nature. At its essence, like all creatures in the world, humanity is inherently blessed a "blissed human."

However, it cannot be overlooked that human bliss is profoundly complex, shaped by the extraordinary interplay between cultural dynamics and biological intricacies. Of particular note is the human brain, a highly advanced organ that operates uniquely at the border of reality and imagination.

This complexity means that human bliss is influenced by a web of internal and external factors—though, as with most phenomena related to humans, distinguishing between the two remains an almost impossible task.

This insight seeks to explore both external and internal factors that contribute to human happiness (or bliss) to understand its foundations and the principles that govern it. By doing so, it aims to highlight the key elements that shape human well-being. Ultimately, twenty essential topics have been identified within this context, which can be called "the foundations of human happiness."

These topics are as follows:

1. **Balanced Environment**
   Living in harmony with both nature and society, fostering balance and sustainability in every aspect of life.
2. **Security**
   Feeling safe—physically, emotionally, and financially—is an essential pillar for peace of mind and human flourishing.
3. **Livelihood and Purposeful Work**
   Engaging in work that not only supports one's life but also carries meaning and benefits others.
4. **Health and Vitality**
   Caring for one's physical and mental well-being, which serves as the groundwork for happiness and progress.
5. **Close-Knit Family**
   The presence of loving, dependable family ties that offer comfort, encouragement, and belonging.
6. **Self-Dignity and Respect**
   Acknowledging one's own worth, while honoring the dignity and rights of others.
7. **Freedom and Individuality**
   The ability to live authentically, express oneself, and follow personal goals without undue restrictions.
8. **Justice and Fairness**
   Living in a society where fairness prevails, rights are safeguarded, and no one is left behind.
9. **Curiosity, Exploration, and Creativity**
   The natural urge to learn, discover new possibilities, and express one's creativity in daily life.
10. **Channeled Aggression, Play, and Sports**
    Channeling energy through positive competition, playful engagement, and physical activity that strengthens both body and mind.
11. **Art**
    Experiencing and creating beauty in its many forms—whether through art, music, or meaningful design.
12. **Restlessness and Drive for Progress**
    The subtle discomfort that inspires individuals to grow, improve, and seek new horizons beyond the familiar.

13. **Connection with Nature and Survival Skills**
    Rediscovering the rhythms of nature and learning essential survival skills—like observation, patience, and resourcefulness—that reconnect us with our roots.
14. **Honesty, Integrity, and Keeping One's Word**
    Living truthfully, acting with moral integrity, and honoring the promises we make.
15. **Awareness of Moral Defilements (The Abysses)**
    Becoming conscious of the ethical pitfalls and harmful tendencies that can pull us away from joy and inner peace.
16. **Tools and Technology**
    Creating and using tools and technologies responsibly to improve life without undermining humanity or nature.
17. **Peace-Seeking, Love, Compassion, and Service**
    Actively promoting peace and kindness, showing compassion, and serving others without expectation and exception.
18. **Wishing Happiness for All Without Exception**
    Cultivating a heartfelt desire for the happiness and well-being of every person, regardless of who they are.
19. **Critical Thinking and Freedom from Dogma**
    Remaining skeptical of rigid beliefs and practicing open, thoughtful inquiry in the search for truth.
20. **Spirituality and Transcendence**
    Seeking meaning beyond the material world, embracing the spiritual dimension of existence.

The twenty foundations discussed each address our requirements for happiness from various perspectives, highlighting different aspects of this concept. Understanding these factors, we can logically conclude that there is a direct relationship between the *intensity of happiness* and the level of achievement of these twenty foundations. In other words, greater access to these factors leads to more complete happiness, while lacking access results in diminished happiness.

While this conclusion seems reasonable, let's consider the perspective of the NUSAWVO school on this matter. The NUSAWVO school acknowledges that the assumptions outlined above are rational and that achieving these twenty foundations is essential for happiness. However, it emphasizes a crucial point regarding *human agency and authority over one's happiness*. According to NUSAWVO, while realizing these foundations is crucial for our happiness, it is not the complete picture. We must find agency and authority in our lives regarding happiness.

This involves consciously taking measures on the "path of happiness."- (see Note 20) This means our happiness is not solely determined by our circumstances, fate, or the authority of others. Instead, we possess the power to direct our lives and make choices that contribute to our happiness.

Clearly, "being present on the path of happiness" is not something separate from the fullness of our existence. Perhaps we are already on this path, but we might not be aware of it, or we lack agency in this regard.

Now, the question is: how can we "actively" and "consciously" be present on the path of happiness?

To answer this question, the NUSAWVO school leaves the position of a neutral observer and humbly offers its unique approach to the concept of "human agency over one's happiness." In doing so, the school temporarily sets aside the question of "how much we have benefited from the twenty foundations of happiness" and deliberately avoids relying on uncertain, ambiguous concepts like luck, fate, and so on—factors often considered relevant to the original question. Instead, it seeks to explore happiness from a broader, more comprehensive perspective.

Note 20- *We believe that happiness is not a "goal" but a "path," a "way," a "journey." We are born onto this path, and our most natural state is to remain on it, continuously growing and being enriched along the way.*

In its progression, the school introduces a fundamental and ancient concept under the theme of "humanity's full-scale nature."- (see Note 21) It asserts that true and enduring happiness can only be achieved when all aspects of human nature are fully activated. This raises a crucial question: What are these aspects?

The NUSAWVO school defines human nature as comprising four main aspects: physical, emotional, mental, and spiritual. When these aspects are activated and balanced harmoniously, human nature moves from potential to action on a broad scale.

Given that these four aspects are general concepts, an important question arises: How can they be practically realized and integrated into everyday life?

In response to this question, the NUSAWVO school offers a simple and natural solution through a "special arrangement" of some of the most fundamental elements of human existence. This arrangement consists of six natural human characteristics that collectively represent the four aspects of "our full-scale nature."

Furthermore, they encapsulate and, in a sense, recreate all the "twenty foundations of our happiness." We refer to these six characteristics as the "six agent gems of human happiness," or, in other words, the core of the "NUSAWVO School" (the meaning of "NUSAWVO" will be explained below).

Note 21 *- Human full-scale nature can also be referred to as human transcendent nature or human sublime nature.*

According to this school, the comprehensiveness, pragmatism, and agency embedded in these six gems enable the harmonization and activation of our nature across all four aspects (and, correspondingly, all elements of the twenty foundations of our happiness). These "active gems of human happiness" are:

a. **Naturalness** (N) – **The Natural State**
b. **Unaltered-Untainted- Untampered-Undefiled**(U)
c. **Seeking the Truth** (S)
d. **Abiding in Truth** (A)
e. **Wishing Happiness for Everyone** (W)
f. **Volunteering to Serve Everyone** (Vo)

As observed, the first letters of these six characteristics can be extracted to form the acronym 'NUSAWVO.' Now, the next question is how to activate these six natural characteristics.

Typically, when discussing a natural human characteristic and aiming to activate it, we must first provide information about it. This information should arise from a natural context. Furthermore, it is essential to explain the biological, sociological, and anthropological evolution of the characteristic, its significance in the current cultural and civilizational landscape, and other relevant factors. Additionally, we must gather the necessary details on how to activate it.

Beyond providing information, is any other action necessary?

Awareness alone is not sufficient; some form of motivation or power is required to enhance awareness, foster acceptance, and cultivate habits and culture—both individually and socially. Various influences can shape this process.

At the individual level, these influences may include encouragement, discipline, control, force, and fear. At the social level, - (see Note 22) they might involve instructions, laws, fines, and punishments.

***Note 22-** While we believe that happiness depends on individual agency, the reciprocal relationship between personal and societal happiness remains undeniable.*

While each of these factors may be effective under certain conditions, and sometimes even necessary, the NUSAWVO school does not consider them suitable for managing its proposed programs. Instead, the school emphasizes the importance of 'love and compassion,' which aligns with the principles of the law of attraction. Among the various forces and motivations shaping human life in daily challenges, this school recognizes only the power of 'love and compassion' as worthy of activation, direction, and organization within the framework suggested by NUSAWVO.

When considering who should channel this power, we can point to parents, individuals, society, culture, and even humanity as a whole. However, it is ultimately the individual who must take responsibility for their own happiness and destiny. This holds true regardless of the support and compassion they may receive from others or society. When the six characteristics described under 'NUSAWVO' are activated through the power of love and compassion, in alignment with the law of attraction, a person's full-scale or transcendent nature is realized. As a result, the individual gains control over their destiny and happiness, effectively guiding their life along what we call 'the path of happiness.' It is important to emphasize that this is not merely an individual pursuit but also a collective one. When members of a society actively pursue the path of happiness, the entire community becomes consciously engaged in this endeavor.

In conclusion, this insight presents proposed solutions for actively achieving happiness under three key headings: 1. "The Full-Scale Nature of Human," 2. "The Six Characteristics Known as NUSAWVO, Essential for Realizing the Full-Scale Nature of Human," and 3. "The Power of Love and Compassion (and the Law of Attraction) as the Driving Force Behind the Six Characteristics of NUSAWVO." Accordingly, we may redefine the term "happy human" as "agent happy human," emphasizing the proactive and intentional role individuals play in realizing their happiness. However, this insight goes beyond merely defining happiness or proposing a solution; it takes an additional step toward completing the process of happiness by connecting it to human awareness and self- awareness.

Inviting happiness into the realm of human consciousness and self-awareness does not require extraordinary preparation or effort. On the contrary, it only requires a few simple reminders. Every Individual who has influence over their destiny—or is somehow on the path to happiness—already possesses an awareness of their happiness within their subconscious mind. However, like many things, this awareness must be emphasized and brought to the forefront. Through such reminders, a person's perspective can be renewed, allowing them to consciously recognize and embrace their happiness.

This irreplaceable gift, referred to as "self-aware happiness," encompasses five noble traits. By understanding and embracing these traits, whether consciously or unconsciously, we allow happiness to enter the realm of our self-awareness. These five traits are:

- Thanksgiving (Appreciation, Gratitude) – The Thanksgiver (T)
- Hope – The Hopeful (H)
- Astonishment and Elation – The Astonished and Elated (A)
- Opportunity Finding and opportunity making – The Opportunity Finder and Maker (O)
- Seeking Meaning – The Seeker of Meaning (S)

By selecting the first letters of these traits, they can be referred to as "THAOS." However, if you prefer a different name, feel free to choose other letters to create any term you like. The choice is yours!

In this manner, the entire proposed solution of this school, in the context of our happiness, is summarized in the following two steps:

- First, with the power of "love and compassion" and the law of attraction, and through the six characteristics known as "NUSAWVO core", the "full-scale nature of humanity" is realized.

On this foundation, a person becomes the agent of their own happiness. In this regard, it is as if the gem of their existence is cleansed and polished, and the things that once drained their energy either fade or disappear. At this point, the individual is prepared for any type of wonder, including the ultimate secret of happiness: "self-awareness of happiness" (which will be discussed below).

• Second, human happiness, in the form of the five noble traits mentioned, is invited into the realm of self-awareness. In this way, human happiness becomes "active self-aware happiness."

After that, it matters little whether we refer to ourselves as "active self-aware happy human," "self-aware happy active human," "happy human," or simply "human." The work is complete, and the secret of happiness—bliss—has been revealed. - (see Note 23)

However, this school designates this excellent and noble state of being as "self-aware active, happy human" or, for instance, "Nusawvo Thaos human."

In continuation, the themes of the "full-scale nature of human," its structure, and mechanisms, including the "six agent gems of human happiness" within the framework of "NUSAWVO," will be explored in the fourth to eleventh insights. This section will also delve into the power that activates these characteristics, namely "love" and "compassion."

Additionally, the twelfth insight will briefly introduce the "quintet Noble traits," referred to as "THAOS".

Note 23- *There is no doubt that, in addition to the items mentioned, other requirements for happiness could be added to this list—requirements that may not have been addressed by this school. We acknowledge this limitation and leave the discovery of these matters to the seekers of truth.*

The symbol of this insight is a circle, representing the "happy human." This circle will encompass the related insights, namely those from the fourth to the twelfth.

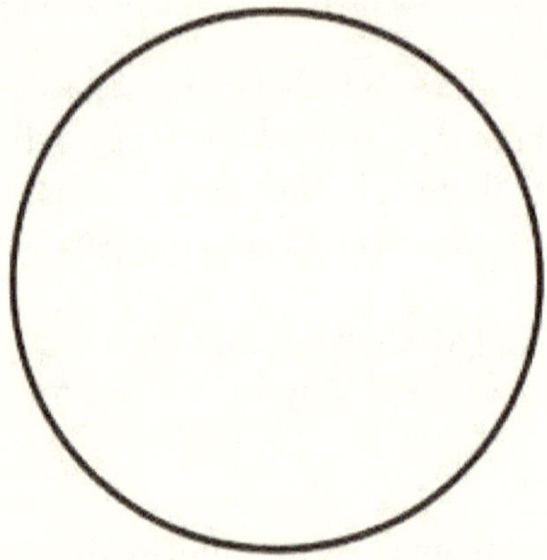

The set of symbols from the first to the third insights of this school is as follows:

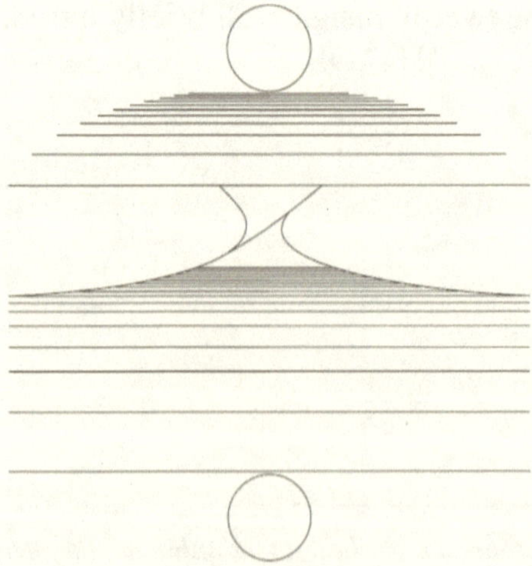

To clarify the connection between the third insight, which focuses on the "happy human," and the related insights, we can outline the following:

a. The fourth insight emphasizes the full-scale nature of humans, interpreted as transcendent human nature.
b. The insights from the fifth to the tenth are collectively referred to as the NUSAWVO core.
c. The eleventh insight highlights the power of love and compassion, along with the law of attraction.
d. Finally, the twelfth insight pertains to self-aware happiness.

This framework illustrates how these insights interrelate.
Below, the interconnection of these insights has been depicted in the figure:

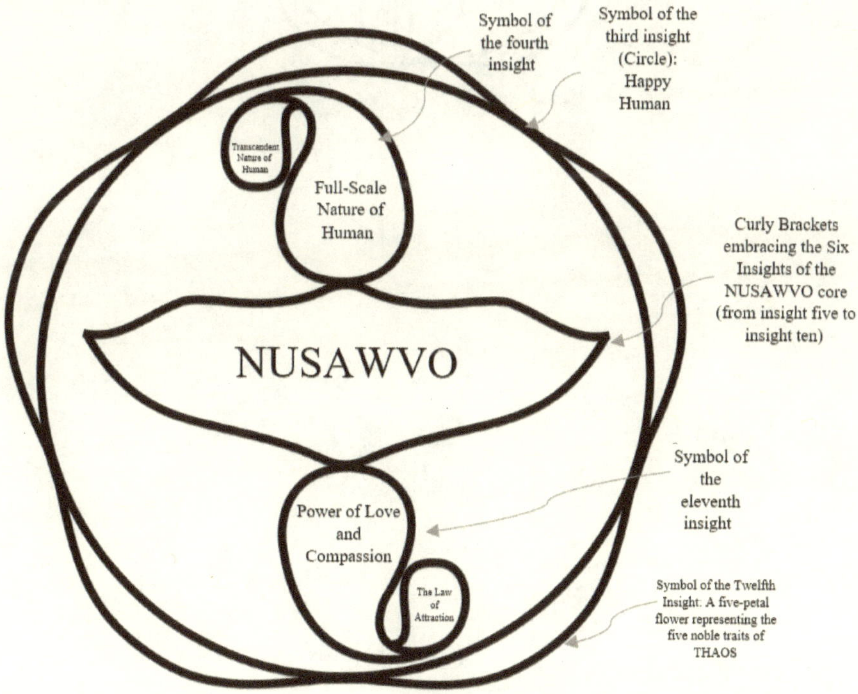

The symbol describe can be represented more simply below, reflecting the true spirit of this school: a smile and a flower. - (see Note 24)

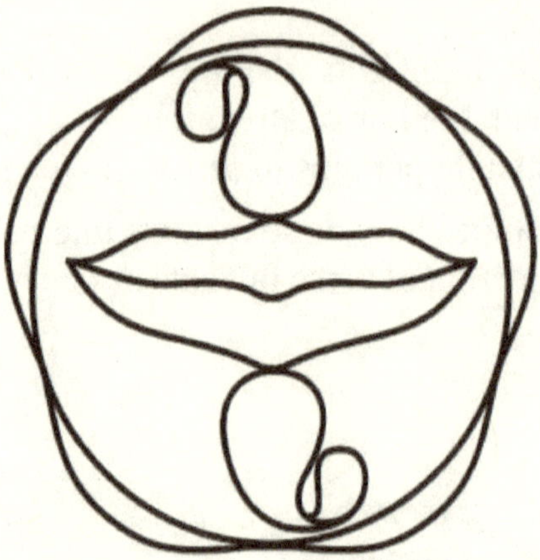

Note 24- *We will never know what a simple smile can do. (Mother Teresa)*

The Fourth Insight: The "Full-Scale Nature" of Human

The fourth insight of this school is dedicated to exploring the full-scale nature of humanity, often referred to as the transcendent nature. This insight introduces the concept of human agency over one's own happiness. In this exploration, we discover that human nature consists of four general aspects: physical, emotional, mental, and spiritual. Achieving genuine human happiness requires the activation of all these aspects. These aspects are broad concepts, implicitly addressed through the twenty factors underlying human happiness, as discussed in the third insight. To fully activate these aspects in a balanced and harmonious way, this school proposes a specific structure based on "six natural human characteristics," known as the "six agent gems of happiness." The abbreviation for these six characteristics is "NUSAWVO." With the power of love and compassion, along with the law of attraction, NUSAWVO offers a practical approach to embodying the four aspects of our complete nature, thereby elevating our happiness to what is termed "active or agent happiness."

"NUSAWVO" is considered the core of the NUSAWVO school.

In the following sections, we will briefly explain the six insights mentioned earlier. Finally, we will discuss the eleventh insight, which focuses on "the power of love and compassion," along with the law of attraction.

The symbol of this insight is shown below

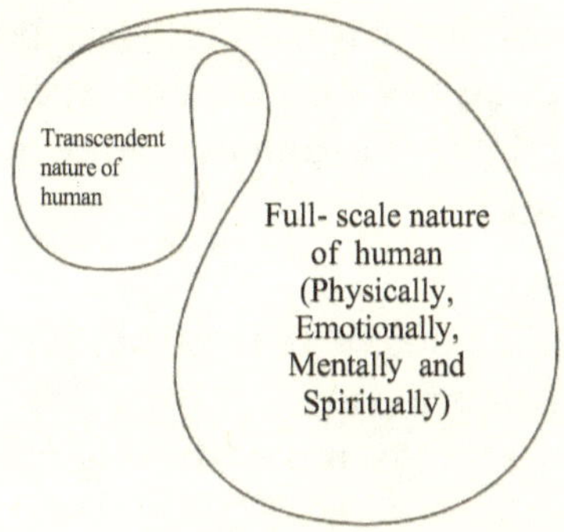

The Fifth Insight: "Naturalness" (N)-The Natural State

The fifth insight, known as "naturalness," focuses on the first element of the "NUSAWVO core." This insight revisits and emphasizes the twenty foundations of happiness introduced in the third insight. But why return to these foundations? By bringing them into awareness, these factors—referred to as the requirements and foundations of happiness—become fully recognized in our perception and consciousness. This recognition not only highlights their importance but also underscores their necessity. It marks the first step toward uncovering the comprehensive foundations of our happiness, or, as we may also call it, the "full-scale nature" (transcendent nature) of our being. Once again, this insight seeks to reveal and redefine the twenty foundations of happiness, particularly in relation to our mind and consciousness. In doing so, it offers a deeper understanding of what it means to be a natural human being. Furthermore, it sets the stage for the next five insights—the core ideas of NUSAWVO—through which these foundations will gradually take shape and manifest in reality. Ultimately, this process leads to the realization of our full-scale nature. The message of this insight is clear: "It is necessary to trust in the noble and peaceful nature of the human being." By understanding the importance of the twenty principles and engaging in personal and social practices to realize them, we can foster the flourishing and growth of our true nature and elevate human society.

Emphasizing "trusting the noble and peaceful nature of the human being" carries an additional message: the pure and unspoiled conscience of a person should not be tarnished by negatively charged or malicious doctrines and beliefs. Such ideologies harm both the individual and their happiness. Maliciousness refers to concepts and practices such as:

- Greed and excessiveness,
- Fear of others and reducing people to mere objects,
- Market-driven anti-ethical behaviors,
- Extremist ideological teachings rooted in dogmatism and hatred, and

- Extremist religious doctrines intertwined with dogmatism, superstition, and hatred.

These destructive ideas do not originate from the sublime nature of human beings. According to the NUSAWVO school, such anomalies stem from "ignorance"—a lack of awareness of the natural foundations of our happiness. This ignorance manifests as "cultural shortcomings" and "cultural deviations," which we will address in more detail later.

While we emphasize belief in the "honorable and noble nature of humanity," we must also humbly warn against the erosion of this noble essence. We must avoid testing our sublime and noble nature in the depths of degradation or treating it frivolously. In our next exploration, we will delve deeper into this important topic.

The symbol of this insight is part of NUSAWVO's smile, as mentioned at the end of the third insight.

The Sixth Insight: "Unaltered-Untainted-Untampered-Undefiled State" (U)

The sixth insight forms the second pillar of the "NUSAWVO" core, focusing on the concept of the "untainted state." To be untainted is to recognize "moral taints" and deliberately avoid them. Defilement encompasses beliefs, speech, and actions that inflict harm upon us or others, obstructing the realization of our noble and complete nature.

Knowledge and awareness of defilements are cultural and social treasures that have emerged from millions of years of human history, shaped by both evolutionary instincts and societal development. This awareness is rooted not only in human efforts and mistakes but also in pre-human natural history. As we will explore, some of this awareness—often referred to as 'taboo'—is influenced by biological predispositions and reinforced by culture. For instance, prohibitions against incest reflect an interplay of evolutionary mechanisms, such as the Westermarck effect, and deeply ingrained cultural values.

The number of defilements typically does not exceed the number of fingers on two hands. Everyone has the right to be aware of these abysses and understand their dangers. In fact, it is virtually impossible to find a culture that sanctifies or encourages vices such as murder, theft, lying, revelry, alcohol and drug abuse, adultery, malice, dogmatism, laziness, and greed. If such a culture does exist, it is likely to be vulnerable and prone to degradation.

It is important to note that moral principles are not learned solely from sermons or speeches. Rather, they are learned through the behaviors and actions of parents, teachers, and community elders. We will explore these ideas in more detail later. Knowledge of defilements is similar to understanding the rules of guidance and driving; it helps us navigate life safely and enjoyably.

This insight, both individually and socially, acknowledges 14 key defilements, as follows:

1. **Murder**: The deliberate taking of another person's life. A severe violation of human rights, causing irreversible harm and deep societal trauma.
2. **Theft**: Stealing or taking something without consent. Theft undermines trust, creates insecurity, and disrupts social and economic systems.
3. **Lies and dishonesty**: Falsifying information or misleading others, which erodes trust and damages relationships. It creates confusion and leads to further harm.
4. **Adultery**: Engaging in romantic or sexual relations outside of a committed partnership. It damages trust, harms individuals emotionally, and destabilizes families and relationships.
5. **Alcohol and drug use**: The consumption of substances that impair judgment, health, and behavior. Abuse of alcohol and drugs can lead to addiction, violence, neglect, and physical harm.
6. **Violence**: The use of physical force to harm or intimidate others. Violence causes direct harm to individuals and society, often leading to long-term psychological and physical damage.
7. **Dogma and imposition of beliefs**: Rigidly adhering to a particular set of beliefs and forcing them upon others. This stifles intellectual freedom, fosters intolerance, and leads to conflict and oppression.
8. **Malice, slander, and foul language**: Intentional harm through harmful speech, spreading false information, or using abusive language. These behaviors degrade individuals, create divisions, and damage reputations.
9. **Laziness and neglect**: Avoiding responsibility and neglecting duties. This leads to stagnation, missed opportunities, and societal inefficiency, affecting both personal and collective progress.

10. **Greed and excessiveness**: An insatiable desire for more, whether in material wealth, power, or pleasure. Greed leads to exploitation, inequality, and harms both individuals and the environment.

11. **Slavery**: The forced labor or ownership of individuals as property, depriving them of their freedom and dignity.

    Slavery is a severe violation of human rights and perpetuates systemic injustice and inequality.

12. **Discrimination**: Treating individuals unfairly based on inherent characteristics such as race, religion, or identity. Discrimination fosters division, perpetuates inequality, and prevents social harmony.

13. **Creation of totalitarian (authoritarian) political systems**: Establishing a political system where power is concentrated in the hands of a single authority or group, suppressing individual freedoms and dissent. Such systems are often oppressive and lead to human rights violations.

14. **Militarism, warmongering, and Colonization**: The promotion of military power and the willingness to engage in war for territorial expansion or dominance. Militarism fuels conflict leads to unnecessary suffering and diverts resources from peaceful development.

In the end, awareness of the defilements empowers us to recognize the abysses along the path to happiness. By avoiding these moral taints and fulfilling our full nature, we gain agency over our own happiness.

The symbol of this insight is embodied in NUSAWVO's smile.

The Seventh Insight: "Seeking the Truth" (S)

The seventh insight, forming the third core component of the NUSAWVO worldview, emphasizes "the seek for truth." Its full title, *"The Seek for Knowledge, Art, and Truth,"* reflects its comprehensive nature. This insight integrates diverse dimensions of understanding, offering a holistic framework that spans the entirety of human knowledge and cognition. The search for truth, as emphasized by this school of thought, underscores the importance of free-thinking, the rejection of dogma, and a relentless drive for inquiry and understanding. It is evident that no knowledge or truth is entirely free from error or approximation. The notion of attaining absolute truth—whether in science, philosophy, or mysticism—remains implausible. This highlights the inherent relativity of all concepts and categories. As a result, the pursuit of knowledge and truth should be regarded as a perpetual and essential responsibility.

Moreover, one of humanity's most profound instincts—the pursuit of art—must remain unconstrained by any individual or extremist ideology. The quest for artistic expression, too, should be recognized as a continuous and self-evident endeavor, encouraging unfiltered creativity and expression. As previously mentioned, the full title of this insight is *"The Seek for Knowledge, Art, and Truth."* This insight emphasizes the necessity of ongoing personal exploration in these areas. At the societal level, it stresses the importance of providing opportunities for research and inquiry into these topics for all members of the community. Ultimately, it highlights the vital role of this pursuit in empowering individuals to exercise their agency and attain true happiness. In this book, both “thruth” and “the truth” are used with intention. The former reflects an ongoing pursuit, the latter, a deeper realization. The reader is invited to perceive through whichever lens resonates more profoundly.

The symbol of this insight is a part of the NUSAWVO smile.

The Eighth Insight: "Abiding in Truth" (A)

The eighth insight, forming the fourth component of the NUSAWVO core, is titled *"Abiding in Truth."* The complete title of this insight is *"Adherence to the Truth" and "Conditional Persistence in Interests, Rights, and Truth."*

Adherence to the Truth refers to a steadfast commitment to truth, correctness, honesty, and truth-oriented principles. This principle sets the beliefs of the NUSAWVO school apart from others that may justify pursuing goals by any means, often overlooking the value of truth. In contrast, within this school of thought, our commitment to truth, correctness, and honesty is fundamental—even when it conflicts with our personal interests or goals.

The second part of this insight, titled **"Conditional Persistence in Interests, Rights, and Truth,"** addresses a crucial issue at the heart of many societal problems. This issue reflects both the excessiveness and one-sidedness in the benefits and rights we claim for ourselves, as well as the rigidity, stubbornness, and dogmatism tied to ideological beliefs. The insight emphasizes the need to:

1. Investigate and critically examine the excessiveness and one-sidedness in the benefits and rights we envision for ourselves.
2. Acknowledge that our beliefs—whether ideological, political, social, scientific, or otherwise—are never free from error or imperfection.

Furthermore, it is essential to approach conflicts with others with care and to act from a place of love and compassion. As stated in the second insight, we must also strive to understand the noble essence of those involved in the conflict, recognizing their inherent dignity and worth. Ultimately, the title *"Abiding in Truth"* encapsulates the role of truth in fulfilling our complete nature and, by extension, in exercising our agency over our happiness.

The symbol of this insight, like other parts of the NUSAWVO core, is a part of the NUSAWVO smile.

The Ninth Insight: "Wishing Happiness for Everyone—Universal Benevolence" (W)

The ninth insight emphasizes *"wishing happiness for everyone"*—a concept referred to as *universal benevolence.* Benevolence is inherent to our true nature, but societal structures, cultural differences, and various relationships—whether familial, professional, political, or international—often obscure this natural inclination. Instead of wishing happiness for all, our desires tend to narrow, focusing on personal, familial, group, ethnic, religious, commercial, or national interests. This narrowing of focus can lead to malicious thoughts and intentions.

At the heart of these disturbances lies *ignorance,* particularly ignorance of the fundamental principles of happiness. As discussed in the fifth insight, this ignorance can be described as *"cultural shortcomings"* or *"cultural deviations,"* terms that essentially convey the same idea and can also be understood as perversions.
In this context, perversion refers not to its narrow sexual definition, but to any significant deviation from the full-scale nature of the human being—especially in terms of awareness, ethics, and conscience. These perversions disrupt the balance of our complete nature, distorting our thoughts and behaviors in ways that prevent us from recognizing our true essence and living in harmony with it.

Here are five examples of these perversions:

1. **Greed and excessiveness** – An insatiable desire for more, often at the expense of others.
2. **Fear of others and objectification** – A failure to recognize the unique essence in each individual, driven by a sense of separation.
3. **Market-driven unethical behaviors** – Practices rooted in ignorance, greed, and excessiveness, driven by profit at the expense of ethics
4. **Narrow-minded ideological views**

   – Rigid and limiting worldviews, born from ignorance and often leading to intolerance.
5. **Extremist religious beliefs** – A product of ignorance, fostering division, hatred, and intolerance

In addition to these larger perversions, there are also *small personal or social displeasures* that, if not addressed, can escalate into more significant problems. These disturbances might seem minor initially but can, over time, lead to serious conflicts or even war. Identifying and addressing their root causes—such as excessiveness, fear, prejudice, and dissatisfaction—is essential for peace.

The principle of collective and universal benevolence—without exception or discrimination and without expectation—serves as a remedy to these disturbances. The NUSAWVO school provides solutions to address the root causes of these conflicts through its 13 insights. To address smaller personal grievances, the school introduces a circular diagram known as the *Attitude Cycle,* which encompasses 10 interconnected topics. This diagram is designed to help individuals uncover the underlying causes of their conflicts and better understand their role in them. By fostering awareness, it encourages a transformative shift in our attitudes, guiding us toward a more harmonious existence.

By manifesting universal benevolence, we can address these distortions in our thinking and behavior, leading to a more peaceful and balanced world.

Ultimately, wishing happiness for everyone is a fundamental and self-evident condition for realizing our full nature and exercising our agency over our happiness.

The symbol of this insight is also part of the *NUSAWVO smile.*

Attitude Cycle

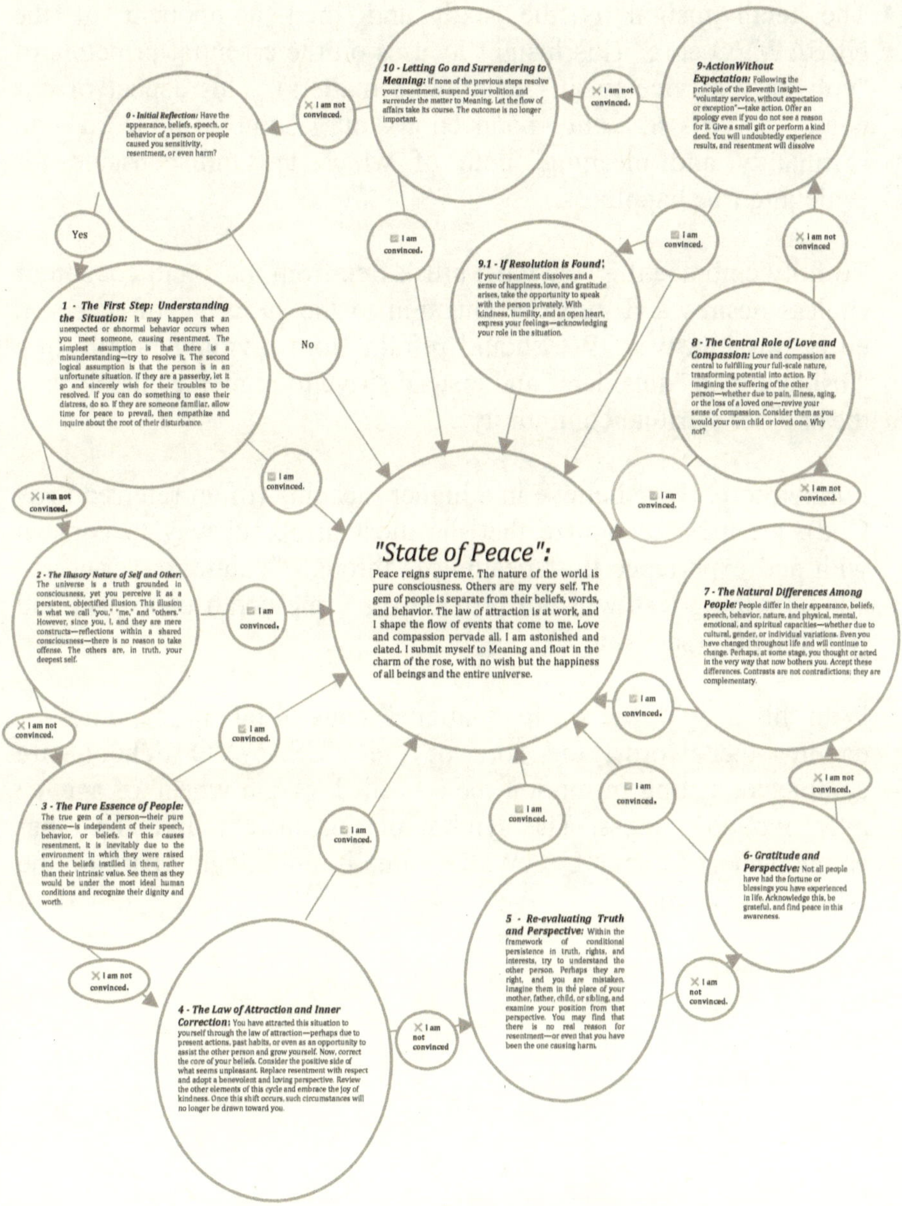

The Tenth Insight: "Voluntary Service to all" (VO)

The tenth insight is the sixth and final component of the NUSAWVO core. This insight focuses on the essential principle of "voluntary service to all" (charitable matters). This aspect of our being serves as our connection—our "umbilical cord"—to spirituality and meaning, both of which are indispensable for achieving true happiness.

To fully embrace this, we must break free from the small cocoon of endless desires and turn our attention to the suffering of others in every possible way. We should refrain from dividing people into "insiders" and "outsiders" and instead strive to immerse ourselves in the broader spiritual community.

Furthermore, if we believe in a higher meaning (often referred to as God), we must recognize that the most profound way to connect with and experience this meaning is through "voluntary service to all." Anything less will not provide the same depth of impact and fulfillment. - (see Note 25)

With this insight, the six foundational gems of human happiness—or, in other words, the core of the NUSAWVO School—are completed, setting in motion the so-called 'magic wheel' of agency over one's happiness. The symbol of this insight, like the other components of the NUSAWVO core, is an integral part of the NUSAWVO smile.

Note 25- *Worship is not in praise or bowing low, But in the service to others, true devotion to show.*
– Saadi

The Eleventh Insight: "Love and Compassion"

The eleventh insight emphasizes the essential concepts of love and compassion. Within the NUSAWVO school, love and compassion— along with the six core elements known as the "NUSAWVO core"— are fundamental for realizing and activating a key principle of the school: the **full-scale nature of human**.

This focus on love and compassion underscores that the NUSAWVO school rejects the use of force, violence, or coercion in its teachings. Furthermore, love and compassion are linked to the **law of attraction**, which interprets the principle of action and reaction. In Hinduism, this is referred to as *Karma*, while in Zoroastrianism, it is known as *Asha* (to some extent). Ultimately, love and compassion represent the emotional and spiritual dimensions, while the law of attraction pertains to the mental and rational aspects. Together, they form a unified principle.

In explaining this insight, it should be noted that the concept of love is simple, and no event in our lives is entirely separate from love. For instance, the reason you and I are communicating right now is rooted in love. If love had been rejected by even one pair of our ancestors— whether going back to our fish ancestors or even earlier, hundreds of millions of years ago—then at least one of us (and most likely both) would not exist.

Love has another facet known as **compassion**. Compassion is closely linked to reason, and when it accompanies love, it brings love into a realm of moderation, gentleness, thoughtfulness, honor, responsibility, and wisdom. Compassion reminds us of the innocence and vulnerability of those who receive love, transforming the magnificent flame of love—which can burn intensely—into a constant, gentle, and noble warmth.

Love and compassion also have another dimension, expressed through the **law of attraction**. This law represents the principle of action and reaction. It suggests that everything that happens to us in life and elicits a response is generally a reflection of our intentions, thoughts, words, and behavior.

The law of attraction functions like a mirror, reflecting reactions that correspond to our actions. This principle operates on the idea that we attract experiences that align with the core of our beliefs, speech, and behavior.

It is crucial to choose the essence of our existence—our core beliefs— carefully and with good intentions. By fostering positive thoughts for ourselves, others, and the entire world (without discrimination or exception), we can ensure that the law of attraction brings happiness and bliss to everyone and the universe as a whole.

In the main discussion of this insight, the power of love, compassion, and the law of attraction will be explored in greater depth.

The symbol of this insight is:

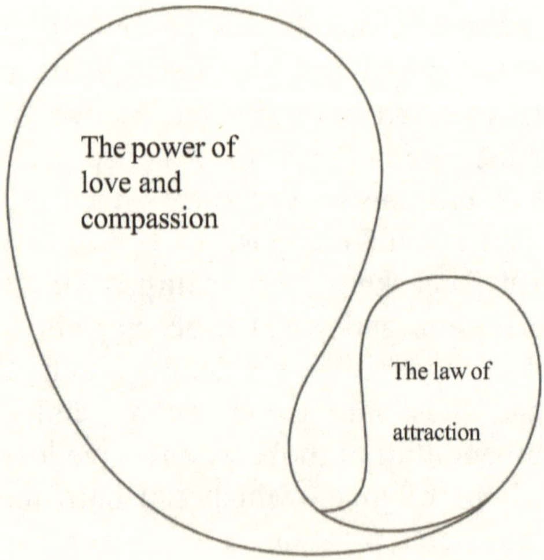

The Twelfth Insight: "Self-Aware Happiness"

From the 4th to the 11th insights, we delved into the theme of **"Agency over Happiness."** This discussion explored our **full-scale nature,** the operational and implementation framework known as the **NUSAWVO core,** and the essential role of **Love and Compassion** as the activating force within this framework.

The twelfth insight transitions to the concept of **self-aware happiness.** Here, we will briefly examine five noble traits that have the potential to elevate our experience of happiness—our bliss—into a state of profound self-awareness.

The First Trait: Thanksgiving (Appreciation, Gratitude, Thankfulness)—The Thanksgiver (T)

Without gratitude and appreciation, a person cannot truly become aware of their happiness or understand its value. In such a state, there is little hope for improving one's circumstances. The NUSAWVO school emphasizes the importance of this noble quality, reminding us of the countless blessings we have received from existence. Even if we may not fully benefit from the 20 elements identified as the **foundations of happiness,** we can still cultivate thankfulness for what we do have and for everything that contributes to our sense of wholeness. Gratitude invites us to acknowledge and honor the abundance around us: being thankful to nature, to the people throughout history, and to the spirit of humanity, whose toil and perseverance over millions of years have paved the way for our comfort, prosperity, security, and modern conveniences. We can express gratitude to **God, to consciousness itself, and to those who have shaped our lives**—our ancestors, families, employers, employees, customers, and countless others. Without this sense of gratitude, we fail to recognize the joy that already exists in our lives (the "half-full" part of the glass) and miss the hope that gratitude inspires for further improvement.

The Second Trait: Hope—The Hopeful (H)

Hope embodies the belief in the possibility of improving circumstances, the anticipation of a better future, and the desire for the continuation of positive outcomes. It recognizes that every crisis or seemingly bitter experience holds the potential for growth and transformation. To truly live is to hope—not only for the best but even for hope itself, for hope is the essence of life.

Life flourishes through hopeful and joyful individuals who, regardless of their physical health or challenges, appreciate the value of their happiness and express gratitude for it. These individuals envision a brighter, more fulfilling future. **Life truly belongs to the hopeful,** and each person, as long as they draw breath, carries the noble responsibility of bearing this radiant torch and passing it on to future generations.

From the perspective of this philosophy, if we view everything as a **manifestation of consciousness**—a force that has shaped our world through persistent imagination—then there is no need for worry or despair. **For those who grasp the profound value and mysteries of life, no crisis represents the end of the road; instead, it becomes a doorway to new possibilities.**

A diagram was once introduced by a wise soul, and to present it here would not lack grace.

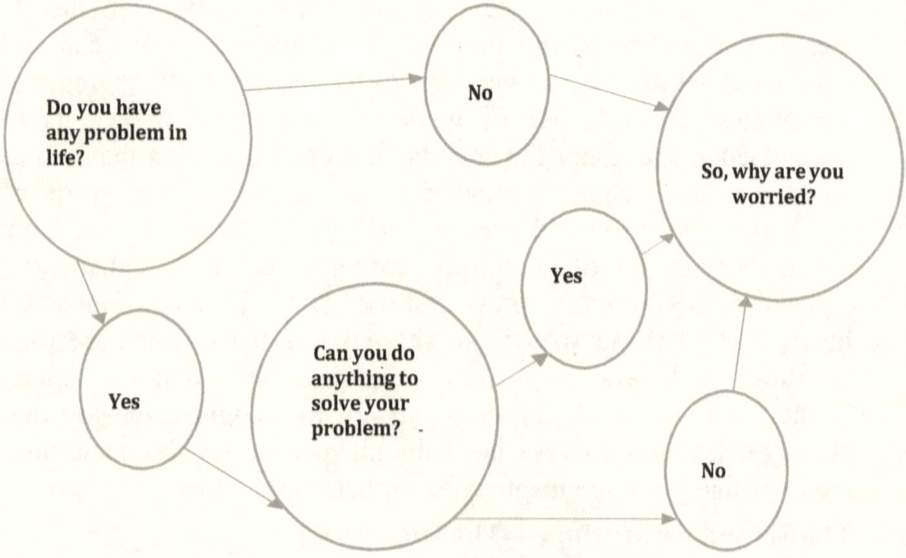

The Third Trait: Astonishment and Elation (A)

Existence itself is awe-inspiring, amazing, and truly astonishing. The very fact of existence—the existence of all things, of ourselves, of our being—is a marvel. Hearing the sound of rain, the ticking of a clock, seeing the shifting shadows, smelling the scent of grass, listening to the birds, feeling the cool breeze of a stream, tasting cherries—these simple experiences remind us of life's wonder. The joy of hugging one another, laughing together, crying together, and the way great music moves us evokes a deep sense of unity and oneness with everything in the world—this is astonishing.

The miracle of existence lies in the awareness that we are here, in this world. We existed when the world was not, and we will continue to exist even when the world ceases to be; yet everything endures. This is deeply astonishing.

Our presence in this world, and the awareness of this presence, is not only astonishing but also blissful and uplifting. It is a natural, spiritual, and pure feeling—a reminder of the profound beauty that surrounds us.

The Fourth Trait: Opportunity Finding and Opportunity Making (O)

A human is both an opportunity finder and an opportunity maker. As far back as we can trace, our ancestors were opportunity finders, even as they journeyed across the oceans as brave single cells billions of years ago. Nothing can truly be deemed a disaster for humans. Every crisis—and what ancient literature often refers to as a "problem"—is, in fact, an opportunity for growth. Without the ability to find opportunities, life could not have persisted, even in its most primitive forms. Our human ancestors would have perished millions of years ago, during the vanishing of the forests, had they not been opportunity finders. Humans are creatures who both find and create opportunities. With the power of reason, imagination, and an unwavering will, a person can transform any situation into a valuable opportunity and leave it for the benefit of others. It is no coincidence that it is said: "For a human, there is no dead end; either they find a way, or they build a way." Entrepreneurship is one of the most prominent manifestations of this trait.

The Fifth Trait: The Seeking of Meaning (Spirituality) – The Seeker of Meaning (S)

Spirituality is one of the four fundamental aspects of our "full-scale nature" and an integral part of our true essence. If it is true that our essence is "consciousness," then we are, by nature, inherently and entirely "spiritual" beings. In other words, there is no separation between us and the rest of existence. This interconnectedness forms the foundation of spirituality within us.

Moreover, our authentic and direct connection with "meaning" (God) represents another vital dimension of our spiritual essence. This connection ensures that no door remains closed as we walk the path of happiness.

It is through the pursuit of meaning and spirituality that our happiness ultimately becomes self-aware. By tearing away the veils of separation, loneliness, and alienation, we transcend feelings of boredom, futility, and confusion. This pursuit brings us closer to the profound harmony that lies at the very heart of existence.

The symbol of this twelfth insight is a flower, widely regarded as one of the most beautiful and serene symbols in the world. It serves as a representation of a self-aware human being. This flower features five petals, each symbolizing one of the five aforementioned noble traits.

This symbol is represented as follows:

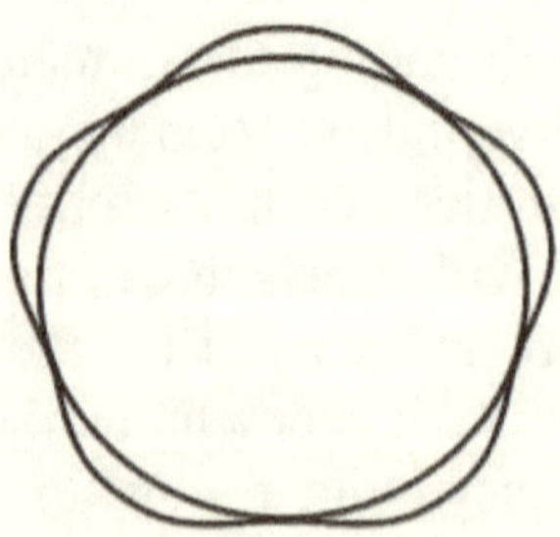

Thirteenth Insight: "Meaning"

The thirteenth insight of this school, much like the first, revolves around the concept of "meaning" (God). However, unlike the first insight, this one introduces no new topics, as all essential themes have already been addressed in the preceding insights. The reappearance of this insight underscores that "meaning" is the foundational pillar of this school of thought, and that's why among the thirteen insights of this school, two are adorned with the name Meaning.

This school begins and ends with "meaning," exploring questions about existence, truth, the full-scale nature of humanity, love, and more. While understanding these concepts is invaluable, it is crucial to acknowledge that they are products of human thought and emotion. As such, they are not immune to error and may lead to uncertain conclusions.

For this reason, the book (or school) both opens and closes with an emphasis on the central belief in "meaning," serving as a reminder to avoid the pitfalls of flawed reasoning. The hope is that this connection to meaning will illuminate and guide us through every stage of life. - (see Note 26)

Lastly, the symmetry between the thirteenth insight—"meaning"—and the number thirteen is seen as a blessing, intended to dispel any superstitions associated with this number. This alignment reinforces the overarching theme of embracing deeper understanding over arbitrary or baseless beliefs.

Note 26- *All that we spoke, but tales of the friend,*
Regrets that linger, never to end."
-Saadi

The symbol of this insight is represented by the circle from the first insight.

And finally, the symbol of the NUSAWVO School will be as follows:

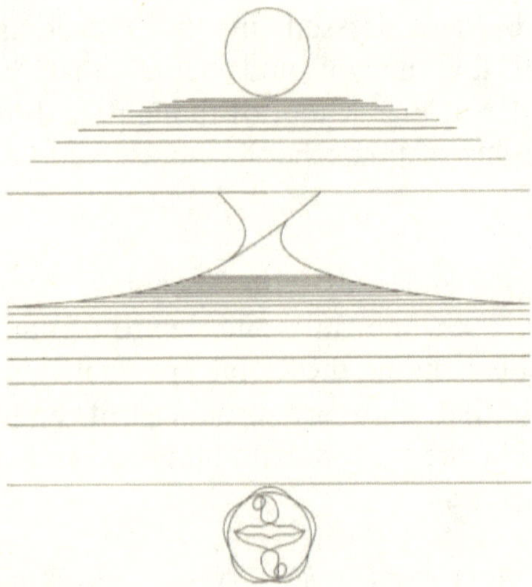

The End of the Extended Abstract

This rather lengthy summary concludes here. We take this moment to humbly pay tribute to the wholeness that, in its various forms and titles—sometimes broad and expansive, sometimes profound and focused—has endowed our existence with continuity and meaning. Among its many gifts, it has granted us the opportunity to write and share this book.

We extend our deepest gratitude to God, our beloved, whose unique and indescribable relationship with us has been the guiding grace and companion on this journey.

We are thankful for existence—or, as we might say, "consciousness"—which, from infinite worlds and combinations of possibilities, has bestowed upon us this chance to manifest as both "intensive" and "extensive" beings of consciousness, to experience life in this extraordinary realm, and to share our dreams with others.

We are grateful to our beautiful planet, our ancient mother, whose generosity has nurtured the birth and evolution of life over billions of years.

We give thanks to humanity, which, through millions of years of relentless effort and boundless hope, has brought us to this era of prosperity and comfort, enabling the exchange of ideas with ease and convenience.

We honor the hardworking and free-thinking human society, whose generosity and magnanimity have created spaces for doubt, free discussion, and the sharing of ideas.

We offer heartfelt thanks to our ancestors, parents, and families, whose sacrifices and contributions have shaped our existence and growth. Without them, none of this would have been possible.

This list of gratitude could extend indefinitely: to our employers, employees, colleagues, friends, neighbors, teachers, and countless others who have enriched our lives.

Finally, we thank you, the reader, for your patience in journeying through these lengthy and sometimes repetitive topics with us.

As we reach the conclusion of this work, we humbly acknowledge that, in essence, we are without substance. We recognize the many mistakes that may reside within these pages and within the framework of this school of thought. We earnestly seek your understanding and forgiveness, reminding ourselves and our readers that no part of this book is beyond error or revision. - (see Note 27)

Note 27- *"A fool who recognizes his own ignorance is, to that extent, wise, but a fool who thinks himself wise is indeed a fool."*
-The Buddha

NUSAWVO has a hymn, as shown below. You can find the full performance under the title "NUSAWVO Hymn" or (Sorood-e Tanjouhpanda) on the World Wide Web. The musical notation for this hymn is as follows:

We wish to see humanity natural,
By the power of love, not by force, not by force. By the power of love, not by force, not by force.

We wish to see humanity pure,
By the power of love, not by force, not by force. By the power of love, not by force, not by force.

We seek the transcendent truth,
By the power of love, not by force, not by force. By the power of love, not by force, not by force.

We long to abide in the truth,
By the power of love, not by force, not by force. By the power of love, not by force, not by force.

We yearn for human happiness,
By the power of love, not by force, not by force. By the power of love, not by force, not by force.

We strive to serve all people,
By the power of love, not by force, not by force. By the power of love, not by force, not by force. By the power of love, not by force, not by force.

We are in love.

NUSAWVO hymn: (Lyrics in Persian)

Sorood-e Tanjouhpanda

2
a re zou da rim pai dar dar ha ghi ghat be ma nim ba ni rou ye eshgh
na ba zour na ba' zour ba ni rou ye eshgh na ba zour na ba zour ma a re
zoo da rim nick bakh ti ye en san ra be bi nim ba ni rou ye eshgh na ba
zour na ba zour ba ni rou ye eshgh na ba zour na ba zour ma a re zou da
rim be khed ma te en san dar a yim ba ni rou ye eshgh
na ba zour na ba zour ba ni rou ye eshgh na ba zour na ba zour
ba ni rou ye esgh na ba zour na ba zour ma a she ghim

The First Insight

"Meaning"

For years, the heart sought a cup to show the world,
Begging from a stranger, the treasure it already held.
A gem beyond the bounds of space and time,
It searched the shore, among the lost, for what was near.

God stood by in every moment,
Yet the heart could not see, and worshiped from afar.

—Hafiz

The world, in awe, agrees with His divinity,
Yet wonders at the depths of His essence.
No bird can soar to the peak of His truth,
Nor can intellect fathom the fullness of His features.

Countless heroes have ridden the path,
Yet, in the end, none have succeeded.

— Saadi

I seek no grace from God through another's art,
Such a bond can tear people apart

– Rumi

If one asks of Him, what can I impart?
The traceless can't be told by a heartless heart

– Saadi

The First Insight: (Meaning)

The world of existence has been described through various themes and concepts such as the universe, cosmos, world, time, space, eternity, pre-existence, being, appearance, consciousness, energy, the circle of existence, truth, reality, nothingness, and illusion. However, the NUSAWVO school of thought posits that beyond these themes and concepts lies a Meaning that transcends the realm of being and existence.

This Meaning exists beyond the reach of human thought and understanding, defining titles, attributes, or actions that might attempt to define it. Yet, paradoxically, it is a profound wonder: people discover and experience it, seek it, love it, turn to it for guidance, take refuge in it, and find peace within it.

One might now ask: Is there any "meaning" behind our speculations and thoughts—something that eludes definition, title, feature, action, or quality? If such a meaning exists, what does it imply? We are confronted with something that does not belong to the world as we know it, something that cannot be grasped through conventional understanding yet can still be experienced. What, then, are we truly discussing? Are we not speaking of a concept that is fictitious, illusory, or even superstitious?

Perhaps the discovery of this "meaning"—and the subsequent experience of it—is the most profound and transformative event in a person's life. However, such an experience is deeply personal, and for this reason, its essence cannot truly be conveyed to others.

This profound and transformative experience—this connection, which could be called the discovery of "meaning"—can arise in any circumstance.

However, it is often said that it manifests most powerfully when one is in deeply unfavorable conditions. When an overwhelming sense of helplessness and despair consumes them—when they stand at the brink of losing their own sense of wholeness—it is precisely at this moment, through accepting their absolute helplessness and surrendering their individuality, - (see Note 28) that a sudden wave of rich, warm emotions unfolds.

***Note 28**-Forget the cleverness, let it slip away,*
For in its grasp, true wisdom cannot stay.
Buy perplexity, let it be your guide,
In mystery's depths, true insight will reside.

Cleverness is ignorance, a fleeting show,
While perplexity reveals what we must know.
In the quiet of the unknown, we find our sight,
For in the dark, we see the truest light.

-They asked Sheikh Abul Hasan: "Where did you see God?"
He replied: "In a place where my ego was no more."

-A man approached Sheikh Mumshad Dinawari and said, "Pray for me."
The Sheikh replied, "Go to God's village, so you need no prayers from Mumshad."
The man asked, "O Sheikh, where is God's village?"
The Sheikh answered, "It is where your ego no longer exists."

-A young man joined the companions of Junaid and, for several days, kept his head bowed in silence, lifting it only for prayer. Then he departed. Junaid sent a disciple after him with a question: "How can a Sufi, described as pure, come to know something that is beyond description? "The disciple caught up with the dervish and asked. The dervish replied: "Become indescribable, and you will find the indescribable. "When Junaid heard this answer, he was overcome by its profundity. For several days, he remained immersed in the greatness of the insight and lamented, "Alas! He was a giant, and I failed to recognize his worth.

Translated from "Tazkerat al-Awliya," a book written by Farid al-Din Attar in Persian.

Trust, love, purity, tenderness, vitality, hope, compassion, and infinite mercy envelop the individual. - (see Note 29)

At this point, a connection to a super-personal source—fundamentally different from anything known before—occurs. This intimate, personal relationship, once experienced, becomes an undeniable reality for the one who has felt it.

This truth transforms their life at every level, guiding them toward a deeper understanding of others and filling their being with humility, purity, and dignity. Most significantly, in the light of this relationship, any form of injustice and violence is perceived as unnatural and unjustified. –(see Note 30)

Note 29- *There are lights in the world that are hidden in the deepest darkness.*
-Gandhi

Note 30- *There is no doubt that ordinary devotional religious faith carries valuable psychological, social, and spiritual functions. However, when we have personally and authentically experienced "meaning," we encounter a pure emotion of unity, love, and compassion—naturally free from any form of hostility or violence. In contrast, if we rely solely on inherited devotional faith, there is a risk of perceiving others in a binary manner: either "with us" or "against us." Under certain conditions and influenced by radical interpretations of faith—labeling others as pagans or outsiders—we may be drawn toward violence and aggression. As the saying goes:* "The person who has strong faith can move a mountain, but if that faith is blind, it may bury the best things beneath it."*It is important to clarify that the NUSAWVO school does not position itself as an alternative to any faith, nor does it seek to challenge or oppose religious belief. Rather, within the framework of this insight, it aims only to illuminate a deeply personal experience—one that can strip away the superficial layers of faith and rejuvenate its essence.*

In truth, the splendor of this love-based, profound experience diminishes—or even erases—the individuality of the person. It suspends their disruptive and unruly will, replacing it with an unlimited, unconditional -(see Note 31) acceptance and an unshakable trust in the "Meaning." This transformation is accompanied by a boundless, all-encompassing love that embraces everything and everyone.

In this process, the allure and glamour of fleeting pleasures fade away, and new perspectives unfold before the individual. A deep desire to serve others awakens within them, and their compassion gains a profound and pure depth.

This process of internal refinement and emancipation transcends a person's background, gently restoring their self-respect and dignity while moving them away from self-centeredness and self-referential views. It imbues their presence in the world with meaning and value, as if granting them a unique responsibility—small though it may be—arising solely from their own being.

Note 31*- Viktor Frankl, the renowned Austrian psychiatrist and Holocaust survivor, addressed the resilience of faith even in the wake of unimaginable atrocities. In his book Man's Search for Ultimate Meaning, he wrote:*

"God is not dead, not even after Auschwitz. For either belief in God is unconditional, or it is not belief at all. If it is unconditional, it will stand and face the fact that six million died in the Nazi Holocaust; if it is not unconditional, then it will fall away if only a single innocent child has to die. Just as a small fire is extinguished by the storm, whereas a large fire is enhanced by it—likewise, a weak faith is weakened by predicament and catastrophes, whereas a strong faith is strengthened by them."

At this point, we may encounter the same confusion as before. On the one hand, those who undergo this experience report vivid, irreplaceable, and deeply transformative emotions—explicitly pleasant and passionately stirring. On the other hand, it is asserted that "meaning" lies beyond our ordinary thoughts, feelings, and perceptions. - (see Note 32) This apparent contradiction might seem perplexing. - (see Note 33)

Yet the truth is that what unfolds in the depths of the experiencer's being is not of the nature of ordinary thoughts or experiences. Rather, it may be a kind of emotional perception—a direct knowledge or a profound connection. - (see Note 34) From this connection, a rich, overflowing relationship emerges, one suffused with love and hope. It is a relationship that brings depth and vitality to the dream we call life.

Note 32*-"He has not allowed minds to define the boundaries of his attributes, yet he has not withheld the ability to perceive Him."*
—Ali Ibn Abi Talib, Nahj al-Balagha

Note 33*-The universe is brimming with contradictions: life and death, being and nonexistence, wave and particle, and more. In the Second Insight, we aim to explore and comprehend these paradoxes.*

Note 34*-It is no coincidence that Christ said that to experience God, one must become like a child again—that is, free from names and signs, returning to the pure experience of a baby and its mother. Similarly, in Hindu culture, the embrace of a couple symbolizes connection and unity with the spirit of the universe. The Upanishads, among the most profound spiritual texts of Hinduism, state:*

"When a man holds his beloved wife in his arms, he becomes unaware of anything within or beyond himself. Likewise, when one is embraced by the conscious soul, they dwell in a state of unawareness of both inner and outer realities."

Nonetheless, despite all these pure, glorious, and jubilant feelings, the NUSAWVO school maintains that, as much as possible (however challenging it may be), one should refrain from attributing verbs, traits, or characteristics to "Meaning." The moment we begin to assign such attributes—even general ones like being a source, a reference, existing, a creator, or superior—we impose definitions and limitations upon it. This is a fundamental error that deprives us of the bounties and blessings of this profound relationship.

Worst of all, whether consciously or unconsciously, people often anthropomorphize "Meaning," projecting their own negative traits onto it. Through self-made interpretations, they not only distance themselves from the blessings of this relationship but also bring harm to themselves and others.

The NUSAWVO school's perspective on "Meaning" is founded upon five key pillars:

1. **Belief in "Meaning"** as a love-based, reciprocal relationship that is free from demands or conditions.
2. **Avoidance of attributing any specific action, attribute, or characteristic** to "Meaning."
3. **Activation of balanced and holistic human natural abilities** through the insights of this school—especially:
    - *The Ninth Insight:* "Wishing happiness for everyone" universally, without exception or discrimination.
    - *The Tenth Insight:* "Voluntary service to all,"- (see Note 35) encompassing universal inclusion, without exception or discrimination and without expectation. This approach naturally enriches the relationship with "Meaning."

Note 35 *– "Worship is not in praise or bowing low, but in the service to others, true devotion to show."*
– Saadi
"If you are not able to serve other people, how can you serve the spirits?"
– Confucius

4. **Complete and unconditional trust in "Meaning,"** involving the suspension of self-centeredness, selfishness, and personal biases (i.e., setting aside the disruptive aspects of personal volition). This includes surrendering the routine course of affairs to "Meaning," seeking its guidance, and taking refuge in it.
5. **Supplications and prayers** (entirely optional) addressed to "Meaning," which may include:
   - Repetition of the *Hymn of 'God"*.
   - A simple ritual of NUSAWVO alongside the repetition of the healing hymn *"O God"* or any other form of optional supplication.
6. (see Note 36)....

In the end, it is evident that the NUSAWVO school does not seek aesthetic exploration in naming "Meaning," particularly when it concerns attributes, features, or verbs. However, in the spirit of sharing a unique experience with all, this school—humbly and wholeheartedly—carries the name *Khoda* (in Persian)- (see Note 37) and *God* (in English) upon the tongue and in the heart.

***Note 36** – At the highest levels, the experience of merging with and living in God—often attributed to saints and perhaps analogous to* Nirvana *in Buddhism—could also be considered part of the NUSAWVO school's principles. However, as such states and levels lie beyond the scope of the capacity of this school, we entrust the exploration and pursuit of these profound truths to dedicated seekers of wisdom.*

***Note 37** – The name* Khoda, *within the rich heritage of the ancient Iranian peoples, has always been associated with peaceful, pleasant, and harmonious thoughts and emotions. It has never invoked as a foundation for violence or enmity and transcends the narrow notions of* disbelief and atheism *or the divisive notion of* with us or against us.

Yet, it is more fitting for seekers of truth to follow their own preferences, as illustrated in the story of Moses and the shepherd. This is a deeply personal matter, shaped by individual taste and experience. We hope that seekers of truth find this approach acceptable and that, in their pursuit of the signs of hope, trust, love, peace, and enlightenment, the distinguished name will manifest in ways that harmonize with the culture, civilization, and language of each community.

"Meaning," and by extension "God," as previously explained, serves as the foundation of the NUSAWVO school of thought. It is both the subject of the first insight and the final insight of this school. Throughout the subsequent insights and appendices of this book, we will revisit this concept, examining it from various perspectives, including those of atheists, agnostics- (see Note 38), and religious extremists.

While the NUSAWVO school does not assign a specific symbol to "Meaning," it does feature a circle—or, more aptly, a ring—at its apex. This ring, akin to a wedding band, symbolizes the reciprocal and loving union between humanity and God. This interpretation and symbolism closely resemble the circle held by the wise elder in the Faravahar symbol, a hallmark of the Zoroastrian tradition.

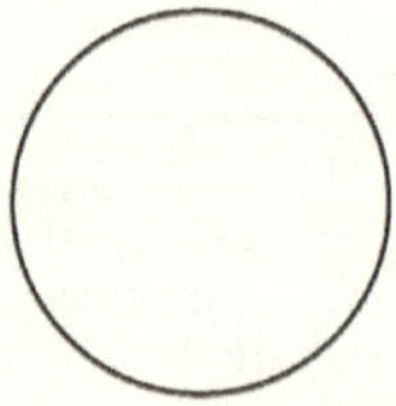

Note 38 – Agnosticism is the epistemological stance that the existence of God is either unknown or inherently unknowable. It reflects a position where a person neither affirms belief in God due to a lack of definitive evidence, nor rejects belief outright, as the truth remains uncertain.

The End of the Topic and the Summary of the First Insight

At the conclusion of this discussion, it is essential to revisit the question posed at the outset. We reflect on the central proposition of this insight: "There is a Meaning hidden beyond existence-being, our thoughts, sensations, and ordinary perceptions—a Meaning to which no title, attribute, characteristic, or action can be ascribed." It seems that despite all the elaborate explanations presented in this insight, this "narrative" remains flawed, self-centered, contradictory, and ultimately unacceptable for several reasons:

1. **First flaw:** The assertion that "there is a meaning hidden beyond existence, being, and beyond our thoughts, feelings, and perceptions" presents an extremely flawed claim. This argument is based on the belief that this school has accurately and intelligently comprehended the concepts of "meaning," "existence and being," and "beyond existence" in every aspect. Consequently, it concludes that there is a "meaning" beyond existence and being (the universe).

   However, it is evident that no one possesses the ability to grasp this with certainty, making this claim exaggerated and unfounded. -(see Note 39)

Note 39 *– Hallaj said: "The person who claims to know him through the truth has actually acknowledged that their own existence is greater than that of the subject of recognition. This is because the one who truly understands something, in that process of knowing, becomes stronger than what they are trying to know. O, so-and-so! There is nothing smaller than a particle in existence, and yet you do not even comprehend that. How can you then grasp the essence of truth, which is undeniably more subtle than a particle?"*

2. **Second flaw:** The statement presents a logical contradiction by suggesting the existence of something beyond the realm of existence that is simultaneously inaccessible to ordinary perception. Logicians argue that anything that can be perceived must exist within the realm of existence. Therefore, if you perceive something, it indicates that it exists within the realm of existence. Conversely, if you cannot perceive it at all, its existence becomes an irrelevant and non-discussable topic. - (see Note 40)
3. **Third flaw:** Even if this "statement," as asserted by this school, is true, it cannot be generalized to all times and all people. For instance, some individuals may reject the concept of "meaning" entirely, while others may perceive it as more concrete, terrestrial, sensible, and perceptible. Some may even embrace the philosophy of the "unity of being," which suggests that everything in the world is merely "meaning." Furthermore, people in the past may have thought differently, and perhaps future superhumans will approach this issue in ways we cannot even imagine today.

Note 40- *This follows Socrates' idea that a true contradiction cannot exist—something cannot both exist and not exist at the same time in the same way. For example, a door cannot be both fully open and fully closed at once. Similarly, if something truly exists, it must, in some way, be perceptible. If it were completely undetectable by any means, then calling it 'existent' would go against the very idea of existence itself. However, some logicians argue for the existence of an independent entity beyond existence itself—one that is still perceivable and which they call 'truth.' Yet, if it is perceivable, does it not, in some way, belong to existence? They insist that this so-called independent 'truth' is merely a construct of logic rather than something real. And so, the paradox unfolds. Welcome to the maze of logic!*

4. **Fourth flaw:** The second part of the discussed statement is even more concerning because it presents contradictory claims. For example, the assertion that "something is beyond our ordinary thoughts, feelings, and perceptions" does not necessarily mean that this thing is devoid of attributes, characteristics, or actions. Even if it is claimed to be beyond ordinary perception, we cannot universally conclude that no one, at any time, could ascribe attributes or actions to it, even if those attributes are the most revered and admirable qualities imaginable. This claim seems excessive and unacceptable, stemming from an illusion of comprehending "meaning" and an assumption of understanding all individuals across all times, as well as other misconceptions.

There is no doubt that the logical and epistemological challenges mentioned above are evident from various perspectives. In a final attempt, we will humbly offer two distinct approaches to address these issues. Ultimately, the choice of the most suitable solution, if any exists, rests with you.

The First Answer

Undoubtedly, "meaning" is knowable because there are no such things as "unknowable" or "inscrutable."- (see Note 41) The very act of discussing "meaning" indicates that some form of knowledge—however limited—has been acquired. However, it is important to recognize that there are different ways of knowing. Rational knowledge and logic are only two of those ways. What might be seen as contradiction or arrogance in this school actually stems from the realm of logical and rational argumentation. Within that context, the objections raised are entirely valid.

This school does not claim that rational and logical knowledge can grasp "meaning." On the contrary, it holds that the best way to understand "meaning" is through what might be metaphorically referred to as "heart knowledge."- (see Note 42)- This form of knowledge does not require logical reasoning or even mystical insight. Heart knowledge does not demand significant effort, especially philosophical sophistication. When the time is right, understanding will naturally come.

Note 41- *In critiquing the concept of the 'unknowable' or 'inscrutable' matter, refer to the views of English philosopher Walter Terence Stace, as discussed in the final section of Appendix No. 2 of this book.*

Note 42- *In philosophical discourse, "heart knowledge" is often contrasted with "head knowledge." While head knowledge pertains to intellectual understanding and rational thought, heart knowledge involves a deeper, more intuitive grasp of concepts, often associated with personal experience and emotional insight. This distinction highlights the multifaceted nature of human understanding, suggesting that some truths are apprehended not solely through logic but through a more holistic, experiential process.*

On the other hand, experience has shown that attempts to logically rationalize the concept of "meaning" and categorize it often led to confusion and frustration. In some cases, this can even unintentionally harm those seeking the truth. It is akin to trying to hold water in your palm: if you grasp it tightly and close your fist with all your strength, the water will slip away, leaving your hand empty. In the process, you might hurt yourself and even startle the friend standing nearby!

To defend ourselves in relation to the discussed statement ("There is a meaning beyond the world..."), we can argue that this approach resembles the metaphor of the finger pointing at the moon. When someone points to the moon with their finger, the focus shifts from the finger to the moon itself. Similarly, the significance of this statement serves primarily as an indication—not something to dwell upon for too long.

It is crucial to acknowledge and emphasize that the mentioned statement lacks a philosophical foundation; instead, it serves as a summary of practical observations from a long journey filled with ups and downs. This process can be outlined as follows:

1. A joyous experience, referred to as "the connection," occurs or is at least perceived in this way.
2. After some time, a fruitless attempt is made to rationally understand that event, resulting only in exhaustion and a sense of aimlessness.
3. Ultimately, by understanding and accepting this statement, the individual finds peace, leading to a renewed relationship with meaning.

The Second Answer

We must admit that the proposed objections are entirely valid, and we have not provided reasonable responses to them. In other words, we must acknowledge that theorizing, making claims, or discussing the meaning and expression of such a statement is both unwise and arrogant. - (see Note 43) Therefore, it is better to refrain from these flawed and sophistical theories and instead humbly remain silent. It is also advisable for you to avoid these unfounded interpretations and focus on more valuable pursuits.

Note 43*- Hallaj addressed one of his students:*
"My son, restrain your heart from contemplating him and your tongue from speaking of his attributes. Instead, dedicate both to unceasing gratitude. For it is a grave sin and an unfathomable arrogance to attempt to fathom his essence, define his attributes, or seek to prove his existence."

-Abdul Wadud bin Saeed bin Abdul Ghani Zahid narrates: "One day, I came into Hallaj's presence and said, 'Guide me to monotheism.' He replied, 'Monotheism lies beyond the realm of words that can describe what God is.' I then asked, 'What, then, is the meaning of "There is no god but God"?' He responded, 'It is a phrase adopted by the common people to distinguish themselves from the people of monotheism. The description of monotheism, as they understand it, is bound by Sharia.' His cheeks flushed, and he continued, 'Shall I summarize it for you?' I said, 'Yes.' He said, 'Whoever speaks of monotheism affirms himself, and whoever speaks of himself falls into polytheism.'

-Ibrahim bin Muhammad Nahrwani recounts:
"I asked Hallaj about monotheism, and he replied: 'Understand that if a servant claims the monotheism of his Lord, he has affirmed himself. And whoever affirms himself has fallen into hidden polytheism.'
Translated from the book "The Martyr of the Truth Path" by Yashar Nouri

The Second Insight

"The Existence, a truth grounded in consciousness appearing objectively as a persistent illusion."

In reflecting upon the nature of existence, we come to understand that consciousness is the very foundation of reality. This insight, although paradoxical in its simplicity, holds a profound implication for our perception of the universe. The following reflections—rooted in both scientific and philosophical discourse—point to the truth that existence is not an independent, objective phenomenon, but is deeply intertwined with consciousness.

Quotations:

1. *"How it is that anything so remarkable as a state of consciousness comes about as a result of irritating nervous tissue is just as unaccountable as the appearance of the djinn when Aladdin rubbed his lamp in the story." – Thomas Henry Huxley*
2. *"The universe is something that happens every once in a while, due to a quantum shift in the vacuum." – Edward Tryon*
3. *"Reality is merely an illusion, albeit a very persistent one." – Albert Einstein*
4. *"The world is made of mind." – Sir Arthur Eddington*
5. *"The world looks more like a big thought than a big machine." – Sir James Jeans*
6. *"This world is a mental image, and the question of existence is but a dream." – Gottfried Wilhelm Leibniz*
7. *"All that we call reality is made of things that cannot be regarded as real." – Niels Bohr*
8. *"What makes you see is not seen; what makes you hear is not heard. What makes thinking, it doesn't think." – Upanishads*
9. *"If you want to find the secrets of the universe, think in terms of energy, frequency, and vibration." – Nikola Tesla*
10. *"The experience of death is nothing but a change of a person's awareness from one holographic realm to another." – Kenneth Ring*
11. *"I regard consciousness as fundamental. I believe that matter is derived from consciousness. We cannot get behind consciousness. Everything that we talk about, everything that we regard as existing, postulates consciousness." – Max Planck*
12. ***Poem by Nouraldin Abdulrehman Jami:***

    "The sophist, blind to wisdom's light,
    Claims the world is fleeting, out of sight,

An illusion, drifting in the breeze,
A transient dream that aims to please.
Yes, the world is shaped by thought,
An image from the mind once caught,
But within this realm, so vast and grand,
Eternity's truth takes its stand.
For behind the veil of transient show,
A timeless reality continues to grow,
In every form, in every hue,
It is the essence, pure and true."

13. *"The apparent material world is a projection of a deeper, non-local level of reality—the implicate order—from which our universe emerges, much like a hologram. Our perception of a separate, material universe is an illusion, and beneath the surface lies a more profound, interconnected reality. Contrary to its materiality and vastness, the universe does not exist independently or for itself; instead, it is part of a much wider and more indescribable reality." – David Boehm*

This second insight draws upon a variety of viewpoints, emphasizing that what we experience as "reality" is deeply intertwined with consciousness. It suggests that existence cannot be fully grasped through the lens of materialism alone, urging us to see beyond the illusion and recognize the underlying unity and interconnectedness of all things.

The Second Insight: The existence, a truth grounded in consciousness appearing objectively as a persistent illusion.

Introduction

What is the origin, essence, and truth of the existence-being, the universe itself?

"Although it appears simple, this question remains one of the greatest mysteries of the human experience."

Why is this question so important? Because within it lie deeper ambiguities, each intricately connected to the truth of our existence and our destiny. In fact, any attempt to answer this question must address three fundamental issues: first, it must uncover and clarify the origin and true essence of the world, as well as the origin and truth of our own existence as an integral part of it.

Second, it must aid in understanding the meaning of life and unraveling the mystery of death. Third, it must provide a logical and solid foundation for answering our most pressing concern: how we should live.

Within the framework of the current insight, we will seek to answer the first topic and lay the groundwork for addressing the subsequent ones.

In this context, we have drawn upon both primary methods of human cognition: 'rational-logical cognition' and 'intuitive-mystical cognition.'

Each of these approaches brings a unique lens through which to explore the truth of existence, and together, they form a holistic understanding that enriches the inquiry.

We have identified three sources - (see Note 44) to guide our exploration. The first and second sources belong to the realm of 'rational-logical' knowledge, while the third draws from the 'intuitive-mystical' domain. The first source encompasses scientific discoveries in fields such as cosmology, paleontology, biology, geology, anthropology, and particularly modern physics (including relativity and quantum mechanics). The second source involves the enlightening discussions found within various philosophical approaches. The third source is rooted in the mystical and intuitive insights of ancient cultures around the world. Our research, conducted through the three mentioned sources, on the foundational question of 'the true nature of the universe,' has ultimately led us to the following hypothesis: *The structure and essence of existence—the universe itself—is a truth grounded in consciousness.* It manifests as a persistent illusion, taking on an objective form as the everyday world we experience. – (see Note 45)

Note 44 *- In this discussion, which represents the second insight, the world is primarily explored from a scientific perspective. However, in the three appendices of this book, we examine this topic in detail from all three perspectives: scientific, philosophical, and mystical.*

Note 45- Consciousness *represents the same fundamental truth that, with some approximations, has been referred to in various traditions by different names: in Hinduism as* ***Brahman****, in Buddhism as* ***Dharmakaya****, among certain Iranian mystics as* ***Jan-e Jahan*** *(the universe itself), in Taoism as* ***Tao****, and among Toltec shamans as* ***Nagual****. While these terms are not direct equivalents, they share overlapping qualities that allow for meaningful dialogue across traditions. It is also conceivable that the concept of the* ***soul****, in its philosophical and religious context, aligns with* ***consciousness****. However, the term "soul" has acquired complex definitions across different religions and schools of thought, often signifying diverse categories. To preserve the sanctity and distinct significance of this term within various traditions, the* ***NUSAWVO*** *school avoids equating it directly with* ***consciousness****.*

If this claim is true, it implies that the familiar world we live in—the entire universe with all its contents, facts, and evolutionary processes, including space-time, events, objects, particles, forces, fields, galaxies, stars, planets, and all their creatures (including human beings with their feelings, emotions, and thoughts)—is part of a much broader truth. This also extends to worlds preceding or succeeding our current one, parallel worlds that may exist, are being created, or will be created in the future, along with all their entities and creatures, including their feelings, emotions, and thoughts. In essence, everything that has existed or potentially exists in the 'world of existence'—all forms of energy in their diverse manifestations—is encompassed within this definition. Their fundamental essence is an infinite, singular, integral, indeterminate, and non-local truth referred to as 'consciousness' or "—"- (see Note 46). According to the nature of each type of world, this consciousness manifests as a 'persistent illusion,' taking an objective form appropriate to that world. For our specific world, with a history spanning 13.8 billion years, this 'persistent illusion' assumes the spatial and temporal characteristics we recognize. It is this objectified form of consciousness that we identify as the world—what we call our everyday reality.

If we tentatively accept the claim—that the true essence of the world is "consciousness," and its objective form is a "persistent illusion"—we are inevitably led to the conclusion of two hypothetical-allegorical realms of existence. These realms can be understood as two sides of the same coin, each offering a different perspective on the nature of reality. Continuing this discussion, we will examine this claim by first defining each of these two allegorical realms in summary. Then, we will analyze and critically assess them through a scientific lens, using more precise categories as we proceed.

Note 46*- During this discussion, we will soon uncover the meaning of "—"*

The Definition of the First Allegorical Realm of Existence - (see Note 47)

The first allegorical realm of the world is the one we experience daily and perceive as our "real world." In this realm, the world appears as a diverse and fragmented collection of phenomena. The universe is composed of scattered objects, components, beings, events, and occurrences—all of which manifest as various forms of energy, such as fields, forces, waves, particles, and so on. These entities undeniably follow causal relationships that align with traditional interpretations, and they are considered to have independent existence outside of our minds. In this realm, our perceptions and sensations are logical, linear, and causal, governed by the mind. It must be stated that within this realm, the reality of things and events is not only non-illusory but is, in fact, the very essence of what is being experienced. Nothing lies hidden behind it. In other words, everything is real, and human beings—including ourselves—are the products of the world's evolution over 13.8 billion years (and, of course, the natural outcome of Earth's evolution over the past 4.6 billion years). Undoubtedly, we are among the most real aspects of this vibrant realm. Furthermore, there is no doubt that our physical, emotional, mental, and spiritual abilities—which are the result of our complex brains and rich human culture—are some of the most remarkable and the most real - (see Note 48) achievements of life's evolution, both biological and social. These abilities, like all aspects of life, are considered to have emerged from the very fabric of this real realm. Therefore, it is clear that no logical and balanced mind can deny the reality of our world and our existence.

---

*Note 47- **In Hinduism**, this realm is referred to as "Maya." **Toltec shamans** describe it as the "Tonal," while **philosophers and mystics** often label it as "appearance," "illusion," the "world of sensations," or "phenomena," among other terms.*

***Note 48- This definition of the 'most real' resonates with Descartes' famous statement, "Cogito, ergo sum" ("I think, therefore I am").** It suggests that the most real aspect of the world is, at the very least, the self—the one who thinks and perceives.*

Now, the only point that needs to be added here is that, despite the established reality of this realm, it is possible that it does not possess independent existence (or, according to some philosophers, true existence). - (see Note 49) Instead, it may be dependent on a deeper truth, often referred to as 'consciousness.' This truth can be allegorically termed the 'second realm.' It is perhaps from this perspective that mystics and many seekers of truth distinguish between the first realm—often called the 'outside' or 'appearance'—and the second realm, which they refer to as the 'inside' or 'being itself.'- (see Note 50)

Note 49- *Just like the eddies, waterfalls, and waves of a river—whose existence depends on the river itself—or the waves of an ocean, the shadows of objects, or even the dreams we experience while sleeping, the first realm may not possess independent existence. Instead, it might be entirely dependent on a deeper truth, often referred to as 'consciousness.'*

Note 50- *David Bohm, one of the greatest theoretical physicists of the 20th century, proposed that the tangible reality of our daily life is, in fact, a kind of illusion or a "holographic world." Beneath this apparent reality lies a deeper, hidden order of existence. Bohm refers to this profound level of reality as the "implicate order," which he also associates with the concept of "consciousness." In contrast, he describes the everyday, perceivable level of existence as the "explicate order." Bohm's idea of a holographic world suggests that every small piece or particle contains all the characteristics and information of the whole. In other words, the entirety of existence is implicated within each part. This perspective extends to the human brain, which Bohm argued possesses a holographic structure, reflecting the same underlying principles. Bohm challenged conventional distinctions between the living and non-living, asserting that such divisions are fundamentally meaningless because animate and inanimate matter are inseparably intertwined. Life, according to Bohm, permeates the entire universe. Even a piece of stone, he suggested, is "alive" in a particular way, as life and intelligence are not confined to living organisms but are present in all matter, energy, space, time, and everything else we have abstracted from the constant movement of reality. He argued that our tendency to perceive these abstractions as separate entities is a misconception. Bohm further posited that the world does not exist by itself or for itself. Instead, it is the expression of something far greater and more indescribable. He viewed the tangible world not as the primary product of this greater reality but as a fleeting shadow, a minor ripple within the vast and dynamic movement of existence. (Inspired from Michael Talbot 'the holographic universe') This perspective invites us to rethink the nature of reality, emphasizing interconnectedness and the profound mystery underlying the universe.*

The Definition of the Second Allegorical Realm of Existence (Basis Realm) - (see Note 51)

In the second allegorical realm, it is possible to experience existence itself in accordance with its essence— "pure consciousness"—which is integral, indeterminate, singular, and beyond time and space. This realm can be referred to as the "inside". It has been interpreted through terms such as "Unity," "Oneness," "the essential connection of all things," "Whole," "Nothingness," "Fullness," "Continuum," "Mind," "Integral," "Implicate Order," "Consciousness," or perhaps even more directly through symbols or silence—represented simply as "——." This silence signifies that the realm defies interpretation through conventional language or symbols.

No form of linear or causal logic governs this realm. Although it can be experienced through perception and knowledge beyond the senses—as countless mystics, seers, and seekers of truth - (see Note 52) have claimed—it is essential to note that such perception does not correspond to everyday, linear understanding. Many describe their experiences implicitly, using terms like "the feeling of cosmic unity" or "the oceanic feeling." Yet, the knowledge of this realm transcends the ordinary framework of thought, making it inherently indescribable. From the perspective of the second realm:

Note 51- *The Chinese, much like the Indians, believe in an ultimate truth that underlies and unites the myriad things and events we perceive. This truth is expressed through three terms: "complete," "all-inclusive," and "whole." Although these terms may appear distinct, they ultimately converge in meaning, pointing toward the same fundamental reality. The Chinese refer to this ultimate truth as "Tao," which literally translates to "the way" or "the path." This concept is drawn from the teachings of Chuang Tzu and further explored in Fritjof Capra's The Tao of Physics.*

Note 52- *Seekers of the second realm are known by various names across different cultures and spiritual traditions: clairvoyant, sage, ascetic, mystic, seer, shaman, Sufi, Zar, guru, yogi, Salek, and many more.*

1. "The one who experiences," or, in philosophical terms, "the subject"
2. "The theme of the experience or knowledge," or "the object"
3. "The audience," or "the one to whom the experience is conveyed"
4. "The conversational tool," i.e., the words used to communicate the experience

All these elements are inseparable and unified. - (see Note 53)

A person who truly experiences the second allegorical realm (or their true self) encounters a profound challenge in expressing that experience. From their perspective, there are no adequate tools (verbal or otherwise) to convey it. Moreover, the experiencer, the object of the experience, and the audience (or questioner) are practically indistinguishable. - (see Note 54)

Note 53 - *"The soul of the soul holds no place, is it strange to say? Yet what the one seeks within our being, where does he stay?"*
— Rumi

Note 54 - *A Zen monk named Doku approached his teacher and asked, "In the search for truth, what mental state must I cultivate to achieve it?" The master replied, "There is no mind to cultivate, no stages to progress through, and no truth for you to attain. "Perplexed, Doku asked, "If there is no mind to train and no truth to find, why do you gather all these monks to study Zen daily and seek guidance from you?" The master responded, "I have no place to gather monks, no tongue to teach them, and no training to offer. How can I guide anyone?" "Confused, Doku asked, "Master, how can you say such things? Are you lying to me?"*
The master replied, "If I have no tongue, how could I speak a lie to you? "Frustrated, Doku said, "I don't understand your words. I don't know what you mean." With a serene smile, the master answered, "I don't understand my own words either."

— Quoted from One Hundred Tales of Zen by Nyogen Senzaki

Thus, the most valid response to the question of what this experience entails is silence. For this reason, the most fitting representation or symbol of this realm is "——." - (see Note 55) Whether it is understood as a subtle definition, a mystery (Koan), - (see Note 56) or an action by the mystic—such as striking the apprentice's back, shouting, - (see Note 57) or even a more intense encounter—these gestures lack concrete meaning in the everyday world (the first realm). However, under special conditions, they may provoke a sudden moment of insight or enlightenment, potentially elevating a person's perception to a higher level.

Note 55 *- To understand or describe anything, something other than it—its opposite—must be present. Imagine writing with black ink on black paper! In such a case, no distinction could be made, and no description would be possible. Now, if we accept that the contents of the world and our mind are one in the second realm, there are no tools of description or analysis available. Words and thoughts cannot exist because each requires at least two components: "Zero and One," "Silence and Sound," "Darkness and Light." As the Upanishads say: "Where the eyes do not function, neither speech nor thought can transcend. We do not know, we do not understand. How can we teach this?"*

Note 56 *- Consider these two Koans:*

- *"What was your original face? Was that the same face you had before you were born from your parents?"*
- *"You can make a sound by clapping your hands together, but what about the sound of one hand?"*

Note 57 *- A monk once approached Bodhidharma, the sage credited with transmitting Zen (Chan) Buddhism to China and said: "My mind is not at peace. Please, calm it for me." Bodhidharma replied: "Bring me your mind, and I will calm it." The monk paused and answered: "But when I search for my mind, I cannot find it." Bodhidharma then said: "There, I have calmed your mind." "Adapted from a Zen kōan in The Gateless Gate (Case 41), attributed to Bodhidharma and his disciple Huike."*

An ambiguity may arise here: If we accept the "existence of the second realm," which, by its very nature, cannot be described or defined, how can this realm be researched, understood, or experienced? Are we not, then, merely discussing a nonsensical idea? In response, it can be argued that such research is indeed possible, and the experience of this realm is, in fact, likely. The second allegorical realm can be explored through two primary methods: the first is the "rational-logical method," and the second is the "intuitive or mystical method." - (see Note 58)

The First Method

The first method, the rational-logical approach, involves conducting scientific, philosophical, and analytical mystical research (distinct from experimental mysticism). In this approach, understanding our true self—"consciousness"—is developed through logic, reasoning, testing, and intellectual processes, including scientific, philosophical, and mystical studies. Since the pursuit of knowledge encompasses all aspects of our world, including both realms, it is inevitable that research in this field must begin with scientific and philosophical knowledge of the first realm (considering its historical context and nature) before extending to scientific, philosophical, and mystical (analytical) exploration of the second realm.

Note 58- *In this context, Fritjof Capra, a renowned American physicist, states: What Eastern mystics regard is a direct experience of the truth that transcends not only rational thought but also emotional perception. Capra further explains that Buddhists refer to the knowledge arising from such experiences as "absolute knowledge," because this knowledge is free from distinctions, abstract separations, and mental classifications, which are always relative and approximate. As the Buddhists teach, this knowledge is a direct experience of truth—one that is without discrimination, classification, or limitation. A full understanding of this truth lies not only at the heart of Eastern mysticism but is also a central feature of every mystical experience. Mystics consistently emphasize that the ultimate truth can never be a subject of cognitive argumentation or proof, nor can it ever be adequately expressed in words, because it lies beyond the realm of the senses and reason. It is a consciousness from which both our expression and understanding emerge.*

Quoted from The Tao of Physics by Fritjof Capra.

The Second Method

The second method involves non-linear cognition, relying on mystical experiences, enlightenment, and intuition. In this approach, the journey begins with meditation, establishing mental silence and experiencing **unaltered states** of consciousness. As one progresses, these experiences deepen through practices such as solitude, seclusion, and extended periods of meditation, - (see Note 59) distancing oneself from the everyday phenomena of life. In these states, the mystic employs various techniques aimed at quieting the mind's internal chatter, slowing the thinking process, and reducing the activity of sensory receptors.

This creates the conditions necessary for accessing deeper layers of the human psyche, including trance-like states that may facilitate indirect encounters with the second realm. In more advanced stages, under the guidance of a mentor and through rigorous exercises, rituals, and prolonged periods of solitude—sometimes lasting decades—the pursuit of experiencing the second realm continues.

The ultimate goal is to reach a direct encounter with this realm. However, these practices are not without risks. Due to the secretive nature of such traditions, they are vulnerable to misuse by fraudulent claimants and incompetent guides. The greatest danger lies in when individuals within such schools are coerced into using substances, which often result in hallucinations and the destruction of both physical and mental health.

Note 59- *Meditation refers to the practice of engaging in activities that quiet the thought process. For instance, one may meditate on the breath or focus on a mantra—a word without meaning—or perform specific body movements done without mental interference, as seen in practices like* Yoga *or* Tai Chi. *In the tradition of Muslim Sufis, meditation takes on various mystical forms, including the repetition of God's name, unique customs and rituals, and sometimes through the special spiritual dance known as* Sama.

The NUSAWVO School's Approach to Exploring the Second Realm (Basis Realm)

To explore the second realm, the NUSAWVO school adopts the first method—the intellectual approach—which includes scientific, philosophical, and analytical mysticism. However, it also embraces certain safe and unassuming modes of meditation. - (see Note 60) These methods are entirely free from the influence of narcotics, and they do not surrender an individual's free will, intellect, or life to any mentor or guide. The NUSAWVO school asserts that the second realm is not separate from our being; rather, it is our true self. Therefore, exhaustive philosophical and mystical efforts are not always necessary to experience it. With proper mental preparation, even reading the introduction of an enlightening book could spark intuition and insight. In an instant, - (see Note 61) the seeker may uncover the secret of the ultimate self, experience oneness with existence, and realize that their true nature is part of the essence of consciousness—beyond time, space, and limitations.

Note 60- *Meditation can take many forms, whether through 'meditation on breathing,' 'meditation on certain words' (mantra), or other techniques. To try a very basic meditation, simply gaze at the center of the smile in the middle of the NUSAWVO symbol for a few minutes with half-closed eyes. Meditate—do not concentrate—without expecting or aiming for anything. (Try to breathe with your belly rather than your chest.) After a few minutes, your mental chatter may naturally subside—provided you are not striving for this outcome. With practice, you will be able to perform this meditation with your eyes closed.*

Note 61- *In the mystical Buddhist-Taoist school of Japan, known as 'Zen,' humor is emphasized because the nature of a joke relies on the sudden perception of humor. A burst of laughter arises only when there is a rapid transfer and immediate understanding of the content of an anecdote. This moment must be instantaneous and cannot be accompanied by the interpretation or rational analysis of the joke. This technique is purposeful and meaningful within the Zen school, facilitating an understanding of enlightenment and the sudden intuition of truth. -In ancient Indian culture, the belief that consciousness (Brahman) is the truth and essence of all things is widely accepted. In this context, it is said: "The one who has not realized Brahman is like a fish that has not found water."*

Further Exploration of the First Realm

An introduction to cosmology, life history, and human history.

Three widely accepted facts in cosmology, which support and confirm each other, aid scientists in understanding the structure and history of our universe. These three facts are: 'the reality of the expansion of the universe,' 'Cosmic Microwave Background Radiation' (the residual radiation from the early stages of the universe), and 'the abundance of elements, such as hydrogen, helium, and trace amounts of lithium, which were formed during the early moments of the universe.'

Based on these facts, nearly all scientists today agree that the universe was not always as it appears. In the past, its components were much closer and more compact. If we trace this back far enough, approximately 13.8 billion years ago, all matter and energy in the universe existed in a hot, dense state, where matter and energy were interchangeable. Due to the infinite nature of the universe, that state of extreme density and energy was not confined to a single point but existed throughout the entire expanse of space-time.

Physicists refer to the moment of the universe's emergence, about 13.8 billion years ago, as the 'Big Bang.' Just as the compression of matter increases its density and temperature, scientists have reached a similar conclusion through calculations and the consideration of all related physical factors. They believe that, at the moment of the universe's birth, the observable universe - (see Note 62) — and even regions beyond it in the infinite expanse—were initially in a state of high-energy density. In that first instant, the temperature was unimaginably high, on the order of billions of billions of degrees. During this time, all forms of energy, - (see Note 63) including the four fundamental forces and fields, as well as particles (or wave-particles), were unified. Only after this initial moment did the universe begin to expand and develop into its distinct forms, with the four fundamental forces - (see Note 64) (gravity, electromagnetism, weak nuclear force, and strong nuclear force) becoming separate.

Some scientists also propose the existence of a fifth fundamental force, though this remains speculative.

Scientists have little to no knowledge about the state of the universe before its emergence, approximately 13.8 billion years ago.

***Note 62- The observable universe** is a crucial topic in cosmology and will be explored in detail in Appendix One of this book. In brief, the 'observable universe' refers to everything that can be detected using the most advanced human-made instruments, such as telescopes and radio telescopes. This vast cosmic sphere contains more than two trillion galaxies, including those similar to the Milky Way, and its number of stars far exceeds the grains of sand on Earth. Its estimated diameter is approximately 27.6 billion light-years. However, as we will discuss in Appendix One, when accounting for the expansion of the universe, this diameter extends to about 93 billion light-years.*

It is important to note that the boundaries of the observable universe do not represent the actual limits of the cosmos. If we were to scale the observable universe down to the size of an electron, the universe beyond it would still be vastly larger. Many cosmologists propose that the universe is truly infinite. This notion is supported by observations indicating that space appears flat in all directions, with no detectable curvature on the largest scales. In fact, it is often more challenging to conceive of a finite universe than an infinite one, as it immediately raises the question: what exists beyond the boundary of a finite cosmos.

***Note 63- Energy and Its Role in the Nature of Reality.** By definition, energy is a property of objects that can be transferred to other objects or transformed into different states and forms, enabling work, phenomena, or effects to occur. In its familiar forms, energy manifests as kinetic energy, heat, potential energy, chemical energy, electrical energy, nuclear energy, radiation, and more. In its denser form, energy appears as conventional matter—a portion of which is well understood, while another portion, including dark matter (discovered only in recent years), remains obscure in its characteristic.*

Energy is quantized, meaning it is transmitted in discrete packets known as quanta. This principle is mathematically represented by Planck's equation:

$$E = h\nu$$

where E is the energy of a quantum, h is Planck's constant, and ν (nu) is the frequency of the corresponding wave.

(continued from Note 63)*-According to the first law of thermodynamics, also known as the law of conservation of energy, energy can neither be created nor destroyed but can be transferred or converted from one form to another. Energy and matter are fundamentally equivalent and interchangeable, as described by Einstein's famous equation:*

$$E=mc^2$$

This relation, tested countless times, demonstrates that the essence of everything in our observable world can ultimately be interpreted as energy. Therefore, the first law of thermodynamics can also be understood as the conservation of both matter and energy. One of the most remarkable applications of this principle is the emergence of material particles from pure energy. This phenomenon occurs in high-energy physics experiments today, where pairs of particles—one with negative charge (electron) and the other with positive charge (positron)—emerge from pure energy. When these pairs annihilate, they convert back into pure energy, highlighting the fundamental reversibility between energy and matter.

Capra's Perspective on Energy and Reality *The renowned Austrian American physicist Fritjof Capra expressed this concept eloquently: "The theory of relativity has shown that mass is not an inherent property of objects or elements, but merely a form of energy. Energy itself is a dynamic quantity, intrinsically linked to action and activity. The fact that a particle's mass is equivalent to a certain amount of energy means that particles cannot be seen as static objects; rather, they must be understood as dynamic patterns—energetic phenomena that take on the appearance of mass."Capra further asserts that in modern physics, classical concepts such as fundamental particles, material objects, and separate, distinguishable entities lose their meaning. Instead, the universe reveals itself as an inseparable, dynamic web of energy patterns. Moreover, the observer is inherently a part of this whole, challenging the traditional distinction between subject and object. In this view, even our familiar notions of space, time, individuality, and causality dissolve, much like the perspectives found in Eastern mystical traditions.*

Whitehead's Philosophical View on Energy. *The English philosopher Alfred North Whitehead also echoed a similar sentiment, stating: "Mass refers to a certain amount of energy, taking into account its dynamic effects." From this interpretation, we see that the fundamental nature of reality, as understood through physics, is energy. However, we may take this idea one step further. Instead of merely defining the universe as "energy," we might propose that its true nature is energy-consciousness—or, to be even more daring, claim that the ultimate essence of the universe is consciousness itself. This profound assertion demands deeper, more meticulous exploration, which we will undertake in the following sections.*

Note 64-The Fundamental Forces. *The fundamental particles of our universe interact in four distinct ways. Each of these interactions is a specific type of force. A region of space where a force exerts its influence is known as a force field. Every particle serves as the source of one or more forces, and any particle capable of generating a particular force field will, in turn, interact with a similar field produced by another particle. Typically, this interaction results in one of two outcomes: either the particles are drawn toward each other (attraction) or repelled (repulsion). The four fundamental forces are as follows:*

- ***Gravitational Force****: This force keeps us anchored to the Earth, governs the motion of planets around stars, and holds celestial bodies together. Without gravity, stars and planets would disintegrate. Gravity acts on all particles with mass and has an infinite range. The theoretical quantum particle associated with gravity is the graviton, although it has yet to be discovered.*
- ***Electromagnetic Force****: Responsible for solar radiation, electricity, and magnetism, this force enables life on Earth and powers modern technology. It causes like charges to repel and opposite charges to attract. Like gravity, the electromagnetic force has an infinite range. Its force-carrying quantum particle is the photon.*
- ***Strong Nuclear Force****: This force binds protons and neutrons together within atomic nuclei, preventing them from flying apart. Without it, atoms would disintegrate, rendering matter unstable. The strong nuclear force has an extremely short range, limited to the nucleus of an atom. Its quantum particle is the gluon.*
- ***Weak Nuclear Force****: This force governs the radioactive decay of unstable atoms. It can either attract or repel depending on specific conditions. Unlike gravity and electromagnetism, its range is confined to the atomic nucleus. The force is mediated by the W and Z bosons.*

These four fundamental forces govern the structure and behavior of the universe at every scale, from subatomic particles to cosmic phenomena.

However, a growing number of physicists, including Steven Weinberg, Edward Tryon, Lawrence Krauss, and Michio Kaku, propose that the universe emerged from a "quantum vacuum" or even from "nothing." Edward Tryon, a renowned American physicist, famously stated: "*The universe is simply one of those things that happens from time to time due to a quantum fluctuation in the vacuum.*" - (see Note 65) Similarly, American cosmologist Alan Guth remarked: *"The universe is the ultimate free lunch!"*

In any case, based on the aforementioned cosmological theories and in light of established principles in geology and biology, the history of our world—within the first allegorical realm of the universe—can be summarized as follows: Approximately 13.8 billion years ago, our universe emerged from a quantum vacuum within an infinite expanse.- (see Note 66) In the very first moment of its existence, supersymmetry, which concealed the quantum vacuum, broke. This symmetry breaking released energy and gave rise to fundamental manifestations, including the four fundamental forces of nature, their associated fields, and subatomic particles (such as electrons, protons, and neutrons). With these elements in place, the universe began its expansion.

Note 65*-It is not assumed that our universe, even within infinite dimensions, is the only existing one. According to the hypothesis known as the "parallel worlds hypothesis" or the "multiverse theory," countless universes may have existed before ours, could exist alongside it, or may emerge afterward—each existing independently, without apparent interaction between them.*

Note 66*-In the context of the origins of the universe, there are differing theories regarding whether space-time was born in an infinite expanse or whether it began as a singular point that subsequently expanded to infinity. One prominent theory, known as* ***cosmic inflation****, supports the latter view, suggesting that the universe rapidly expanded to cosmic scales in a fraction of a second after the Big Bang. Importantly, this inflationary expansion is distinct from the gradual expansion of the universe observed today, which continues at a much slower rate. A key point to consider, regardless of which model we adopt, is the framework provided by* ***Einstein's theory of relativity****, which posits that space and time are fundamentally intertwined. According to this understanding, before the universe's birth, there was no "before" in the traditional sense, because there was no space or time as we know them. Therefore, in either scenario—whether space-time existed in an infinite state or began from a singularity—the "Big Bang" occurred within the context of an infinite, evolving space-time continuum.*

With the passage of time and the cooling of the early universe, within the first few minutes, the first atomic nuclei emerged. This included about 75% hydrogen, 25% helium, and a trace amount of lithium. After approximately 380,000 years, as the temperature dropped to around 3,000 Kelvin, electrons combined with hydrogen, helium, and lithium nuclei to form neutral atoms. At this point, the universe became transparent, allowing photons to travel freely without being scattered by free electrons. This radiation, known as the cosmic microwave background radiation (CMB), was later detected by advanced satellites such as PLANCK, WMAP, and COBE.

A few hundred million years later, gravity caused clouds of hydrogen and helium to collapse and fragment. As the fragments condensed, they gave rise to the first stars. Some of these stars, several times the mass of the Sun, created heavy elements up to iron through nuclear fusion. Others, with much greater mass, underwent nuclear fusion at a faster rate, producing elements up to iron more quickly. These massive stars eventually collapsed under their own gravity, and during the collapse, they fused even heavier elements. The stars then exploded in a spectacular event known as a supernova, dispersing all the elements they had created into interstellar space. These supernova explosions are detectable throughout the universe, including in our own galaxy.

Our Sun and the planets, including Earth, emerged around 4.6 billion years ago from these enriched interstellar clouds. For a long time, due to extreme temperatures and other factors, Earth was inhospitable to life. However, over millions of years, the planet gradually cooled, allowing the formation of oceans. Under the influence of various forms of energy—such as heat, lightning, and volcanic activity—and driven by the chemical tendencies of certain elements, particularly carbon, which exhibit a high affinity for forming complex compounds, the first forms of life emerged. These processes likely took hundreds of millions of years. In fact, the oldest single-celled fossils, found in deep layers of the Earth's crust, are estimated to be about 4.1 billion years old.

After the emergence of the first unicellular organisms, life evolved under the pressures of fundamental forces—self-preservation and self-regeneration—as well as evolutionary mechanisms such as random genetic mutations and natural selection. Multicellularity likely emerged around 1.6–2.1 billion years ago, eventually giving rise to more complex life forms. However, another billion years passed before large organisms colonized dry land, around 500–550 million years ago.

During this period, life diversified into various forms, traditionally classified into five kingdoms: Monera (bacteria), Protista, Fungi, Plantae, and Animalia. Among animals, two major groups emerged: invertebrates (including sponges, corals, worms, mollusks, and arthropods) and vertebrates (including fish, amphibians, reptiles, birds, and mammals). Evolutionarily, vertebrates are more closely related to us.

Humans belong to the mammalian class, which first emerged around 200 million years ago. However, mammals remained relatively small and ecologically constrained until the Cretaceous-Paleogene mass extinction (~66 million years ago), which eliminated the non-avian dinosaurs and allowed mammals to diversify into various ecological niches. Genetic and fossil evidence suggests that humans, bonobos, and chimpanzees share a common ancestor that lived approximately 6–7 million years ago. Over time, evolutionary pressures—shaped by environmental changes and adaptive needs—led our ancestors to diverge from other great apes, setting the stage for human evolution.

Throughout this evolutionary process, numerous hominin species and subspecies emerged, adapted, and eventually disappeared. However, it was around 300,000 years ago that anatomically modern humans (Homo sapiens) first appeared in Africa. During this time, early Homo sapiens coexisted and interbred with other hominin species, such as Neanderthals (Homo neanderthalensis) and Denisovans, contributing to the genetic diversity of modern human populations. Over time, Homo sapiens became the sole surviving human species, expanding across the planet and adapting to diverse environments. Today, all human populations, regardless of geographic or cultural differences, are descendants of this single species.

After exploring the brief history of the universe and life within the first realm of existence—our observable reality—we now turn our attention to examining the records and insights concerning the second realm of existence.

Further Exploration of the Second Realm

As previously mentioned, many physicists propose that our universe originated from a quantum vacuum—a state that, in some interpretations, can be described as emerging from "nothing." Now, after 13.8 billion years since the universe's formation, we seek to understand, from a physical perspective, the current state of this universe that arose from the quantum vacuum and the transformations it has undergone.

A widely accepted view among physicists is that the total balance of the universe's content is effectively zero. But why zero? Consider the following:

1. Total Energy Balance:

- Cosmological calculations suggest that the sum of the universe's matter and energy equals zero. While the total positive energy—encompassing all matter, including stars, planets, galaxies, radiation, and dark matter—is immensely large, the negative energy stored in gravitational fields counterbalances it exactly.
- Applying Einstein's relativity equations, particularly the mass-energy equivalence formula $E = mc^2$ leads to the conclusion that the net energy of the universe is effectively zero.

2. Total Electric Charge:

- The total sum of positive and negative electric charges in the universe appears to be zero, implying a fundamental charge neutrality at a cosmic scale.

3. Total Angular Momentum:

- Observational data suggests that the universe as a whole does not exhibit net rotation—the sum of rotational motion from all cosmic structures, including galaxies, remains balanced at zero.

Thus, we observe a universe that has emerged from the quantum vacuum, where the total net content of matter and energy sums to zero. In other words, a universe that originated from "nothing" in a physical sense and maintains a balance that suggests a profound underlying symmetry.

Quantum - (see Note 67) Physics and the Second Realm

To further explore the second realm, we turn to quantum physics, which provides profound insights into the nature of the universe. The principles of quantum mechanics reveal deep fundamental truths, including:

***Note 67**- Quantum" refers to the smallest discrete unit of a physical quantity that can exist independently. For example, a **photon** is the quantum of **electromagnetic radiation**, meaning it is the smallest indivisible unit of light. More generally, in quantum mechanics, energy, charge, and other physical properties are quantized, meaning they exist in discrete, rather than continuous, values. By analogy, just as a single egg represents the smallest whole unit in the 'world of eggs,' a photon represents the fundamental quantum unit of light.*

1. **Wave-Particle Duality:**
   • Every component of the universe exhibits a dual nature, behaving both as a particle and a wave, depending on the method of observation.

2. **The Wave Function: Superposition and Instantaneous Collapse:**
   • At the quantum level, any single wave-particle exists in a state of superposition—simultaneously occupying multiple potential states—until it interacts with its environment. The moment this interaction occurs, the entire wave-particle collapses instantaneously, regardless of how vast its spread may be—even if it extends across billions of light-years. This collapse defies the limitations of light-speed transmission, occurring as a unified whole without delay.
   • Furthermore, when such particles become entangled and then move apart—no matter how great the distance between them—they remain connected in a way that transcends space and time, without requiring the transfer of information. This is discussed in the next topic (*Quantum Entanglement*).

3. **Quantum Entanglement:**
   • When two or more particles become entangled, their states remain intrinsically linked, regardless of distance. Changes in one particle instantly affect the other, without any information transfer or force application, suggesting a non-local connection that defies classical physics.

4. **Heisenberg's Uncertainty Principle:**
   • At the subatomic level, it is fundamentally impossible to precisely measure certain pairs of physical properties simultaneously—such as a particle's position and momentum. The more accurately one property is measured, the more uncertain the other becomes, reflecting the

fundamental limits of knowledge imposed by quantum mechanics.

5. **Quantization of Physical Properties:**

• At the atomic and subatomic scale, nature is inherently quantized—meaning matter and energy exist in discrete packets (quanta) rather than continuous values.

6. **Quantum Leap: -(see Note 68)**

• Electrons within atoms do not transition gradually between energy levels; instead, they undergo instantaneous quantum jumps, disappearing from one orbital and reappearing in another without traversing the space in between. This phenomenon fundamentally alters the wave function of the system.

In the following sections, we will explore this concept further by examining key cases and interpretations that reinforce this perspective.

Note 68- *A quantum leap refers to the instantaneous transition of an electron between discrete energy levels (orbitals) within an atom. These orbitals are quantized and completely distinct, meaning there is no gradual movement between them. When an electron jumps from a higher to a lower energy level, it emits a photon with a specific energy corresponding to a precise wavelength of light. This produces a unique spectral signature—akin to a barcode—for each element. This principle forms the foundation of* ***spectroscopy****, a fundamental scientific technique that allows the identification of elements not only on Earth but also in the Sun and across the entire universe.*

1. Wave-Particle Duality - (see Note 69)

Wave-particle duality is a foundational principle of quantum physics that reveals the dual nature of all quantum entities. Particles, such as electrons or photons, exhibit both wave-like and particle-like characteristics. As they propagate through space, these quantum entities can behave like waves. However, when they interact with another particle or field—particularly during energy exchanges—their wave function collapses, and their particle nature emerges. In the absence of interactions that cause collapse, quantum entities continue to evolve as probability waves, displaying their inherent wave-like properties.

This duality is universal, meaning it governs every physical and chemical interaction, whether within the fiery core of a star, deep in Earth's oceans, or in the neural networks of the human brain. These interactions unfold within the framework of wave-particle behavior - **(see Note 70)**, described mathematically by a wave function. The wave function dictates how quantum entities behave, interact, and ultimately determine the structure of reality.

***Note 69- Everything is Waves, Not Particles** - In modern physics, the classical notion of solid "particles" is no longer an accurate representation of reality. Instead, everything fundamentally exists as waves. These waves are typically in an extended, spread-out state but can appear localized under certain conditions. The key property of these waves is their **momentum**—which, in relativistic terms, is related to their mass and velocity, or equivalently, their energy. In other words, what we perceive as a "particle" is actually a wave behaving in a specific manner. Regardless of whether this wave is propagating freely through space or appearing localized as an electron, proton, or other subatomic components, it remains a wave at its core. Consider a simple example: when you place your hand on a surface (such as a desk), the electron waves (the wave functions of electrons) of the desk's atoms repel the electron waves (the wave functions of electrons) of your hand's atoms due to the **Pauli exclusion principle** and electrostatic repulsion. This means that what we perceive as "solid" objects making contact is actually the interaction of overlapping energy waves—specifically, negatively charged electron wave functions repelling each other. In reality, the classical idea of physical "matter" as rigid, tangible substance does not exist. Instead, all that exists are fluctuating energy waves interacting according to the laws of quantum mechanics. The sensation of touch you experience is nothing more than an*

(continued from Note 69)- *electrical impulse carried as a wave through the atoms of your nervous system and eventually interpreted in the wave-like atoms of your brain.*

The Wave Nature of Perception: *The process of perception—whether vision, smell, taste, or touch—also follows this wave-based framework.*

Vision: *Light consists of electromagnetic waves (photons), which behave as wave-particles. When photons strike the surface of an object, some are absorbed, and others are reflected. Those reflected photons enter your eyes and are absorbed by photoreceptor cells in your retina. Upon absorption, the photons cease to exist as discrete particles; instead, their energy is transferred to electrons within the retinal cells, exciting them. This excitation propagates as an electrical signal—still fundamentally wave-like—through the neurons of your optic nerve until it reaches the wave-structured neurons of your brain, where it is processed into the experience of sight.*

Hearing, Smell, and Taste: *The process is similar for other senses. Sound waves, chemical molecules carrying odors, and taste compounds all interact with specialized receptors in your sensory organs. These receptors convert incoming stimuli into electrical impulses, which travel as wave signals through your nervous system, eventually reaching your brain. The brain, itself composed of wave-like atomic structures, interprets these signals into conscious perception. Thus, at the most fundamental level,* ***everything we perceive—whether physical objects, sensory experiences, or even thoughts—is composed of waves and vibrations, ultimately reducible to energy.***

The Universe as Energy-Consciousness: *Because everything in the universe is an expression of waves and energy, the fundamental nature of reality can be understood as* ***energy-consciousness****—a dynamic interplay of vibrations that encode and transmit information. As Nikola Tesla, one of the pioneers of electromagnetic theory, famously said:*

"If you want to understand the secrets of the universe, think in terms of energy, vibration, and frequency."

Note 70-Wave Nature of Atoms and Chemical Bonding

If atoms and their subatomic components were not wave-like in nature—if they behaved as classical solid particles governed solely by Newtonian mechanics—then every atomic collision would lead to catastrophic fragmentation. Just as two colliding planetary systems would break apart into smaller celestial bodies, classical "solid" atoms would shatter upon impact, making the formation of stable matter impossible. However, in reality, atoms interact seamlessly and form stable

(continued from Note 70)*-structures through various types of bonding—covalent, ionic (electrovalent), and metallic bonds—which are fundamentally governed by quantum wave mechanics. For instance, in a covalent bond, two atoms partially overlap their electron orbitals, effectively sharing their electrons. Since electrons exist as quantum waves, their wave functions extend into a common space between the atoms. This shared electronic wave function stabilizes the atomic assembly, forming molecules like hydrogen gas (H_2), where two hydrogen atoms bond through overlapping orbitals. In an ionic (electrovalent) bond, an electron wave transitions from the orbital of one atom to another, resulting in charge separation. The atom that loses an electron becomes positively charged (cation), while the atom that gains an electron becomes negatively charged (anion). The resulting electrostatic attraction between these oppositely charged ions stabilizes the bond, as seen in sodium chloride (NaCl, table salt). These remarkable interactions arise primarily from the wave nature of atomic components and secondarily from the inherent self-organizing properties of matter at its most fundamental level.*

High-Energy Collisions and Particle Creation

While atoms naturally bond and interact due to their wave-like behavior, extreme conditions—such as those created in particle colliders—allow scientists to observe how subatomic components behave under immense energy inputs. In high-energy collisions, particles do not merely "shatter" like macroscopic objects; rather, they undergo transformations dictated by quantum field interactions. Inside a particle collider, charged particles (such as protons) are accelerated to near-light speeds and smashed together at enormous energies. Under these extreme conditions:

1. *The colliding particles interact with the energy supplied by the accelerator itself.*
2. *The kinetic energy from the collision is converted into mass-energy, leading to the creation of new particles, as described by Einstein's equation, $E = mc^2$.*
3. *These newly formed particles are not merely fragments of the original ones, but rather emergent manifestations of energy interacting with quantum fields.*

This process reveals a fundamental truth: new particles arise from energy rather than being simple byproducts of disintegration. The ability to generate particles in this way is a core principle of modern quantum field theory (QFT), which describes the behavior of subatomic entities as excitations in fundamental energy fields.

2. The Wave Function: Superposition and Instantaneous Collapse

Each quantum particle is associated with a wave function, which mathematically describes the probability of the particle's position in space. This wave function is inherently non-local, meaning it extends across space, with different regions representing the probability of finding the particle in various locations. However, when the quantum system interacts with another particle or an external field with sufficient energy, a remarkable event takes place:

- **Instantaneous Collapse:** Regardless of the distance over which the wave function spreads, it collapses to a specific point in space where the quantum entity appears as a particle. This collapse happens instantaneously, regardless of distance, and does not require any transfer of energy or information.
- **Non-local Action:** The entire wave function "knows" about the interaction as a whole, which leads to what physicist Albert Einstein famously described as "spooky action at a distance."

This non-local, instantaneous behavior has been experimentally confirmed, including a 2014 study at Griffith University in Australia, demonstrating the collapse of the wave function independent of spatial distance. This phenomenon also underpins quantum entanglement, which we will explore in the next section. Furthermore, Wheeler's thought experiments, which will be discussed further in Appendix 1 of this book, offer additional insights into the nature of this non-locality and the instantaneous collapse, expanding on the foundational ideas presented here.

Consciousness and the Dual Realms of Reality

The wave-particles' ability to respond to their environment, interact intelligently, and instantly recognize changes in their system hints at a deeper, intrinsic intelligence embedded in all matters.

This suggests that consciousness is not confined to complex biological systems but is instead a fundamental aspect of reality itself, manifesting at different levels:

1. **The First Realm (Energy-Consciousness):**

   This realm encompasses the observable domain of quantum entities—subatomic particles, forces, and fields—that interact through wave-particle behavior. These entities constantly process information from their surroundings, behaving in ways that suggest a form of awareness. In this realm, consciousness governs physical phenomena.

2. **The Second Realm (Non-Local Consciousness):**

   This deeper layer of consciousness transcends space and time. It allows wave-particles to recognize their wholeness and interact across vast distances without the need for energy transfer or information exchange. This realm reflects the non-local nature of quantum mechanics, where phenomena occur beyond conventional physical limitations.

These two realms, akin to two sides of the same coin, operate in tandem, driving the emergence, evolution, and continuity of existence.

This framework proposes that consciousness is not merely a byproduct of biological complexity but rather the very essence of reality.

Experimental Evidence

The intelligent behavior of wave-particles has been rigorously tested in laboratories worldwide. One of the most famous demonstrations of wave-particle duality is the double-slit experiment, - (see Note 71) which consistently shows that particles can behave as both waves and particles, depending on how they are observed (measured). A detailed exploration of these groundbreaking experiments will be provided in **Appendix 1**.

Furthermore, the instantaneous collapse of the wave function—occurring independently of spatial distance and the speed of light—was directly demonstrated in a 2014 experiment at Griffith University in Australia, lending further experimental validation to these profound quantum phenomena.

Note 71- *In the double-slit experiment, the wave-particle duality of subatomic particles—including photons, electrons, protons, and neutrons—has been well established. This behavior has also been observed in larger systems, such as macroscopic atoms and molecules, with experiments demonstrating quantum interference in molecules containing up to approximately 800 atoms. However, for even larger molecules, the de Broglie wavelength becomes extremely small, making it practically infeasible to construct a double-slit apparatus with slits of the appropriate dimensions and perform the experiment under controlled conditions.*

3- Quantum Entanglement

Quantum entanglement refers to a phenomenon where two or more subatomic particles become correlated in such a way that their quantum states remain interconnected, even when separated by vast distances. When entangled, these particles share a common **wave function**, meaning they exist as a unified system rather than as independent entities.

A key characteristic of entanglement is its **non-local** and **instantaneous correlation**: regardless of the distance between them—whether a kilometer or millions of light-years—any change in one particle's measured quantum state (such as polarization, spin, or charge) corresponds instantaneously to a complementary change in the other. This occurs without any detectable transfer of information, energy, or force between them, violating classical expectations of locality.

Albert Einstein famously referred to this as **"spooky action at a distance"** because it seemed to imply faster-than-light communication, which contradicted relativity. He and his collaborators, Podolsky and Rosen, proposed the **EPR paradox**, arguing that quantum mechanics must be incomplete and that some hidden variables might predetermine these correlations. However, starting in the late 20th century, **experimental tests—most notably those based on Bell's Theorem—have consistently confirmed the validity of quantum entanglement and ruled out local hidden variables**. These experiments demonstrated that the correlations between entangled particles are genuinely **random**, refuting Einstein's belief that the particles "already knew" their properties in advance.

Implications for Reality and Consciousness

The deeper significance of **spooky action at a distance** is that the only plausible explanation for such an instantaneous connection—independent of space, time, and the speed of light—is that entangled particles **never truly separate**. Instead, they remain part of a **single,**

indivisible quantum system. This suggests that physical reality at its most fundamental level is not composed of independent objects interacting across space but rather of a **continuous, interconnected whole**, akin to a unified field of consciousness.

From a **cosmological perspective**, since all matter in the universe originated from the same quantum state at the beginning of time, it follows that all particles were once entangled. This raises the possibility that the universe itself remains deeply interconnected at a fundamental level.

In the **NUSAWVO school**, this leads to the recognition of **two levels of contact (or two aspects of consciousness) among things**:

1. **The first level** corresponds to **traditional causality and classical reasoning**—where objects interact through force, energy, and information exchange. This is what we allegorically call **"the first realm"** in NUSAWVO thought.
2. **The second level** is a more fundamental **non-local connection**, beyond conventional physical interactions. It does not involve the transmission of force, energy, or information but rather a **pure, instantaneous, and integral unity**. In NUSAWVO thought, we refer to this as **"the second realm" or "basis realm."**

Thus, quantum entanglement not only challenges classical physics but also provides a conceptual bridge between physics and the deeper metaphysical insights of the **NUSAWVO worldview**, pointing toward a universe where **consciousness and reality are fundamentally unified**.

4- The Principle of Uncertainty

As mentioned, The **Uncertainty Principle**, formulated by **Werner Heisenberg**, states that at the **subatomic level**, certain pairs of interrelated properties—such as **position and momentum** (or equivalently, energy and time)—**cannot be simultaneously**

determined with absolute precision. The more precisely one property is measured, the less precisely the other can be known.

This fundamental limit is **not** due to imperfections in measurement devices but is an inherent property of **quantum mechanics**. It arises from the **wave-particle duality** of matter: particles at the quantum level exhibit both **localized particle-like properties** and **distributed wave-like behavior**.

To measure any property of a quantum system, an external influence—such as an applied field or an interaction with another particle—must be introduced. However, this act of measurement **disturbs the system**, causing the **collapse of the wave function** and altering the complementary property. For example, if a particle's **position** is precisely determined, its **momentum (velocity times mass)** becomes fundamentally indeterminate. This is not merely a limitation of technology but a fundamental characteristic of **quantum reality**.

Implications for the Nature of Reality

At its core, the **uncertainty principle** reveals that the **foundation of the universe is intrinsically indeterminate** at the quantum level. The wave-like nature of reality suggests that existence itself is **probabilistic and non-deterministic**, challenging classical notions of a fully predictable, mechanistic universe.

This principle has guided modern **scientific inquiries into the origin of the universe**, as the fundamental **quantum fluctuations** present at the earliest moments of cosmic history may have played a crucial role in the birth of the cosmos.

Beyond physics, this inherent **indeterminacy** suggests that the universe is governed by **a deeper, non-causal truth**—one that is neither bound by classical cause-and-effect nor reducible to mechanical interactions.

In the **NUSAWVO school**, this deeper reality aligns with what we call the **Second Realm of Consciousness**—a level of existence that is not governed by forces, energy transfer, or information exchange but instead reflects a **unified, integral, and non-local reality**. It represents a **pure state of being**, beyond conventional causality, where the fundamental nature of existence is expressed in an interconnected form.

Thus, **the uncertainty principle is not merely a constraint on measurement but a window into the deeper essence of reality itself**—one where consciousness, existence, and quantum phenomena are fundamentally intertwined.

5- Quantization of Matter and Energy

The quantum nature of matter and energy, as described by quantum physics, demonstrates that all physical quantities in the universe exist in discrete "packets" (quanta) at the atomic scale.

Historically, the phenomena of **black body radiation** and the **photoelectric effect** played a crucial role in shaping this understanding. In the late 19th and early 20th centuries, these experimental observations led physicists—most notably **Max Planck** and later **Albert Einstein**—to conclude that energy is not transmitted continuously but rather in discrete amounts. Each of these energy packets is called a **quantum**, with the quantum of light later named the **photon**.

Since energy exhibits wave-like behavior, it possesses a frequency (vibration). The energy of each quantum (E) is given by **Planck's equation**:

$E=h\nu$ where **h** is Planck's constant, and **ν** (nu) is the frequency of the associated wave. This equation also has a wavelength-dependent form:

$E=hc/\lambda$

where **c** is the speed of light, and **λ** (lambda) is the wavelength of the electromagnetic wave.

In 1924, **Louis de Broglie** extended this quantum framework by proposing that matter, like energy, also exhibits wave-like properties. By combining the principles of relativity and quantum physics, he argued that all matter and energy are fundamentally intertwined under **Einstein's mass-energy equivalence**:

$$E = mc^2$$

This led him to formulate the **de Broglie wavelength equation**:

$$\lambda = h/p$$

where **p** is momentum (mass × velocity), and **h** is Planck's constant. This equation confirmed that **all** components of the universe—including atoms, subatomic particles, and energy quanta like photons—exhibit **wave-particle duality**.

Thus, the universe, at its most fundamental level, is quantized and possesses an inherent wave-like nature. The presence of **Planck's constant (h)** in these equations signifies the discreteness of all matter and energy, reinforcing the reality of quantization at the most basic level.

The quantized nature of the physical universe has been experimentally verified and applied across industries, from semiconductor technology to quantum computing, since the early 20th century.

The Question of Consciousness and Quantization

A key question arises: **Is the second realm (Consciousness) also quantized?**

Observations from quantum physics, such as **quantum entanglement** and the **wave-particle duality**, suggest that the behavior of the second realm does not conform to classical notions of force, data, or energy transfer. Since quantization implies discrete, measurable units, the second realm—pure consciousness—does not appear to exhibit such characteristics. Instead, it manifests as **integral, undefined, and impossible to reduce to discrete parts**.

However, when consciousness interacts with the first realm (the physical universe), it appears to take on the properties of energy. In this manifested form—what we might call **Energy-Consciousness**—it does exhibit quantized features. This suggests that while the **manifested** aspect of consciousness may align with the principles of quantum mechanics, its **pure essence** remains beyond quantization.

General Summary of the Discussions on the First and Second Allegorical Realms and Their Relationship

At the outset of this insight, we proposed that the structure and nature of the universe is fundamentally grounded in **consciousness**, which manifests as a persistent illusion in our everyday experience. Allegorically, we referred to this **manifested** reality—our conventional world—as the **"first realm,"** while we identified the **true essence** of existence, i.e., **consciousness**, as the **"second realm"**, or more precisely, the **"basis realm."** To reinforce and elaborate on this initial claim, we will first redefine these two realms and then explore the nature of their relationship through reasoned speculation.

Redefining the First Allegorical Realm

The world we experience—the **first realm**—exhibits a **wave-particle duality**, as described by the **de Broglie relation**:

$\lambda=h/p$

where **λ** (lambda) is the wavelength, **h** is Planck's constant, and **p** is momentum. Ultimately, this realm is composed of **energy**, as also implied by **Einstein's mass-energy equivalence**:

$E = mc^2$

Energy in this realm manifests in various forms, including **force, fields, waves, and wave-particles**. In quantum mechanics, these **wave-particles** (such as electrons and photons) are not merely inert entities; rather, they exhibit behavior that suggests an intrinsic level of **awareness** or responsiveness to their surroundings.

Quantum Consciousness in the First Realm

According to quantum physics, **wave-particles** seem to "know" their own nature and environment, responding in **precise and measured ways**. This is evident in phenomena such as:

- **Quantum superposition** – where the system exists not as a definite particle, but as a **broadly spread wave function** that encapsulates all possible states. Upon **interfering with any other wave-particle and exchanging energy**, including any form of **measurement or detection**, this **wave function collapses**, -instantly- manifesting a definite outcome in a specific location. Throughout this process, the wave function exhibits an **intrinsic awareness** of its own wholeness and boundaries, responding instantly and **non-locally** to any constraints imposed upon it.
- **Quantum entanglement** – where particles maintain an instantaneous connection, regardless of distance, independent of energy or information transfer.

These observations suggest the presence of a **non-local** level of interaction—a form of interconnectedness that transcends conventional time and space constraints. Such behavior implies that at the fundamental level, the **building blocks** of the first realm exhibit properties best described as **"energy-consciousness."**

Redefining the Second Realm (Basis Realm)

Based on the hypotheses presented, we proposed that the true nature of the universe—its fundamental being—belongs to the **second realm**, which we identify as **"consciousness."** Since this realm encompasses everything, including the **first realm**, it is inherently **indescribable**. That is, there is no external reference point from which to define it, as **definition requires contrast with something outside of what is being described**. - (see Note 72)

However, as this description does not provide a practical framework for analyzing the relationship between the two realms, we must attempt to **understand consciousness** at least in its **manifestations** within the first realm. In conventional terms, **consciousness** is often defined as:

"A system's knowledge and awareness of itself (both internally and externally) and its measured response to environmental stimuli."

While this is a broad definition, further refinement is needed to capture the complexity of the subject.

Note 72- *The Dao that can be expressed is not the eternal Dao.*
- Lao Tzu
- "The Dao does not exhaust itself in what is greatest, nor is it ever absent from what is least; and therefore, it is to be found complete and diffused in all things. How wide is its universal comprehension! How deep is its unfathomableness!"
-Zhuangzi

David Chalmers and the Problem of Consciousness

David Chalmers, an Australian philosopher and leading expert on consciousness studies, classifies the **problem of consciousness** into two categories:

1. **The "easy problems"** – These include the cognitive and behavioral functions of awareness, which can be studied using conventional scientific methods.
2. **The "hard problem"** – This is the fundamental question of how and why subjective experience (qualia) arises from physical processes.
   The **hard problem- (see Note 73)** is not just about whether consciousness can emerge from unconscious systems (such as neural activity) but **whether purely physical processes, no matter how complex, can ever explain the subjective experience of being**. Chalmers argues that this mystery remains unresolved and may take centuries to solve.

The Easy Problems of Consciousness

To illustrate the **easy problems**, Chalmers points to various cognitive and behavioral functions associated with awareness, including:

1. The ability to **identify, categorize, and respond** to environmental stimuli.
2. The **integration of information** by a cognitive system.
3. The ability of a system to **access its internal states**.
4. The capacity to **focus attention**.
5. The **measured control of behavior**.
6. **Reportable mental states** (self-reflective awareness).
7. The distinction between **wakefulness and sleep**

Note 73- The ***"hard problem"*** *reflects the same truth we aimed to elucidate in this section—namely, the second insight of the NUSAWVO school—framed under the allegorical title of the Second Realm.*

While these **do not explain subjective experience itself**, they provide a framework for studying consciousness in terms of its observable effects. Depending on which of these aspects we emphasize, **different degrees of consciousness** can be assigned to various systems—ranging from elementary particles like **photons and electrons** to complex beings such as **human minds and the astonishing concept of self-awareness**. - (see Note 74)

Note 74- Self-awareness refers to an entity, system, or organism's ability to recognize itself as distinct from its surrounding environment and other entities. It involves not only the recognition of one's existence but also an awareness of this recognition. According to current research, self-awareness is not exclusive to humans. At least nine other species have demonstrated self-awareness, meaning they recognize themselves as distinct beings and are aware of this awareness. One widely used method for assessing self-awareness is the ***mirror test****. In this experiment, the subject is temporarily anesthetized, and a mark is placed on a part of its body that cannot be seen without a mirror (e.g., the forehead). Upon regaining consciousness, the subject is given access to a mirror. If it uses the mirror to investigate or touch the marked area, this suggests self-recognition and, by extension, self-awareness. Humans typically develop self-awareness around* ***18 months of age****. Other species that have passed the mirror test include* ***gorillas, orangutans, chimpanzees, rhesus macaques, elephants, bottlenose dolphins, killer whales, and magpies****. However, some researchers argue that self-awareness may exist in other forms beyond what the mirror test can measure.*

Refining Our Definition of Consciousness

For our discussion, the **essence of Chalmers' first five criteria** aligns closely with our initial definition:

"A system's knowledge and awareness of itself (both internally and externally) and its measured response to environmental stimuli."

This definition remains **broad enough** to encompass different levels of awareness, yet **precise enough** to differentiate between entities of varying complexity within the **first realm**.

Well! While we have made considerable efforts, we must admit that we have not expanded the concept of awareness (whether from the first or second realm) beyond its general definition. Therefore, we must acknowledge the limits of our findings and, as we proceed, make some assumptions regarding the relationship between the first and second realms, along with some speculative reasoning.

Investigating the Relationship Between the First and Second Realms

Based on the documentation presented in previous discussions of this insight, we can assert that the foundation of our world, within the First Realm, is the composition of "consciousness" and "energy." This suggests a causal (and perhaps supra-causal) relationship between consciousness and energy, which has driven the emergence and evolution of our universe (existence) within the First Realm. The key question now arises: Which is more fundamental—**consciousness** or **energy**? In other words, which one is the origin of the other? It is clear that energy cannot be the origin of consciousness. Energy without consciousness is meaningless and contentless. Even if energy were to exist independently of consciousness, it would still be unconscious and ignorant.

How could an unconscious and ignorant phenomenon produce any action- (see Note 75), let alone give rise to consciousness or a world of such complexity with its fundamental parameters- (see Note 76)—let alone the creation of a single "wave-particle" with such remarkable intelligence- (see Note 77) and abilities? Therefore, energy is not the origin of consciousness. On the contrary, **consciousness** is the origin and source of everything, including energy. Consciousness is a profound and genuine truth, fully aware of its own wholeness- (see Note 78). As noted in quantum physics, it is consciousness that oversees the most fundamental aspect of the universe: the knowledge and awareness of "wave-particles" (the building blocks of our world—photons, electrons, protons, atoms, molecules, etc.).

Note 75- Thomas Henry Huxley*, a renowned English biologist and contemporary of Charles Darwin, as well as one of the prominent advocates of the theory of evolution, once remarked:* "How it is that anything so remarkable as a state of consciousness comes about as a result of irritating nervous tissue is just as unaccountable as the appearance of the djinn when Aladdin rubbed his lamp in the story."

Note 76- Our universe *is governed by six fundamental parameters, commonly referred to as:* ***epsilon (ε), number (N), omega (Ω), lambda (Λ), Q****, and the* ***D number****. Even the slightest deviation in the values of any of these parameters would lead to a very different evolutionary path for the universe. This would significantly affect the development of the cosmos since the Big Bang and, in many cases, would prevent the formation of life as we know it. The fact that our universe operates with these precise parameters is one of the compelling reasons behind the hypothesis of the* ***multiverse****. According to this theory, an infinite number of parallel universes could exist, each with different values for these fundamental constants. Among them, only the universes where these parameters align in a similar way to our own would have the conditions necessary to develop life. In those universes, time could unfold sufficiently to give rise to self-aware intelligence, such as human consciousness. Therefore, we find ourselves living in one of these universes, with the appropriate parameters to support life.*

Note 77- ***Intelligence means****: "The ability to acquire and apply knowledge and skills."*

Note 78- Max Planck*, the renowned German physicist and father of quantum physics, once said:* "I regard consciousness as fundamental. I believe that matter is derived from consciousness. We cannot get behind consciousness. Everything that we talk about, everything that we regard as existing, postulates consciousness."

Consciousness grants them the ability to be aware of their characteristics, integrity, and surroundings, while enabling them to respond intelligently to their environment.

Furthermore, this consciousness exists as an unbroken, whole, and indeterminate truth across all realms, including our universe. It is present in quantum phenomena such as "instantaneous collapse of the wave function" and "quantum entanglement," continuously operating without the need for energy, force, or information transfer. Now, cosmologists' findings about the origin of the universe—based on the **"quantum vacuum"** or **"nothing"**—become meaningful. (Please pay attention: the origin of the universe from **"nothing"**.) Notably, the energy content (matter-energy) of the universe is currently calculated to be **zero**. (Please pay attention: the current balance of the universe's content is equal to **"zero"**.) This paradox suggests that energy and its manifestations (waves, particles, etc.) do not have an inherent essence—or at least not an authentic one. Rather, their existence depends on something beyond them: **pure consciousness**, which, as cosmologists suggest, corresponds to the "quantum void" or, more simply, "nothing." However, this **"quantum void"** is not truly a "vacuum" or "nothing." It is a **pervading, fertile**, and **refreshing** force. This provides the most fitting description of the true nature of the universe—**pure consciousness**. - (see Note 79)

Note 79*-If the simple yet profound idea that* ***"the unity of the world takes the form of truth called consciousness"*** *is true, then we have overcome the long-standing and unsuccessful challenge of the* ***"dual perception of mind and matter."*** *This dualistic framework, formally introduced by* ***René Descartes*** *during the* ***European Enlightenment****, shaped modern thought but also deepened the perceived divide between subjective experience and objective reality. However, the roots of this distinction stretch back to* ***ancient Greek philosophy****, where thinkers like* ***Plato and Aristotle*** *laid the groundwork for later debates on the nature of existence. Descartes, in his quest to establish a secure foundation for knowledge,* ***boldly revived and redefined*** *this dichotomy in an era dominated by medieval scholasticism. His intellectual courage in challenging religious and philosophical traditions placed him at great personal and professional risk, yet it also set the stage for the development of* ***modern science and rationalism****. The* ***NUSAWVO***

Now, let us revisit the entire story of existence—its manifestations and the foundation of our claim—this time through the lens of the **centrality and authority of consciousness**.

From the points briefly mentioned earlier, we find that the fundamental nature of reality is a **truth of the kind of "consciousness."** This **consciousness is beyond time and space**, an indivisible, indefinable, and holistic foundation that underlies everything. All that exists—every entity in the universe—is, in its essence, **consciousness itself**. Humans, too, are **manifestations** of this consciousness. However, the **separation, individuality, and independence** that we perceive in the everyday world—the "**first realm**"—is an **illusion**. Just as **waterfalls, ripples, and whirlpools in a river** appear to have distinct identities yet remain inseparable from the river itself, so is our perception of separateness within consciousness. These illusions are **powerful, tangible, and even consequential**, yet they ultimately depend on the underlying unity of existence.

(Continued from Note 79)- School *does not deny the reality of the everyday world, often referred to as* ***"the first realm of the universe."*** *It fully acknowledges this reality but asserts that its* ***continuity and existence are ultimately sustained by a fundamental truth—consciousness itself.*** *In this view,* ***consciousness is not a byproduct of material interactions but the primary and original essence of all that exists****.*

Thus, NUSAWVO ***transcends the traditional debates*** *between materialism and idealism, as well as the* ***mind-matter dualism*** *that has shaped Western philosophy. Instead, it posits that both mind and matter arise from and are unified in a single, underlying reality:* ***a truth called "consciousness."***

Now, if we attempt—using the **language of the first realm**—to poetically describe the relationship between this realm and the **"second realm"** , including the **emergence of the physical world**, we might propose the following **non-scientific but illustrative assumption**: perhaps **consciousness**, through the **projection- (see Note 80) of what we call "energy,"** has assumed a new state—**"consciousness-energy"**—from which space, time, and the material world have emerged- (see Note 81).

Note 80*- We have borrowed this enigmatic allegorical concept of 'projection' from the mystical teachings of Kabbalah, which describe "creation as an illusory projection of God's transcendent aspects". However, it is important to emphasize that, as outlined in the first 'Insight,' the NUSAWVO school places the concept of 'meaning' (God) beyond the confines of our mental determinations. Instead, it directs its focus on the exploration of "consciousness" as the central subject of its speculations."*

Note 81*- We remain unaware of the various manifestations, combinations, and other states of consciousness. In other words, we do not know whether there are other universes—distinct from our own—that have arisen based on different manifestations and combinations of consciousness. Additionally, we cannot ascertain what the first realm is like in these hypothetical, allegorical realms. Is it capable of generating life with unique characteristics specific to those realms? While some mystics claim to have visited such realms, the validity of these assertions remains uncertain. To explore this further, one can refer to the works of Carlos Castaneda and Paul Twichell.*

In this framework, the birth of the universe, commonly interpreted today as the Big Bang, occurred 13.8 billion years ago—though, from the perspective of pure consciousness, it might still be the same moment today. At the first instant of this event, **the fundamental laws of physics came into existence**, bringing with them **energy, quantum fields, fundamental forces, and the first wave-particle interactions**.

From that moment, the well-documented history of cosmic evolution unfolded:

- The universe expanded and cooled, leading to the formation of **hydrogen clouds**, which later condensed into **stars and galaxies**.
- Over billions of years, **supernova explosions** enriched the cosmos with heavier elements, laying the groundwork for the formation of planetary systems.
- Around **4.6 billion years ago**, our **solar system** emerged, including the **Sun and Earth**.
- Earth's early environment—consisting of **oceans, atmospheric gases, and energy sources like underwater volcanoes**—became a vast **natural laboratory**.
- Within this primordial environment, **chemical evolution** took place over hundreds of millions of years, driven by **organic compounds**, some formed on Earth and others delivered by meteorites.

- The first signs of **life** appeared approximately **4.1 billion years ago**. Over time, through the forces of **evolution- (see Note 82)—random genetic mutations and natural selection**—simple molecular structures gave rise to increasingly complex forms.

- Around **1.5 billion years ago, protocells** emerged.
- Another **billion years later**, around **550 million years ago**, certain life forms evolved mechanisms to survive in dry environments.
- Throughout this evolutionary journey, more complex organisms developed **nervous systems and primitive brains**, allowing for **rudimentary forms of awareness**.
- At the level of **particle-waves, atoms, and molecules**, a primitive kind of intelligence existed from the very beginning, influencing the emergence of self-organizing life.

Finally, with the **continuous expansion of neural complexity**, particularly in **ten key species, including humans**, **awareness evolved into self-awareness**, marking a significant milestone in the unfolding story of consciousness in the first allegorical realm of existence.

Note 82- *As previously mentioned, (if not interpreted as dogmatism), the process of life's evolution is a non-purposeful mechanism, meaning no scientific evidence has been found to support the idea that evolution operates with a specific purpose. When we say evolution is non-purposeful, we do not imply that it is a random, disorganized process. On the contrary, the same intelligence and awareness found in wave-particles— the system's knowledge and awareness of itself (both internally and externally) and its measured response to environmental stimuli— have gradually evolved, giving rise to more advanced forms of life. These life forms carry out vital processes to maintain their essence and ensure reproduction, such as adapting to survive in challenging environmental conditions. Furthermore, the goal of life is not necessarily to become more complex or sophisticated. Whenever survival demands, evolution has even led to reverse processes that simplify organs, as seen in certain evolutionary instances (e.g., intestinal worms, where some body parts were lost during evolution). Essentially,*

***(continued from Note 82)-** the aimlessness of life means there is no predetermined program or ultimate goal, such as the creation of human beings. Life's unfolding was not directed toward a specific end goal, like the emergence of humans. Humans, alongside nine other species (including magpies), that have reached self-awareness, are evolutionary marvels. However, there is no evidence to suggest that a designer exists behind the scenes guiding evolution toward human beings, human intelligence, or even self-awareness. If such a designer did exist, why would it take 13.8 billion years for these processes to unfold? Why wouldn't humans or human-like intelligence appear in a faster or more direct process?*

Consider a simple example: Suppose we travel back 100 million years. Despite the 13.7 billion years since the birth of the universe, there was no trace of humans or even human-like beings. The only mammals at the time were small, mouse-like creatures barely surviving beneath the feet of dinosaurs. If an alien were to arrive on Earth in those conditions, how could they deduce that the purpose of life was human production? It can be imagined that a person's desire to view themselves as the center of the universe—believing that the entire existence and the cosmos were created for their sake—is primarily a result of a false sense of separation from their true essence, which is "consciousness". This desire stems from feelings of loneliness and fear of the surrounding environment. However, by embracing the simple yet profound realization that everything is consciousness, this need for conceptual separation is eliminated.

Moreover, there is a common misinterpretation of the concept of God, which also arises from isolation and the false sense of loneliness. Many attribute to God the task of creating the world and designing it with the purpose of bringing about the existence of isolated human beings, even through evolutionary processes. It is clear that such ideas reflect human-centered thinking, where we project our own desires and expectations onto the concept of divinity. This anthropocentric view effectively "imprisons" God within our individual thoughts and constantly reminds Him of His supposed duties. This way of thinking can prevent individuals from seeking knowledge and truth. It also causes harm to the individual's deep connection with "meaning," as it restricts the true understanding of existence.

Countless books have been written about the real mechanisms of evolution since Charles Darwin's time. One of the most up-to-date and comprehensive works on this subject is The Blind Watchmaker *by Richard Dawkin.*

The journey of human evolution, encompassing the development of our remarkable intelligence and self-awareness, is also inseparable from natural processes. Our emergence as human or proto-human ancestors over the past few million years marks a pivotal point in this evolutionary narrative. Through adaptive changes, we have developed brains more advanced than those of our close relatives—bonobos, chimpanzees, and orangutans. Coupled with the sophisticated cultures we have cultivated, our consciousness not only attains self-awareness but also delves into the mysteries of existence, seeking to understand our origins and the nature of reality.

This evolution of consciousness, along with the intricate tapestry of experiences we call reality, is governed by persistent physical laws. - (see Note 83) It can be likened to a grand feast in which we, as guests, partake—an ephemeral yet profound journey that we navigate, immersed in what we call life- (see Note 84).

Intriguingly, it is conceivable that we manifest in various forms of consciousness across multiple parallel realities, simultaneously engaging in different experiences. Yet, in our present state, we remain unaware of these alternate existences, with the rigid laws that separate our worlds serving as fundamental aspects of this cosmic mystery. To elucidate how humans, as facets of a unified reality called "consciousness," perceive the world, we can turn to the insights of physicist David Bohm. He posited that consciousness is an internal relationship with the whole:

Note 83- *Reality is merely an illusion, albeit a very persistent one.*

- Einstein

Note 84-*O heedless ones, the tangible form is nothing*
And this nine-tiered firmament is also nothing
Be happy, for in the abode of being and corruption
We are bound for a moment, and that too is nothing

-Omar Khayyam

"We are internally related to everything, not [just] externally related. Consciousness is an internal relationship to the whole; we take in the whole, and we act toward the whole. Whatever we have taken in determines basically what we are."

From this perspective, the world—composed of waves, particles, fields, and forces—interacts with our sensory receptors and brain cells, which themselves are part of this fundamental reality. Within the field of our focused consciousness, these interactions are interpreted into sensory experiences such as smells, sounds, tastes, colors, and the perception of physicality, giving rise to the tangible world we experience. As the Kalahari people say: *"There is a dream dreaming us. It is a good dream. The only trouble is we live it badly."*

Neurophysiologist Karl Pribram presented a perspective aligned with this view, suggesting that our perception of reality may be a constructed experience:

"For the first time in three hundred years, science is admitting spiritual values into its explorations."

Both Bohm and Pribram observed that mystical experiences—such as feelings of cosmic unity or oneness with the universe—bear a striking resemblance to the concept of the *implicate order*—in the NUSAWVO school: *The Second Realm*—a deeper, underlying reality. This suggests that mystics may have the ability to transcend ordinary perception—*The First Realm* in NUSAWVO thought—and access a more profound dimension of existence. Gordon Globus, a professor of psychiatry and philosophy, proposed that within this implicate order—*The Second Realm* in NUSAWVO thought—lies an infinity of potential realities. Since the holographic structure that the human brain employs to construct everyday reality is the same mechanism it uses to generate dreams and unaltered states of consciousness (such as mystical experiences), it follows that these three types of reality—waking experience, dreams, and mystical states—are fundamentally interconnected. In a reality where consciousness is fundamental, phenomena such as the mind's influence on random events, telepathy, telekinesis, and levitation can be viewed as minor manipulations within the field of consciousness. Near-death experiences, and perhaps even death itself, may represent shifts within this field. As Kenneth Ring suggested, "The experience of death is nothing but a change of a person's awareness from one holographic realm to another." Don Juan Matus, a Yaqui shaman and mentor to anthropologist Carlos Castaneda, emphasized the fluid nature of our existence:

"We are perceivers. We are an awareness; we are not objects; we have no solidity. We are boundless." Bob Toben, in his book "Space-Time and Beyond," explores the relationship between consciousness and reality, suggesting that our perception is a blend of multiple worlds: "Paths are opening that will lead to the unification of science and spirituality."

Our understanding of the essence and truth of existence—how it impacts us and what it reveals in response to the great question of "how to live"—will be explored in greater depth in future insights of the NUSAWVO School. However, we will briefly outline some key aspects here:

- When you realize that the entirety of existence is manifesting in every word you are reading right now, that eternity has already enveloped us, that each passing moment is eternity itself, and that what we call individuality is but a fleeting dream, an expression, or a reflection of an indivisible, infinite whole—whose truest name is "pure consciousness"—then your personal past, present, and future, the grand story of "your extraordinary and singular self," will begin to seem ironic, even illusory.
- And when you grasp the inherent unity of all things—including your oneness with everything outside of you, from other people to the living and inanimate world—you will arrive at the core of self-awareness and the essence of spirituality. This essence is not abstract but tangible: humility, empathy, loyalty, love, and compassion. To see the world as one is to feel the pain of others as your own and to dedicate yourself to healing it.

- Then, when you perceive that everything in this realm is an exciting game—one that could be harmonious and joyful for all players—you will no longer be bound by the illusions of competition and the relentless pursuit of control. The anxieties born from these attachments—fear of failure, envy, anger, despair—will naturally give way to higher, more luminous states: cooperation, acceptance, contentment, humility, and hope. And because you now understand that you play the central role in this dream, that you alone shape its beauty, your life will be infused with meaning. Instead of drifting in boredom, depression, or aimlessness, you will embrace joy, purposeful action, and unwavering dedication. Solitude will no longer be emptiness but a temple—a space for reading enlightening texts, meditating, and nurturing creativity. It will be a workshop where your artistic expressions and constructive contributions take form, where acts of goodwill and voluntary service become the natural extension of your being.
- Then, when the circle of your existence extends beyond yourself—when you dissolve from a mere wave into the vastness of the ocean—you will see past superficial appearances and social facades. Human relationships will acquire a depth and meaning that transcends fleeting interactions, revealing the unspoken truth that connects all beings.
- And since realizing your true nature grants you an understanding of immortality- (see Note 85), you will come to see that death—whether your own or that of your loved ones—is not an end but a shift in awareness. As Kenneth Ring stated, *"The experience of death is nothing but a change of a person's awareness from one holographic realm to another."*

***Note 85**- When one has realized that which is soundless, intangible, formless, unchanging, tasteless, odorless, unwavering, beginningless, and endless, that which is infinite and perfectly stable, then one is freed from the jaws of death.*
-Katha Upanishad

- In other words, death is merely the suspension—perhaps even voluntary- (see Note 86)—of one's concentrated vigilance and the shedding of all worldly concerns, a return to the timeless ocean of consciousness beyond space and time. Once this deepest mystery is unraveled, your sorrow will no longer be rooted in despair but in reverence, and healing will come more naturally. For you will understand that those who surrender their transient roles—whether as a wave, a rapid, or a waterfall in the great river of being—do not vanish. Rather, they become more present, more authentic—not as a part but as the whole itself. Like the river, ever fresh and ever flowing, they remain- (see Note 87).

Note 86- *Roger Penrose, a renowned English physicist, and Stuart Hameroff, an American anesthesiologist, proposed a hypothesis called "Orchestrated Objective Reduction" (Orch-OR). They suggest that structures within the protein microtubules of brain cells are capable of quantum computations and information processing. According to their hypothesis, during a near-death experience or at the moment of death, the quantum information stored in these microtubules may persist outside the brain, potentially continuing the individual's consciousness independently of the physical body. However, this hypothesis remains unproven and is still under investigation. Furthermore, even if the theory were valid, it does not necessarily imply that individuals would retain a strong desire to preserve their intellectual knowledge from this world after death—especially given the profound transformation such an event might entail.*

Note 87- *Lest these expressions and interpretations, which may seem comforting, be misconstrued in a way that diminishes the dreadful reality of suicide—an act regarded as equivalent to murder in the NUSAWVO school—leading to confusion and destruction. Here, we must issue a stark warning: not only can one misunderstand NUSAWVO's message, but the message itself, if taken at face value, might appear crude, naive, or even foolish. Beware of the raw and unrefined nature of NUSAWVO's words*

- And then, when this truth matures within you into absolute faith—when you recognize that we are truly one, not just in origin but in every moment and every circumstance—you will no longer perceive others as "other." You will not merely treat them as you wish to be treated; rather, the very concept of "otherness" will dissolve. Your mind, yourself, your inner being will no longer be separate from theirs. Not only you, not only humanity, but all living things, the entire universe—it will all be you.
- With this realization, your spiritual understanding will deepen, and your heart will overflow with joy, peace, sincerity, and dignity. You will begin to shine. And in that radiance, you will grasp the profound wisdom embedded in the ancient spiritual traditions of humanity: the law of *Asha* (to some extent) in Zoroastrianism, the law of karma- (see Note 88) in Hinduism and Buddhism, and countless others. These laws are not external decrees but natural truths arising from the unity of all things. Every thought, every action, no matter how small, ripples through the infinite field of pure consciousness—our truest self—and inevitably returns to us. Rumi, in his wisdom, expressed this eternal principle:

Note 88- *In* The Seven Spiritual Laws of Success, *Deepak Chopra defines* karma *as both action and its consequence—both cause and effect. Every action generates a force that inevitably returns to us in kind. The law of karma is not mysterious; it is commonly understood through the saying,* "You reap what you sow." *If we seek happiness in our lives, we must consciously plant the seeds of happiness.*

At its core, karma *implies conscious choice. We are, by nature, unlimited choosers, and in every moment of our existence—within the infinite field of pure potentiality—we have access to an endless array of choices. Some of these choices are made consciously, while others unfold unconsciously. However, the most effective way to understand and harness the law of karma is to cultivate conscious awareness of the choices we make in each moment.*

This world's a mountain, high and steep,
Our deeds the echoes that it keeps.
The calls we send, both dark and bright,
Return to us in sound and light.

- Knowing this, you will awaken to the immense power of your thoughts, intentions, and actions—what modern literature calls the *law of attraction.* You will realize that it is you who shapes your reality. Your environment, your experiences, and your future are not separate from you; they are you. From that moment on, you will make choices with a deeper awareness. And with every decision, you will ask yourself: *Will this choice bring happiness and well-being to myself and all others—without exception?* This single question will illuminate the path ahead, leading you to its simplest answer: cultivate positive thoughts, speak kind words, and perform righteous deeds. It is no coincidence that the prophet Zoroaster distilled his teachings into three guiding principles: *good thoughts, good words, and good deeds.*
- And finally, when your knowledge of the universe merges with an intimate relationship with the Divine, you will find that no problem in this world remains unsolvable. Obstacles will transform into opportunities, and in such a state, nothing but hope, empathy, wonder, love, creativity, and joy will endure. All else—fear, doubt, suffering—will be impermanent, mere shadows against the eternal light of your awakened being.

End of Discussion

From a physical standpoint, Newtonian mechanics—and to some extent, relativity—describe the characteristics of the first realm of existence, the domain of classical, macroscopic reality. In contrast, quantum physics reveals the properties of the second realm, a domain where the conventional laws of space, time, and causality break down, giving rise to phenomena such as superposition, entanglement, and wave-particle duality. The nature of existence, examined through these perspectives, will be discussed in detail within the first appendix, where these three branches of physics are explored in relation to both realms of existence.

From a philosophical perspective, the ideas of idealist philosophers—those who uphold the primacy of consciousness or mind as the fundamental reality—resonate, to varying degrees, with the principles of the NUSAWVO school. These include ancient and modern thinkers such as Parmenides, Socrates, Plato, Descartes, Leibniz, Berkeley, Kant, Hegel, Schelling, Fichte, Schopenhauer, as well as certain Buddhist, Hindu, and Chinese philosophers like Laozi. Many idealists argue that the essence of existence transcends its empirical, everyday manifestation. They often assert that "consciousness" is the true reality, the substratum underlying all appearances. However, certain idealists, such as Spinoza, conceive of "God" as the only absolute reality, distinct from the allegorical realms of existence. Spinoza's monism identifies God with the totality of existence, negating any reality beyond it. The second appendix will explore both realms of existence from various philosophical perspectives. From the viewpoint of mysticism and Sufism, most traditions regard the perceived world—the first realm—as illusory or insubstantial, while affirming the reality of a higher, transcendent domain: the second realm. However, many mystics also emphasize that beyond all realms, nothing truly exists except the divine essence itself. In this view, the world is not separate from God but is rather an "appearance," a "projection," a "radiance," or a "glimpse" of the divine. This perspective will be further examined in Appendix No. 3, where the nature of existence is analyzed from a mystical standpoint.

A Critical Clarification

Before concluding this discussion, an essential point must be emphasized.

While the NUSAWVO school asserts that the true essence of existence is "consciousness," it firmly rejects any interpretation equating "consciousness" with "meaning"—or, in other words, with "God." This distinction is clearly articulated in the foundational vision of NUSAWVO. As emphasized there, "meaning" transcends existence, being, and all allegorical realms of reality. It lies beyond all descriptions, definitions, characteristics, limitations, and attributes—whether positive or negative.

Thus, although NUSAWVO upholds "consciousness" as the fundamental essence of both realms of existence, it unequivocally asserts that "meaning," or "God," is beyond all human interpretations, subjective conceptions, and conceptual boundaries. Any attempt to conflate the two is a fundamental misunderstanding of the NUSAWVO worldview.

The second insight is symbolized by a Möbius surface, as illustrated below:

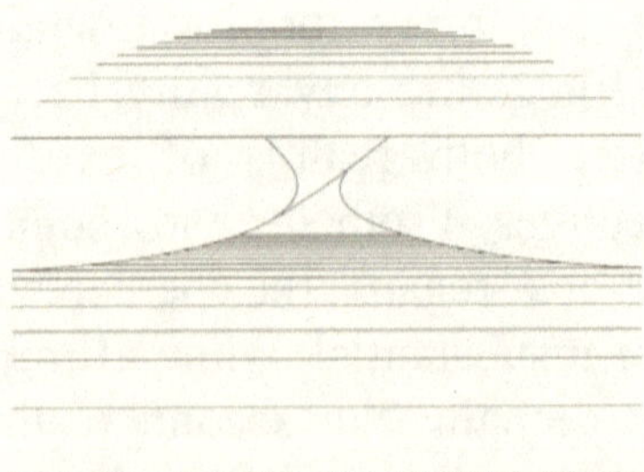

This image consists of two surfaces, representing the first and second realms—or the outside and inside. However, as we know, a Möbius surface has only one continuous side, meaning it lacks a distinct front and back or an inside and outside. By moving along its surface, one inevitably returns to the starting point. - (see Note 89) Thus, choosing the Möbius surface as a symbol of the world serves as an allusion to both the first and second realms, ultimately signifying their unity and inseparability.

The symbol of the second insight, considering the first insight, is depicted in this form:

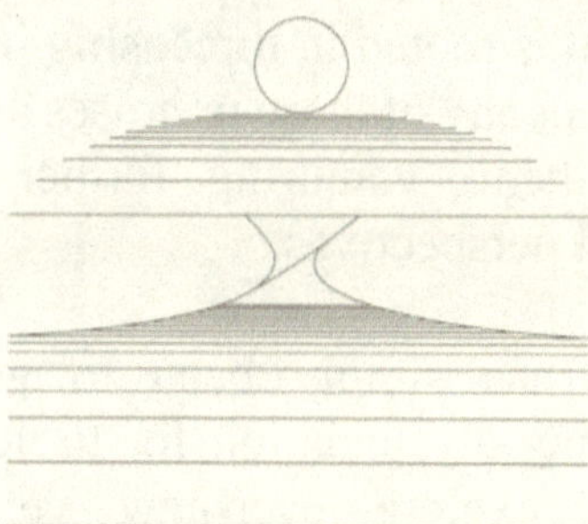

Note 89- *"The outer nature of things reflects their inner nature. The small is equivalent to the large."*
(Hermes Trismegistus)

- *"This is identical to That, and That is identical to This. This and That transcend contradiction, for they are both manifestations of the same existence—the essence of the Tao. Only this existence, at the center of all, gives rise to transformation."*
(Chuang Tzu)

- *"As above, so below; as within, so without."*
(Emerald Tablet, circa 3000 BCE)

Summary of the Second Insight

With this, the second insight comes to a close. Within the framework of this insight, and in our pursuit of understanding the universe—existence itself—we have proposed certain hypotheses, foremost among them the assertion that the essence and ultimate truth of being are of the nature of **consciousness**.

Regardless of the validity of this claim, and setting aside the arguments presented in its support, the very spirit of the **NUSAWVO School** is rooted in **unceasing inquiry**. Therefore, we urge you **not** to accept any of these premises as absolute conclusions but to transcend them—venturing further to cultivate **richer, broader** intellectual perspectives.

Our purpose here has simply been to **gather and organize knowledge**, providing a foundation for tireless seekers of truth—those destined to explore ever-new yet ancient realms of understanding. It is evident that the pristine and boundless landscapes of thought **have no final frontier**. The task now falls upon you—to venture forth with your **brilliant, lucid, and profound** mind, expanding and enriching these insights.

Additionally, we encourage you to refer to the **first, second, and third appendices** of this book, where you may find clarifications to some of your uncertainties.

In the next discussion—the **Third Insight**—we will delve into the **social and biological history of humankind**, unraveling the mysteries that govern **our happiness and bliss**.

The Third Insight

"The Happy Human"

"Happiness is not a possession to be prized, it is a quality of thought, a state of mind."

-Daphne du Maurier.

"We are here on Earth to fart around, and don't let anybody tell you any different."

-Kurt Vonnegut.

"When life is sweet, say thank you and celebrate. And when life is bitter, say thank you and grow."

-Shauna Niequist

"There is no way to happiness; happiness is the way."

-Dr. Wayne Dyer.

"Life is like riding a bicycle. To keep your balance, you must keep moving."

-Albert Einstein.

"It all depends on how we look at things, and not how they are in themselves."

-Carl Gustav Jung.

"Comfort, leisure, and happiness lie in the way we perceive life, not in life itself. It is our perspective that shapes the world we see."

-Masiha Barzegar

"Your situation changes when you change your actions, and your actions change when you transform your beliefs."

-Randy Cage

"To live happily, we must create favorable circumstances rather than wait for them."

-Francis Bacon

"Man is nothing else but what he makes of himself."

-Jean-Paul Sartre

"The world in man found its vast expanse,
And man, a cosmos in mortal dance."

-Mahmoud Shabestari

"I am not what happened to me, I am what I choose to become."

-Carl Gustav Jung

The Third Insight: (Happy Human)

The Third Insight: (Blissed, Self-Aware Agent Human)

Like all **beings** in the world, humans are born happy. Happiness is inherent in life itself—it is the very essence of existence. Every being finds its existence within this essence. Happiness does not require interpretation or justification; life itself is happiness. Life is a miracle, a unique and precious gift. There is no child, no living being, that does not revel in its joy, and no being that can truly reject the wonder of life. Happiness is woven into the fabric of existence.

However, humans are unique among beings—they possess the ability to think, to be aware of their happiness, and to share it with others through the power of intellect and culture. This ability allows humans to make happiness self-aware and enduring.

The third insight of the NUSAWVO school seeks to understand **humanity** and articulate its perspective on human nature. Like other humanist traditions, NUSAWVO recognizes **human beings** in their completeness—as noble, social, intimate, and peaceful, possessing a rich and beautiful nature, and inherently happy.

From this perspective, no definition is more fitting for a person than that of a **"happy human."** Thus, the NUSAWVO school encapsulates its view of humanity in two simple yet profound words: **"Happy Human."** - **(see Note 90)** However, since happiness is a broad and multifaceted concept, we must clarify two key questions: First, what exactly do we mean by happiness? And second, is happiness merely a subjective, fleeting feeling, or does it have an objective foundation tied to external reality? To answer this, as in all inquiries, we must turn to our evolutionary history and seek the roots of our natural happiness there.

Note 90- *Defining human as a 'Happy Human' stands in direct contrast to interpretations that portray humanity as inherently unhappy, burdened by sin, externally driven, or destined to wander without direction.*

A Review of Human Evolutionary History: The Bipedal Hunter

Our pre-human history stretches back millions of years to a time when our ancestors had not yet diverged from the evolutionary path of bonobos and chimpanzees. During this period, they inhabited the dense forests of Africa. It is not without reason that the memory of our **"mother forest"**—with its joy and relative safety—still resides in our collective unconscious. Even today, when we encounter forests or the color green, we instinctively feel a sense of freedom and joy, a deep-rooted connection to our origins.

However, the course of fate shifted around five million years ago. Geological changes led to the drying of the once-lush region of East Africa, leaving a group of our ancestors in a rapidly changing landscape. As the forests gradually receded, vast open plains took their place. Without the shelter and security of the trees, these early beings faced new threats and crises. Now vulnerable to predators, they had to adapt swiftly to survive. As omnivores, much like their closest relatives, they relied on hunting small animals and gathering seeds—resources that were often scarce.

These challenges, coupled with constant environmental dangers, could have easily led to their extinction. (see Note 91)- **Yet genetic diversity played a crucial role.** A fortunate mutation within the genetic pool allowed a small subset of these ancestors to survive, marking a permanent departure from the evolutionary path of bonobos and chimpanzees. - (see Note 92)

Note 91- *It's astonishing, but so far, 99 percent of all species that have ever existed on Earth are now extinct, leaving us as part of the surviving one percent.*

Note 92- *Even today, the genetic difference between humans and chimpanzees is a mere one percent. Humans, bonobos, and chimpanzees share roughly 99 percent of our genetic makeup.*

This genetic advantage brought with it a few key developments. The first was the ability to walk upright—albeit in a form only slightly more efficient than that of chimpanzees and bonobos. Even this small shift proved invaluable, allowing our ancestors to use their hands more freely for defense and tool use. Stones, sticks, and bones became accessible weapons, easier to carry and wield against threats.

Over time, two other critical adaptations emerged. Natural selection favored individuals whose thumbs were positioned more advantageously than their index fingers, granting them a stronger grip for holding objects. - (see Note 93) Additionally, those with feet more structurally similar to modern humans became better suited for running on two legs. These genetic traits enhanced their ability to survive in the harsh conditions of the plains, ensuring that their offspring inherited these beneficial qualities.

Standing on two legs, a hallmark of humanity's evolutionary path, unlocked a new realm of possibilities. The ability to walk upright freed the hands for tool use and complex manipulation. Over time, changes in diet and biomechanics refined skull structure, indirectly facilitating brain expansion. While brain growth was influenced by numerous factors—such as metabolic shifts, social complexity, and environmental challenges—the evolutionary transition to bipedalism played a foundational role in shaping the trajectory of human cognitive development.

Note 93- *If you ever get the chance to examine the hands of bonobos, chimpanzees, or even gorillas, you'll be struck by how closely their hands resemble those of humans.*

The Role of Hunting in Human Evolution

Through the combined forces of selective pressures—driven primarily by the transition to bipedalism and subtle shifts in thumb and big toe positioning—our ancestors' survival capabilities were greatly enhanced. Within this evolutionary framework, any advantageous genetic mutation or social structure that influenced biological evolution was naturally selected and preserved.

Now, let us turn our attention to the open plains. The scarcity of plant-based food sources in this environment made hunting a vital condition for survival. Under these circumstances, any genetic mutation that enhanced cognitive abilities—particularly those aiding in strategy, coordination, and social bonding—was favored by natural selection. These changes not only improved group cohesion and hunting efficiency but also conferred a survival advantage in competition with other predators. Consequently, there is no doubt that hunting played a pivotal role in shaping human intelligence and adaptability. It honed skills such as planning, deception, and tactical reasoning while fostering resilience, courage, ambush tactics, and overall proficiency in the hunt.

Hunting also contributed to the emergence of social structures by necessitating division of labor and cooperation among group members. This, in turn, reinforced male solidarity and reduced interpersonal aggression. The high-risk, cooperative nature of hunting, warfare, and collective defense required long-term alliances, fostering trust and minimizing intra-group hostility—traits that distinguish human social behavior from that of many primates.

Meanwhile, women played a crucial role in early human societies as mothers, gatherers, and later, as the first domesticators of animals. Their cooperative efforts in food gathering and childcare were instrumental to human evolution. The mother-child bond formed the foundation of a stable social network, much like that observed in great apes. The extension of childhood—enabled by maternal care—enhanced offspring survival and strengthened emotional ties between siblings. This suggests that the earliest familial structures likely developed around the mother, eventually leading to the formation of the nuclear family—one of the fundamental pillars of human nature.

It is important to emphasize that evolutionary processes, particularly genetic mutations and natural selection, were the primary forces driving these changes. The environmental pressures of the plains ensured that individuals with advantageous traits had higher chances of survival and reproduction, passing their genes to future generations.

Returning to the significance of hunting, it must be recognized as a decisive factor in human evolution. Beyond reinforcing preexisting abilities, hunting catalyzed the development of new cognitive and social capacities. It not only shaped human interactions with the environment but also fundamentally transformed relationships among humans themselves.

Hunting on the open plains made our ancestors more strategic and perceptive, sharpening their ability to detect subtle environmental cues that could be interpreted as signs or messages. As intelligence and cognitive flexibility increased, early humans learned to outmaneuver not only their prey but also competing carnivores.

Even before humans emerged, omnivorous great apes occasionally hunted small animals. However, with early humans, hunting evolved from an opportunistic behavior into a central, organized activity. Over time, they transitioned from hunting small prey to medium-sized animals and eventually to large, dangerous game. Hunting methods shifted from mere caution and stealth to the strategic use of tracking, ambushes, and traps. Likewise, weapons evolved from crude implements to more specialized and refined tools. This growing mastery over hunting reinforced what would become one of humanity's most crucial strategic abilities: the capacity to plan for the future.

Hunting also accelerated and intensified the dynamic interaction of four key factors: the brain, hands, feet, and tools. This interplay not only influenced human evolution but also exerted evolutionary pressure on hunted animals, shaping predator-prey relationships. As hunting techniques advanced, so did technological innovation in all areas of life, including the construction of more sophisticated shelters. The discovery and control of fire—a revolutionary milestone—further influenced these evolutionary dynamics. Cooking meat made food easier to chew and digest, reducing the need for large teeth and powerful jaws. This, in turn, decreased mechanical pressure on the skull, allowing for changes in cranial structure. Additionally, since cooked food required less energy to digest, humans were no longer constrained by long, carnivore-like sleep cycles for digestion. This increased their active hours, further enhancing their cognitive and social abilities. Current estimates suggest that the control of fire occurred between 1 to 2 million years ago, profoundly shaping the trajectory of human evolution.

The Emergence of Language

Language is an extraordinarily complex phenomenon. - (see Note 94) For our ancestors to develop the ability to speak—even in its most primitive form—the following evolutionary processes likely took place:

Note 94*-Animals communicate with one another and express emotions through a variety of signals that function as messages. These are not limited to vocalizations, such as birdsong, but also include visual cues—gestures, facial expressions, and body postures—as well as chemical communication through pheromones and other scent-based compounds. The evolution of human language is deeply rooted in these fundamental communication methods. Early forms of communication among our ancestors likely relied on gestures, facial expressions, and vocal signals before gradually evolving into structured speech. Even plants engage in communication. They release volatile organic compounds (VOCs) to warn neighboring plants of threats, such as herbivore attacks, prompting them to strengthen their defenses. Some plants also utilize underground fungal networks (mycorrhizal networks) to exchange chemical signals, influencing growth and enhancing resistance to environmental stressors.*

1. **Social and Cognitive Pressures:**
    - The increasing complexity of social structures, cooperative hunting, and division of labor created strong selective pressure for more advanced communication. The ability to produce and understand structured vocalizations became a significant evolutionary advantage, facilitating coordination and information sharing.
2. **Anatomical Adaptations:**
    - Over time, cumulative genetic mutations led to changes in the larynx, pharynx, and vocal tract, enabling early humans to produce a wider range of sounds. Unlike other primates, the human larynx gradually descended, allowing for greater vocal flexibility—though at the cost of increased choking risk.
3. **Neurological Evolution:**
    - Genetic mutations and neural reorganization contributed to the expansion of brain regions responsible for language processing, particularly Broca's and Wernicke's areas. These changes allowed for the cognitive ability to structure and interpret complex sounds, enabling not just speech production but also comprehension and symbolic thinking.

These evolutionary processes were gradual, likely unfolding over hundreds of thousands of years. Language, even in its earliest form, was more than just a tool for communication—it became the foundation of cultural transmission. It allowed early societies to accumulate and pass down knowledge, skills, and traditions across generations.

Current evidence suggests that the cognitive and anatomical prerequisites for language began developing at least 1.5 to 2 million years ago, possibly with Homo erectus. However, more structured forms of proto-language, capable of conveying complex information, likely emerged between 500,000 and 800,000 years ago, particularly with species such as **Homo heidelbergensis**, considered a possible common ancestor of both **Homo sapiens** and Neanderthals.

Culture and Human Evolution

The gradual advancement of collective hunting, food-sharing, and resource transportation led early human societies toward increasingly complex social structures. The success of these activities required a flexible and sophisticated communication system, which likely drove the evolution of phonetic abilities and, eventually, structured language.

Hunting demanded the ability to identify and categorize different animals, landscapes, and environmental cues. This necessity fostered the development of cognitive skills, memory, and symbolic thinking—laying the foundation of early culture. However, this culture did not emerge from a vacuum; it evolved from the social behaviors and learned traditions of our pre-human ancestors and even other primates. - (see Note 95)

Note 95- Culture in Animal and Human Societies: *Culture is not unique to human societies; it manifests in various forms among mammals, primates (especially great apes), and many bird species. In these groups, culture refers to shared behaviors, practices, and knowledge passed down from one generation to another. For most animals, particularly mammals and birds, young individuals must learn essential survival skills to interact successfully with their peers and adapt to their environments. While certain behaviors may be instinctual, the capacity for learning and the transmission of knowledge often resides within the social group's collective memory, not just the individual's genetic instinct. For instance, behaviors like singing (in birds), nesting, group cooperation (such as social hierarchies), hunting strategies, and defense mechanisms are culturally transmitted within groups. In great apes, such as chimpanzees and orangutans, young members observe and learn behaviors related to tool use, social bonding, and problem-solving from older individuals in their group. This learned knowledge is vital for survival and social integration. In humans, these same principles apply, but culture has evolved to be vastly more complex. Human culture encompasses not only practical survival skills but also highly developed*

Given these advancements, the earliest human groups were not mere herds like those of baboons or great apes. Rather, they functioned as primitive societies characterized by cooperative hunting, food division, defense strategies, early familial structures, and shelter-building. None of these organizational complexities could have existed without the emergence of an efficient cultural system, supported by at least a rudimentary form of language.

Culture, once established, became a driver of further complexity. In other words, biological evolution initially gave rise to cognitive and social abilities that allowed for cultural development. However, culture, in turn, began to shape subsequent human evolution. This does not mean that culture replaced genetic evolution; rather, genetic adaptations continued to refine the brain's ability to store, process, and transmit cultural knowledge. From this point onward, culture acted as a "supra-genetic" system—an independent yet biologically anchored force guiding human society.

The Role of Culture in Human Survival

As instinct-driven behaviors gradually receded, culture took on an increasing role in regulating human actions, shaping social norms, technological innovations, and survival strategies.

(continued from Note 95)- *systems of communication, moral values, art, technology, and abstract reasoning. The shift from simple learned behaviors to intricate cognitive and social structures marks a qualitative change in the evolution of culture. Human culture transcends mere survival and becomes deeply intertwined with symbolic thinking and creative expression. Thus, while culture exists in many species, humans have taken it to a new level, developing an expansive and dynamic system that shapes not only individual survival but also societal development and innovation.*

The French anthropologist Edgar Morin - (see Note 96)- aptly noted that culture is so integral to human existence that if infants were raised in isolation, they would lack the ability to form even a simple society—unlike chimpanzees, whose social instincts allow them to function in groups without cultural transmission. - (see Note 97) Ancient human cultures were highly sophisticated. They encompassed not only practical survival knowledge but also technological expertise in toolmaking, strategic hunting techniques such as tracking and trapping, and social customs that governed group interactions. Culture also functioned as a living encyclopedia, preserving and transmitting critical knowledge about the environment, seasonal patterns, medicinal plants, poisons, food preparation, surgical techniques, and child-rearing. Unlike genetic evolution, which operates over long timescales, cultural knowledge could be updated and passed down within a single generation, providing a crucial adaptive advantage. A remarkable feature of culture is its ability to persist beyond its original environment. Early hunter-gatherer societies, despite originating in open plains, could adapt to new habitats—such as forests—while still retaining their complex cultural structures. This flexibility enabled early humans to spread across diverse ecosystems without losing accumulated knowledge.

Note 96*-The comprehensive research conducted by this researcher has impacted several sections of this chapter.*

Note 97*-French scientist Axel Kahn remarked in this context: "Our behaviors and actions are both 100% innate and 100% acquired. If I did not have human genes, I would not possess a human brain, and therefore, no form of cultural training could make me human in the cognitive and behavioral sense. Conversely, if I have human genes and a human brain but am not raised within human culture, the result is the same. This is evident in cases of so-called 'feral children' raised by animals, who, despite their biological humanity, exhibit non-human behavioral patterns due to the absence of cultural influence."*

Culture as a Biological Driver

At a certain stage, culture did more than just reflect human evolution—it began to shape biological change. Cultural pressures favored genetic mutations that enhanced cognitive complexity, particularly in areas related to learning, problem-solving, and communication. When the brain reached a saturation point—where further complexity could not arise through instinct alone—culture created the conditions for advantageous genetic mutations to take hold. This interaction between culture and biology led to:

- Increased cortical neuron density in the brain.
- Stronger neural connectivity between previously independent brain regions.
- The emergence of new cognitive centers supporting advanced reasoning, memory, and abstract thought.

This process helps explain the dramatic increase in brain volume, from approximately 500 cubic centimeters in early hominins to 1,500 cubic centimeters in Neanderthals and Homo sapiens. However, brain evolution was not merely a matter of size—it also involved greater efficiency, connectivity, and plasticity. - (see Note 98)

***Note* 98-** *From an evolutionary perspective, the human brain can be divided into three main regions:*

1. ***The Brainstem (Reptilian Brain)*** *– This ancient structure, inherited from reptiles, regulates basic survival functions such as reproduction, nutrition, territorial defense, and instinctive social behaviors.*
2. ***The Limbic System (Midbrain)*** *– Evolved from early mammals, this region is primarily responsible for emotions, memory processing, and social bonding. It plays a key role in regulating emotional responses and adaptive behaviors.*
3. ***The Neocortex (New Brain)*** *– The most recent evolutionary development, particularly advanced in primates and highly expanded in Homo sapiens. The neocortex, especially the prefrontal cortex, is the center of higher-order cognitive functions such as reasoning, abstract thinking, problem-solving, and language.*

The "Rejuvenation" of Human Intelligence

One of the most profound evolutionary developments in humans is the prolonged period of learning and cognitive flexibility throughout life. Unlike most animals, whose brains reach maturity and then stabilize, the human brain continues to develop, adapt, and restructure itself well into adulthood and even old age. This phenomenon can be called "the rejuvenation of the human species"—the ability to retain youthful cognitive traits such as curiosity, playfulness, emotional sensitivity, and adaptability long past childhood. While other primates stabilize into fixed social and behavioral patterns in adulthood, humans remain lifelong learners, continuously modifying their understanding of the world and transmitting innovations to future generations. - (see Note 99) As a result, knowledge is not just preserved across generations—it is actively expanded, refined, and reinterpreted, fueling an ever-growing cultural legacy. This process ensures that even the elderly, far from being mere knowledge holders, remain active contributors to innovation and adaptation within their societies.

Conclusion

Culture was not merely a product of human evolution—it became an evolutionary force in its own right. By enabling knowledge transmission, fostering cognitive complexity, and even influencing biological traits, culture established a feedback loop that propelled humanity toward greater intelligence, adaptability, and survival success. The ability to pass on knowledge and modify behavior through learning—rather than waiting for genetic evolution—was the ultimate advantage that set humans apart from all other species.

(continued from Note 98)- *In modern humans, the neocortex has expanded significantly, covering most of the brain's surface. This expansion is associated with our capacity for logic, self-awareness, and complex decision-making.*

Note 99-*French scientist Blaise Pascal expressed this idea with the following insight: "All people across all eras should be regarded as a single, ever-evolving human being—one that remains fundamentally the same yet continuously learns and progresses."*

The Evolutionary Role of the Nuclear Family

Unlike birds and many other animal species, long-term pair bonding is not the fundamental social unit among primates and monkeys. In primate societies, particularly among monkeys, stable family structures are generally absent. Where familial bonds do exist, they rarely coexist with broader societal structures. The emergence of the nuclear family alongside the development of human society is a uniquely human phenomenon, playing a crucial role in our biological and social evolution.

It is believed that the formation of the nuclear family followed a process influenced by both biological and cultural changes. With the advent of bipedalism in early human ancestors, face-to-face interaction became physically possible, transforming courtship dynamics.

This shift facilitated deeper emotional and social connections between partners, leading to long-term pair bonding. Over time, evolutionary pressures reinforced the role of sexual attraction and intimacy beyond mere reproductive cycles, a phenomenon also observed in bonobos, our closest relatives in terms of genetics and social behavior. Unlike many primates, humans and bonobos exhibit continuous sexual receptivity, which strengthens long-term pair bonding and enhances social cohesion.

From Pair Bonding to Family Formation

As stable mating relationships became more common, they fostered emotional intimacy between partners. This closeness, in turn, extended to offspring, leading to the psychological recognition of paternal investment before formal hereditary lineage was even established. In early human groups, males began to develop protective and supportive roles toward their offspring—a pattern reinforced by evolutionary advantages, as children with both parental supports had higher survival rates.

The emergence of paternal care thus marked a critical step in the formation of the nuclear family. - (see Note 100) The nuclear family evolved as a father-mother-child unit, but for this structure to persist, parental bonds had to extend beyond childhood. Unlike in most primate species, where offspring often sever ties upon maturity, human children maintain lifelong connections with their parents. This continuity solidified the family institution, ensuring intergenerational support and cooperation, which became foundational for human societies.

The Role of Kinship, Marriage Rules, and Incest Prohibition

The family, however, does not exist in isolation; it interacts with the larger social system. Traditionally, the father belongs to the male peer group, the mother to the female peer group, and the child transitions through different stages of social categorization—from childhood to adulthood. To integrate families into larger societal structures, systems of kinship organization and sexual norms emerged, particularly the prohibition of incest and the practice of exogamy (marrying outside one's immediate family group).

Note 100- *It is possible that, in the early stages of human social development, the responsibility of caring for a sister's children may have initially fallen to the uncle, within the context of incestuous norms. However, as paternal involvement became more prominent, this role gradually shifted to the father, marking a significant step in the evolution of family structures. As a matter of fact, in some societies, including certain Polynesian cultures, the role of the uncle in caring for nieces and nephews can still be significant. This form of kinship structure is often called avuncular care, where the maternal uncle plays a prominent role in the upbringing and protection of the children, especially in the absence or partial presence of the biological father. This pattern is common in various parts of the world, including parts of the Pacific Islands and other indigenous societies, where extended family members, particularly uncles, take on parental roles. The traditional roles in these societies can differ greatly from the nuclear family model, reflecting a more flexible or communal approach to family and child-rearing. In some cases, this structure still exists today, though modern influences can change the dynamic over time.*

There is no doubt that the prohibition of incest and its direct consequence—marriage between members of different groups—played a decisive role in human biological evolution. Unlike many primate societies, where dominant males monopolize reproduction, early human societies developed systems that distributed reproductive opportunities more evenly among males. This shift created a genetic and social balance, increasing genetic diversity while stabilizing social structures.

Preventing Speciation and Strengthening Human Evolution

One of the most profound effects of these marriage rules was their role in preventing the fragmentation of the human species. In most animal populations, reproductive isolation can lead to the emergence of new species over time. However, by enforcing exogamy and preventing inbreeding, early human societies maintained a genetically unified species, even as they spread across the globe. This effect became especially crucial following the emergence of *Homo sapiens* approximately 300,000 years ago, as it helped sustain a single, interbreeding human lineage rather than divergent subspecies. Furthermore, the nuclear family and its sexual regulations had a direct impact on the spread of genetic mutations. In species where dominant males control harems, genetic traits may remain confined to small groups.

However, in human societies where men were prohibited from mating with their daughters and had to seek partners from other groups, genetic variations were more widely distributed. This mechanism accelerated genetic adaptation and played a fundamental role in shaping the modern human genome.

The Nuclear Family as an Evolutionary Catalyst

Thus, the emergence of the nuclear family, along with rules regulating sexual relationships, had far-reaching evolutionary consequences:

- It prevented the division of the human species, maintaining a single interbreeding lineage.
- It increased genetic diversity, promoting adaptability.
- It provided emotional stability by balancing gender dynamics and reinforcing long-term pair bonding.
- It prolonged childhood and adolescence, allowing for extended learning and cognitive development, which contributed to human intelligence.
- It facilitated human expansion across the globe, ensuring social cohesion in increasingly complex societies.

In this way, the nuclear family was not just a social construct—it was a biological and evolutionary milestone that played an unparalleled role in the emergence of modern humans.

Human Concern – Attention to the Future – Spirituality

Worry originates from the primal instinct of fear, a phenomenon not unique to humans but found across the animal kingdom. Fear serves as an adaptive response to perceived threats, heightening awareness and triggering protective behaviors. However, in humans, fear has evolved beyond immediate survival concerns, encompassing more abstract and existential anxieties. As human societies developed culture, they gained the ability to mitigate many natural threats, which reduced concerns about environmental dangers. Yet, this shift gave rise to a new set of anxieties, grounded in self-awareness and cognitive advancement. As human consciousness expanded, so did the distinction between the "self" and the "external world," leading to new fears, such as loneliness, alienation, and a sense of disconnection from the universe. With the rise of metaphysical beliefs, like animism, the world was often seen as unpredictable and even hostile. Furthermore, the awareness of time and the anticipation of the future introduced a new layer of anxiety. Unlike other animals, humans have the capacity to foresee uncertainties, recognizing the inherent unpredictability of life. This awareness, however, often comes without evolutionary preparation to cope with these realizations.

The Expansion of Human Concerns

While early humans' primary concern was undoubtedly survival, additional existential fears gradually emerged over time. These included:

- Separation from the natural world and its perceived dangers
- Natural disasters, such as earthquakes and volcanic eruptions
- Imaginary entities, as seen in animistic beliefs
- Dreams and altered states of consciousness
- Violating social taboos, later framed as "sin"
- The inevitability of death

Awareness of Death and Its Psychological Impact

Although recognition of death is not exclusive to humans—many animals, especially great apes, exhibit emotional distress when companions or relatives die— (see Note 101) human cognitive and cultural development has led to a more complex understanding of death.

Note 101- *Not only great apes but also elephants, giraffes, and several species of corvids, such as ravens and crows, exhibit an awareness of death. They often remain with their recently deceased companion, showing signs of mourning and distress, which suggests a deep emotional response to loss.*

This awareness of mortality is not merely a biological response but a deeply philosophical and psychological realization.

Archaeological evidence suggests that early humans, such as Neanderthals, were aware of death. Burial sites dating back over 100,000 years show bodies arranged in fetal positions—possibly symbolizing a return to birth or reflecting a belief in cyclical existence—often accompanied by weapons, tools, and provisions. These practices indicate early forms of spirituality, as humans began contemplating meanings beyond survival.

The fear of death typically emerges in childhood, around the ages of 6 to 8, when children start to grasp the concept of mortality and its implications for personal identity. This growing awareness profoundly shaped human belief systems, leading to the creation of rituals, myths, and, later, organized religions.

Art, Abstraction, and the Search for Meaning

Artistic expressions, dating back at least 70,000 years, provide compelling evidence of humanity's capacity for abstraction and symbolic thinking. The emergence of art suggests that early humans sought ways to engage with the unknown, using creativity as a tool to process their existential fears. These early symbolic representations likely played a crucial role in the development of spiritual and religious thought, offering psychological relief from anxieties about life's uncertainties.

The Neuroscience of Worry and the Complexity of the Human Brain

Anxiety has a physiological basis, closely linked to the complexity of the human brain. Cognitive neuroscience shows that the human brain is wired for predictive processing, constantly anticipating potential outcomes and striving to resolve uncertainty. While this ability is advantageous for survival, it also makes humans susceptible to persistent restlessness, curiosity, and existential questioning.

The human brain's high degree of neural plasticity and its constant drive to construct meaning contribute to a perpetual state of cognitive disequilibrium. Unlike simpler nervous systems that follow more predictable patterns, the human brain thrives at the intersection of rationality, imagination, and uncertainty. This trait, while fostering creativity and innovation, also makes lasting peace of mind a rare experience.

Ultimately, worry—stemming from the brain's unparalleled complexity—fuels humanity's unending quest for knowledge, understanding, and existential resolution. This relentless curiosity has driven humans to explore all dimensions of existence, from the physical world to the profound mysteries of consciousness and beyond. - (see Note 102)

Note 102*-Maybe this search for meaning and spirituality serves as our umbilical cord, linking the deepest layers of our existence to the fundamental truth of the universe—a truth embedded in the very nature of consciousness. And perhaps, on the other side of this connection, lies divinity itself.*

Conclusion of Human Evolutionary History

As explored throughout this text, human evolution has been shaped by a complex interplay of biological, cultural, social, and spiritual factors. Edgar Morin aptly encapsulates this complexity when he states:

"Becoming human is neither solely biological evolution nor purely cultural, social, or spiritual evolution; rather, it is a multifaceted process arising from the interaction of genetic, biological, cultural, social, and spiritual forces."

With the emergence of Homo sapiens approximately 300,000 years ago, the role of biological evolution in shaping human behavior began to diminish. While evolutionary processes continue at a genetic level, their influence on human life has become secondary to cultural evolution. Humanity has entered a new paradigm, one driven not by biological mutations or natural selection, but by the evolution of culture. Although cultural behaviors have roots in primate - (see Note 103) social structures, they have undergone profound transformations over millions of years, acquiring distinctively human characteristics.

Note 103- *Primates are a diverse order of mammals that includes monkeys, apes, and humans. They are distinguished by forward-facing eyes, which provide binocular vision and enhance depth perception. Another defining trait is their opposable thumbs—or, in some species, opposable big toes—which enable precise grasping and manipulation of objects. Primates also possess relatively large brains in proportion to their body size, fostering complex social structures, problem-solving abilities, and advanced learning. This intelligence allows them to acquire experience and transmit knowledge across generations. Additionally, all primates exhibit extended parental care, with mothers—or, in some cases, other group members—imparting essential survival skills. This prolonged dependency is particularly pronounced in great apes and humans, where cultural learning plays a vital role in development.*

The impact of culture on shaping human nature is so profound that Morin further asserts:

"Human nature is nothing but formable raw material, shaped entirely by history and culture."

Through cultural evolution, humans have transcended the biological constraints that govern the behavior of other species. While most organisms remain bound by genetic predispositions and instinctive behaviors, humans have, for the first time in nature, broken free from the confines of genetic determinism. Through the accumulation of knowledge, traditions, and symbolic thought, humans have gained the ability to modify, suppress, or even override their innate instincts, transforming them into ethical, moral, and socially regulated behaviors.

This evolutionary journey has led to the development of self-awareness, self-confidence, and self-determination. Humans do not merely recognize themselves as thinking beings; they also understand that they are responsible for their actions. This capacity for self-reflection allows for moral judgment, ethical responsibility, and the ability to envision and construct the future. Unlike any other species, humans actively shape their destiny—not solely through genetic inheritance, but through deliberate choices, innovations, and social structures that redefine their existence.

Thus, our brief review of human evolutionary history concludes. While there remains much more to explore, this foundational understanding provides the necessary context for our discussion on human happiness. In the appendices of this book, we will delve deeper into one of the most transformative periods of human history—the last 10,000 years, known as the history of civilization.

Joy and Happiness: Instinct, Meaning, and Evolution

With a brief review of human evolution and the foundations of human nature, we now return to the central questions raised at the outset:

1. What do we mean by happiness (or bliss)?
2. Is happiness purely a subjective, intellectual matter—dependent on one's mindset and attitude toward life—or is it also influenced by external, objective factors?

To address these questions, we must first define joy, as it represents a more fundamental and universal experience.

Joy: A Natural and Instinctive Emotion

Joy is an innate, biologically driven emotion that arises in response to pleasurable stimuli. The fulfillment of basic physiological and psychological needs—such as nourishment, safety, social bonding, and reproduction—triggers joy. This response is common to all animals, including humans, and is deeply rooted in the evolutionary mechanisms that support survival and well-being.

However, joy is a temporary and fluctuating state. Its intensity diminishes over time as the brain habituates to positive experiences. This phenomenon, known as *hedonic adaptation*, suggests that people (and likely other animals) quickly return to a baseline level of well-being after pleasurable or rewarding experiences.

Happiness (Bliss): A Deeper, More Enduring State

In contrast, happiness represents a deeper, more stable emotional state that extends beyond momentary pleasure. It is closely linked to life satisfaction, meaning, and psychological well-being. Unlike transient joy, happiness involves an ongoing balance between one's beliefs, values, goals, and lived experiences.

Importantly, happiness is not simply the accumulation of joyful experiences. Psychological studies indicate that while short-term pleasures contribute to well-being, lasting happiness often depends on deeper factors such as purpose, personal growth, social connections, and a sense of meaning. This aligns with Aristotle's concept of *eudaimonia*, which suggests that true well-being arises from a meaningful and virtuous life rather than fleeting pleasures.

A key distinction is that happiness often transcends immediate desires and pleasures—sometimes even conflicting with them. For example, individuals may sacrifice short-term joy (e.g., indulging in comfort) for long-term fulfillment (e.g., achieving personal or professional goals).

However, happiness cannot be completely detached from biological, psychological, and social needs. Research in neuroscience and psychology shows that chronic deprivation of basic needs—such as social connection, autonomy, or security—negatively affects well-being, indicating that while happiness is shaped by subjective interpretation, it is also rooted in objective realities.

Happiness and Joy in Other Living Beings

For non-human animals, happiness and joy take on different meanings depending on their evolutionary complexity. Most animals live primarily in the present moment, with little evidence of deep regret about the past or abstract concerns about the future. Their focus on future events is largely instinct-driven, often seasonal, and linked to survival and reproductive cycles.

For many animals, the satisfaction of basic instinctual and physiological needs (e.g., food, safety, social bonds) results in a form of well-being that is functionally similar to happiness in humans.

This suggests that joy and happiness in non-human animals are largely equivalent, with little distinction between immediate pleasure and overall life satisfaction. - (see Note 104)

However, in highly social and cognitively advanced species, such as great apes, elephants, and certain corvids (e.g., ravens), research suggests more complex emotional experiences. These animals exhibit behaviors indicative of empathy, grief, frustration, anticipation, and emotional bonding—emotions that may serve as early evolutionary precursors to the human distinction between transient joy and enduring happiness.

Thus, in our closest relatives, the great apes, we begin to see the emergence of emotions that go beyond simple joy. These include mood states influenced by social context, past experiences, and long-term goals—an essential evolutionary step toward the uniquely human experience of happiness as an abstract, long-term phenomenon.

Conclusion

From an evolutionary and psychological perspective, joy is a universal, instinctive emotional response, while happiness is a complex, higher-order experience shaped by meaning, long-term well-being, and social factors. In humans, the distinction between the two is pronounced due to our self-awareness, capacity for reflection, and ability to construct meaning beyond immediate gratification.

***Note 104**- In animal societies, particularly among our close relatives, primates, rigid and hierarchical social structures are common, often leading to significant disparities in well-being and widespread unhappiness among individuals. In these societies, higher-ranking members enjoy greater freedoms and fewer restrictions in pursuing their desires and pleasures, including reproductive*

Now, let us turn our focus to humans. As previously mentioned, with the evolution and increasing complexity of the brain—coupled with the emergence of culture and distinct human self-awareness—attention to both the past and the future begins to develop. This, in turn, gives rise to abstract intellectual and spiritual activities. Consequently, the concepts of joy and happiness (or bliss) start to diverge in significant ways. While joy remains an instinctive and momentary emotional response, happiness evolves into a more abstract and intellectual experience, becoming intertwined with various psychological factors such as regret over the past, anxiety about the future, and existential concerns, including the awareness of mortality.

(continued from Note 104)- *opportunities. In contrast, lower-ranking individuals face strict obedience, deprivation, subjugation, and heightened stress, often manifesting as anxiety and distress. These social inequalities directly impact overall happiness and well-being within these groups. This phenomenon is especially evident in primates such as baboons, macaques, and chimpanzees, where clear hierarchies dictate differential access to resources and opportunities, shaping vastly unequal life experiences. In contrast, less hierarchical and more decentralized social structures—such as those seen in some forest-dwelling primates—tend to mitigate these inequalities. However, in more competitive and rigidly stratified environments, like the male-dominated societies of desert-dwelling baboons, disparities in access to resources, including mates, become much more pronounced and oppressive.*

Interestingly, social mobility can moderate some of these inequalities. Younger individuals often ascend the ranks as they mature, while older individuals may lose status and face marginalization. However, age alone does not determine one's social trajectory—individuals navigate their position within the hierarchy differently throughout their lives, experiencing both upward and downward mobility. Primate hierarchies are often more complex than they initially appear, consisting of multiple levels, sublevels, and even semi-strata within a group. These nuanced systems reflect varying degrees of dominance and influence, demonstrating the intricate nature of primate social structures. Notably, the nuclear family unit, as understood in human societies, has not yet fully formed in these groups. Instead, social bonds are primarily centered around broader group dynamics and relationships that extend beyond the immediate family.

As a result, even when all external conditions for joy and pleasure are present, a person may still fail to perceive themselves as happy.- (see Note 105) Conversely, due to the dynamic and adaptive nature of human beings—marked by an ability to create opportunities and engage in spiritual reflection—one may experience happiness even in the absence of many conventional sources of joy.

This underscores a crucial distinction: while external factors that induce joy contribute to happiness and are often considered necessary for it, they are not sufficient on their own. The presence of joy does not automatically translate into happiness, nor does its absence necessarily prevent it. In the sections that follow, we will revisit these ideas, synthesizing our understanding of happiness to arrive at clearer insights. For now, we pause this exploration. To introduce variety and deepen our understanding, we will now examine several perspectives on human happiness.

***Note 105**-One of the most well-known examples related to this discussion is the story of Buddha. Siddhartha Gautama, later known as the Buddha, was a prince who lived around 2,500 years ago in the region that is now Nepal, but was then part of Ancient India. He was raised in luxury within his father's palace, shielded from the harsh realities of life. However, upon encountering the inevitable human experiences of illness, aging, and death, a profound transformation occurred within him. The comfort and pleasure of palace life lost their meaning, and he could no longer find true happiness in material abundance alone. Driven by a deep existential quest, Siddhartha left his home and renounced his royal status to seek the truth about suffering and the nature of existence. He practiced extreme asceticism and meditation for several years, withdrawing from worldly pleasures. Eventually, after deep contemplation under the Bodhi tree, he attained enlightenment—awakening to a profound understanding of suffering and the path to liberation. Following this realization, he developed a structured path to achieving inner peace and lasting happiness, which became the foundation of his teachings. This path, known as the Middle Way, emphasized balance between indulgence and extreme self-denial. His teachings later formed the basis of what is now called Buddhism, a spiritual tradition that continues to influence millions around the world.*

The First Point of View: *"We are here on Earth to fart around. Don't let anybody tell you any different."*

— Kurt Vonnegut

This perspective, while humorous and irreverent, promotes a natural, uncomplicated approach to life. It asserts that life, as a unique opportunity, should be embraced, encouraging us to enjoy the fleeting moments we have. Vonnegut's simple, fluid style challenges any viewpoint that suggests life is solely about suffering or enduring hardship. He dismisses the notion that we are victims of some divine or cosmic punishment and instead champions the idea that life, in all its absurdity and imperfection, should be enjoyed.

The Second Point of View: Ali Shariati, from his book *Desert*, offers a different perspective:

"People are not all on the same level and have a distance from each other, but they are all of the same gender. Whoever in this life, under the sky, gets jubilated with something has animal and plant stupidity; I don't know why there is a wave of hateful and ugly stupidity in every jubilation, every belly laugh, every finger-snapping, every feeling of jubilation. I don't know why the jubilant and plump faces in my eyes, to the point of vomiting, look disgusting and indecent. Really, may God give a bit of a chance. What chance? Stupidity! Oh, what a blessing this stupidity is! What a capital it is. Everyone's happiness is as great as his benefit from this blessing, and that's it. This is the only secret of human happiness in life, and all the rest is nonsense words and sophistry. How hard and sad is the fate of a person whom that nature cannot fool? How bitter is the fruit of the insight tree!"

At first glance, Shariati's words may seem harsh, anti-nature, or even offensive, appearing as the reflections of someone jaded, melancholic, or overly serious, placing more value on mourning and sorrow than on joy. However, dismissing Shariati based solely on this excerpt would be a mistake.

Shariati, a distinguished intellectual with a doctorate in sociology from Sorbonne University, was much more than a melancholic thinker. As a professor, writer, and social activist, he captured the minds and hearts of thousands of young Iranians in the 1960s and 70s. In his personal life, he was known for being balanced, lively, and socially engaged, with a supportive family. The sharp, seemingly cynical language he uses should not be interpreted literally. Instead, Shariati's critique targets superficial, self-indulgent lifestyles—a way of living marked by ignorance and thoughtlessness. His call is not to reject joy outright, but to awaken individuals to a deeper understanding of life, urging them to be aware, care for others, and embrace spirituality. He challenges the apathy that comes from mindless indulgence and encourages people to live more meaningfully.

Shariati's aim is not to dismiss the beauty of life's pleasures but to highlight the deeper layers of human experience. His words serve as a shock to the system—a wake-up call to live consciously and with purpose, not just for momentary pleasures. He awakens the conscience and encourages individuals to confront their true selves. May a bucket of cold-water splash over you, be my guest!

Despite the severity of Shariati's message, his viewpoint is not in opposition to the first perspective. Instead, it complements it. Both perspectives encourage a fuller appreciation of life, though from different angles. Vonnegut invites us to laugh and live freely, while Shariati urges us to enjoy life with moral integrity and awareness. In modern terms, Shariati's call is to enjoy life, but to do so consciously and with integrity.

Conclusion: Both perspectives, when examined with care, offer valuable insights into human happiness. While Vonnegut champions the idea of embracing life's absurdities and simple joys, Shariati calls us to live more intentionally and mindfully. Together, these perspectives suggest that true happiness may lie in finding balance—enjoying the fleeting moments of joy while remaining grounded in values, awareness, and deeper purpose.

The Third Point of View:

The idea that a person is born for happiness is a truth that nature demonstrates unequivocally. Joy is scattered across the earth, and every living being is a manifestation of that joy. Existence itself is an expression of happiness. To exist is to be joyful. It is the joy of tasting fruit, which, as it ripens, turns into a melody or a bird's song. This drive toward joy and fulfillment causes the plant to grow, fills the hive with honey, and brings a sense of goodness to the human heart. All creatures partake in happiness, each according to their ability to experience it, depending on how their senses perceive it.

Gratitude is a profound and deeply cherished emotion, so necessary and sweet that even the slightest gesture of it fills my heart with appreciation. It teaches me to seek joy in everything that comes my way. Gratitude opens the heart, and through it, I recognize that my connection to the universe, to life itself, is a continuous act of creation.

When I awaken each morning, I am filled with wonder and surprise at my very existence. This sense of amazement remains with me throughout the day. But I do not want happiness that is built upon the suffering of others. I do not want wealth if it comes at the expense of another's deprivation. If my clothing requires the nakedness of another, I would choose to go without. There are places on earth where poverty, suffering, and torment are so immense that a person who feels happiness cannot think of them without shame.

However, someone who has not experienced happiness for themselves cannot truly work for the happiness of others. My happiness is inherently tied to the happiness of others. I need the happiness of others for my own to be complete. In the Bible, the concept of striving for happiness is praised, and I continue to praise it today. The first message that was delivered to us through Christ was the word of happiness. A true Christian is one who finds joy in the simplest of things—even in plain water. The misinterpretation of this message created a religion based on suffering and sorrow as sacred. Jesus Christ said, "Come to me, all you who are weary and burdened, and I will give you rest." This was misread as a call to suffer before approaching Christ, but the true message is different.

As I once wrote, "He who is happy and thoughtful is truly strong." Happiness rooted in ignorance is meaningless to me. The first words of Jesus, "Blessed are those who mourn," are often misunderstood. Those who interpret this as an encouragement to mourn are gravely mistaken. True joy encompasses both sorrow and happiness—it is an integration of all the facets of existence.
— From a reflection inspired by the spirit of *André Gide, Fruits of the Earth.*

The Fourth Point of View:

We are meant to enjoy the simple pleasures of life: food, pleasant smells, colors, beautiful clothes, music, games, entertainment—things that all humans can partake in without causing harm to anyone. We should embrace life's gifts and savor them as fully as we can. Moreover, we should connect with others, fostering relationships that bind us together, for anything that unites people is inherently good. Sharing our happiness with others should be a priority, as it enriches both ourselves and those around us. It is through this shared joy that we can truly experience oneness with nature.
Thus, let us embrace one another, millions of people, and come together in our shared human experience.

— Romain Rolland, Inner Journey

Key Elements of the Happy Human

Building upon the enlightening perspectives we've reviewed and the explanations we've provided regarding happiness and its relationship with joy, we now return to our central question. Our goal is to establish a clearer definition of happiness and delve deeper into its intricate relationship with the complex external and internal factors that influence it. To achieve this, it is helpful to examine the foundational aspects of our nature and seek clues within it. While human nature is undeniably complex and multi-dimensional, this complexity contains an inner core—an essence—that we can uncover and examine directly, independent of the various cultural lenses through which we view it. After all, human culture differs significantly from one society to another, and even from one individual to another.

However, despite these cultural variations, there is a universal thread that runs through all of us—something fundamental, sincere, and self-contained. This is what we might call the "natural gem of humanity," a shared and simple essence that transcends the diversities of culture and experience. In the following, we will attempt to draw closer to this common inner gem shared by all human beings, free from existing differences, and explore the foundations of our happiness within it. These foundations must be discovered at the intersections where external objective factors and internal mental and emotional factors allegorically meet, weaving together a delicate network of our thoughts, emotions, and beliefs. At these intersections, they form concepts and shape the wholeness of our existence as human beings. According to the philosophy of this school, the common inner gem that resides within all of us—forming "our true nature," "our humanity," and "our happiness"—can be understood through 20 key aspects.

From 20 different windows or perspectives, we can examine this gem. These 20 aspects can be referred to as the "necessities and requirements of happiness," or more simply, the "foundations of happiness." The titles are as follows:

The 20 Foundations of Human Happiness

1. Balanced Environment: Living in harmony with both nature and society, fostering balance and sustainability in every aspect of life.

2. Security: Feeling safe—physically, emotionally, and financially—is an essential pillar for peace of mind and human flourishing.

3. Livelihood and Purposeful Work: Engaging in work that not only supports one's life but also carries meaning and benefits others.

4. Health and Vitality: Caring for one's physical and mental well-being, which serves as the groundwork for happiness and progress.

5. Close-Knit Family: The presence of loving, dependable family ties that offer comfort, encouragement, and belonging.

6. Self-Dignity and Respect: Acknowledging one's own worthwhile honoring the dignity and rights of others.

7. Freedom and Individuality: The ability to live authentically, express oneself, and follow personal goals without undue restrictions.

8. Justice and Fairness: Living in a society where fairness prevails, rights are safeguarded, and no one is left behind.

9. Curiosity, Exploration, and Creativity: The natural urge to learn, discover new possibilities, and express one's creativity in daily life.

10. Channeled Aggression, Play, and Sports: Channeling energy through positive competition, playful engagement, and physical activity that strengthens both body and mind.

11. Art: Experiencing and creating beauty in its many forms—whether through art, music, or meaningful design.

12. Restlessness and Drive for Progress: The subtle discomfort that inspires individuals to grow, improve, and seek new horizons beyond the familiar.

13. Connection with Nature and Survival Skills: Rediscovering the rhythms of nature and learning essential survival skills—like observation, patience, and resourcefulness—that reconnect us with our roots.

14. Honesty, Integrity, and Keeping One's Word: Living truthfully, acting with moral integrity, and honoring the promises we make.

15. Awareness of Moral Defilements (The Abysses): Becoming conscious of the ethical pitfalls and harmful tendencies that can pull us away from joy and inner peace.

16. Tools and Technology: Creating and using tools and technologies responsibly to improve life without undermining humanity or nature.

17. Peace-Seeking, Love, Compassion, and Service: Actively promoting peace and kindness, showing compassion, and serving others without expectation and exception.

18. Wishing Happiness for All Without Exception: Cultivating a heartfelt desire for the happiness and well-being of every person, regardless of who they are.

19. Critical Thinking and Freedom from Dogma: Remaining skeptical of rigid beliefs and practicing open, thoughtful inquiry in the search for truth.

20. Spirituality and Transcendence: Seeking meaning beyond the material world, embracing the spiritual dimension of existence.

The position of these categories is, to some extent, aligned with the natural course of a child's development. Of course, this list can be rearranged in any other order. In the following sections, we will elaborate on the foundations mentioned above.

1. **Balanced Environment**

Contrary to simplistic assumptions, humans are deeply interconnected with their environment on multiple levels—not only biologically, but also ecologically and psychologically. Our bodies, for instance, host trillions of microorganisms, collectively known as the human microbiome, which plays a vital role in our health. It is estimated that around 3% of our body weight (roughly 2 kilograms in a 70-kilogram adult) consists of bacteria living within or on us.

These microbes begin their essential work from the moment of birth. For example, as we pass through the birth canal, we are exposed to beneficial bacteria that help develop our immune system, reducing the risk of conditions such as asthma. Similarly, our gut microbiota—comprising thousands of bacterial species—helps digest food, synthesize vitamins, and regulate metabolism. Disruptions to this microbial balance, such as through excessive antibiotic use, can have severe consequences, including digestive disorders, weakened immunity, and even mental health issues.

Additionally, about 500 species of bacteria reside on our skin, contributing to its health and protection. Though these microbial interactions occur on a microscopic scale—an aspect of our existence we rarely notice—on a larger scale, life on Earth operates within complex ecological cycles. Plants, animals, bacteria, fungi, and even viruses interact to maintain the planet's balance. We, as humans, are merely one component of this vast system, albeit an influential one. The preservation of environmental balance is directly linked to human well-being. A degraded, imbalanced, or sterile environment would inevitably undermine both our physical health and psychological stability. In the long run, no human being can thrive in a world where natural ecosystems are disrupted, biodiversity is lost, and essential life-supporting processes are compromised. Understanding and respecting this intricate relationship between nature and human life is crucial for both our survival and our happiness.

2. Security

Security is undeniably one of the fundamental prerequisites for happiness. In the absence of security, fear, anxiety, and uncertainty take hold, leaving individuals and societies vulnerable to harm, instability, and even destruction. Under such conditions, the very idea of happiness may seem like a distant luxury—an ironic dream. However, history shows that humans have always found ways to adapt to insecurity, much like our ancestors, who for millions of years navigated the dangers of an unpredictable world. Even today, when faced with insecurity, individuals and communities often seek glimmers of hope, resilience, and practical solutions to mitigate risks and create opportunities for improvement. Even in extreme circumstances—where insecurity is absolute, life hangs by a thread, and no apparent hope remains—some individuals find strength in spiritual beliefs and a sense of meaning. For them, surrendering to a higher purpose, whether it be God, fate, or the intrinsic flow of life, allows them to transcend fear and embrace existence with peace. In such cases, the external outcome becomes secondary to an internal state of acceptance and trust. At a broader level, governments and policymakers bear the responsibility of ensuring security. Nationally, this involves protecting citizens' rights from aggression, coercion, and violence through just and progressive laws, fair enforcement, and social policies that reduce the root causes of crime and instability. A well-functioning social safety net, economic opportunities, and education can significantly diminish insecurity at its source. Internationally, peace and security are best achieved through diplomacy, conflict prevention, and responsible defense strategies, rather than warmongering or aggression. In essence, while security is a fundamental condition for happiness, individuals can cultivate inner resilience in the face of insecurity, and societies must actively work to create structures that uphold peace, stability, and justice for all.

3. Livelihood and Purposeful Work

Humans are inherently social beings, deriving pleasure, meaning, and fulfillment from work—especially when it involves collaboration, intellectual challenges, and shared problem-solving. There is no doubt that work has played a crucial role in shaping humanity, both biologically and culturally. Therefore, entrepreneurship and job creation should be a priority for authorities, policymakers, and capable individuals with good intentions and visionary leadership. Without work, a person gradually loses balance, dignity, and even physical and mental health. Naturally, this deterioration extends to their livelihood, undermining the foundation of their life and happiness, while making them more vulnerable to hardship and despair. While social security systems are essential pillars of a fair and stable society, they are not substitutes for the intrinsic value of purposeful work. In this sense, all discussions about employment, economic stability, and entrepreneurship ultimately revolve around a fundamental human need: livelihood. By "livelihood," we refer to the essentials of life—food, housing, and basic necessities. However, securing these necessities is not merely a matter of economic policy; it also depends on individual attitudes toward work, effort, and perseverance. No economic system can force people to work and be productive if they lack a sense of purpose and hope for the future. Though external factors such as religious beliefs, cultural norms, and political-economic systems may influence people's motivation, the individual ultimately holds the key. Reflecting on the struggles and resilience of our ancestors—who labored tirelessly for survival over millions of years—reveals that, despite the difficulties of our present conditions, they pale in comparison to what our forebears endured. Their hard work bore fruit, and we are still here today because of it. Thus, idleness has a cure, poverty has a solution, and the future belongs to those who are hopeful and determined. As Victor Hugo wisely said, "The only word written on every man's forehead is hope." Similarly, another thinker remarked, "For humans, there is no dead end; either they find a way, or they make one."

4. Health and Vitality

Health is one of the fundamental pillars of human happiness. Without it, joy loses its vibrancy and can even become meaningless. However, humans have a remarkable ability to adapt to physical limitations, and in some cases, we witness extraordinary resilience in individuals with severe disabilities—an inspiring testament to the human spirit. Among them, the legendary physicist Stephen Hawking stands out. Despite being completely paralyzed, he revolutionized our understanding of the universe, proving that even in the absence of physical health, the mind can flourish and contribute immensely to humanity. Nevertheless, while individual perseverance is admirable, public health must remain a top priority for governments and societies. Governments are expected to take proactive measures to promote well-being, including:

- Developing and expanding vaccination programs to prevent life-threatening diseases
- Implementing large-scale disease prevention initiatives
- Providing accessible and free healthcare services, which is an essential component of a happy and just society

In developed countries, comprehensive social security systems act as safety nets, protecting low-income populations from hunger, malnutrition, and inadequate medical care. These policies significantly enhance public well-being and, in turn, contribute to greater social stability and happiness. In contrast, in less developed nations, corruption and poor management often hinder healthcare accessibility. However, with effective governance and transparency, essential health services can gradually become available to all social classes. Beyond governmental responsibilities, families also play a crucial role in ensuring health and safety. By fostering a culture of hygiene, proper nutrition, and preventive care, they directly contribute to their own well-being and, consequently, to a happier society. Ultimately, the strong link between health and happiness cannot be overstated. A healthy life enables individuals to experience joy fully, pursue their goals, and engage meaningfully with others—all of which are vital components of human fulfillment.

5. Close-Knit Family

Humans are inherently creatures of love—we are drawn to deep, meaningful connections, and the family serves as the first and most fundamental source of love, support, and belonging. A person thrives when they grow up in a nurturing, happy, and safe family environment, benefiting from unconditional love and emotional security. As they mature, they, in turn, seek love, build their own families, and pass down this essential human experience to future generations. It is the responsibility of culture and society to cherish and uphold this natural and fundamental value. A strong family structure is not merely a personal or sentimental matter—it is one of the pillars of human development and fulfillment. Beyond the immediate household, strong social bonds—between extended family, relatives, and close friends—play an essential role in human happiness. Engaging in social gatherings, shared celebrations, and collective activities fosters mental well-being and enhances social and communication skills. A harmonious and supportive family environment also creates the foundation for healthy relationships between children of different genders, allowing them to develop mutual respect, emotional intelligence, and a sense of connection that extends beyond their own household. Ultimately, vibrant, balanced, and loving families are among the most powerful contributors to human happiness—shaping individuals, strengthening societies, and ensuring that love and care continue to flow through generations.

6. Self-Dignity and Respect

Human dignity is fundamental to mental and emotional well-being. A person's self-esteem and psychological balance are deeply influenced by the respect they receive from their family and society. In an environment where individuals—especially children—are treated with respect, encouragement, and support, they develop a strong sense of self-worth, which serves as a foundation for a fulfilling and confident life. Conversely, in families and societies that belittle, criticize, or demean individuals, a toxic and oppressive culture emerges—one that breeds insecurity, emotional distress, and long-term psychological harm. When disrespect and humiliation become normalized, they erode not only the individual's happiness but also the collective well-being of the society. This issue becomes even more concerning in societies where extreme ideological or religious beliefs take precedence over human dignity. In such environments, intrusive control and moral policing may be seen as virtues, leading to violations of personal freedom, dignity, and respect. When people's autonomy and individuality are compromised under the pretext of ideology, it inevitably results in the suppression of happiness and the degradation of human value. A society that nurtures dignity and respect fosters happiness. When individuals are treated with fairness, kindness, and recognition, they not only thrive personally but also contribute to a healthier, more compassionate world. - (see Note 106)

Note 106- *In her poem Children Learn What They Live, American writer and family counselor Dorothy Law Nolte (1954) beautifully expresses how a child's environment shapes their development:*

If children live with criticism,
They learn to condemn.
If children live with hostility,
They learn to fight.
If children live with fear,
They learn to be apprehensive.
If children live with pity,
They learn to feel sorry for themselves.

Understanding human freedom and individuality is crucial in shaping political and economic systems. Over time, people will inevitably resist any system—particularly ideological ones—that seeks to reduce them to mere followers or confine them in an ant-like existence of servitude. It appears that free and democratic systems, which allow for dynamic social, political, and economic mobility, as well as the rotation and renewal of power, are more aligned with human nature. However, these systems are not without flaws, and addressing their shortcomings is essential for their sustainability.

(continued from Note 106)-
If Children live with ridicule
They learn to feel shy.
If children live with jealousy,
They learn to feel envy.
If children live with shame,
They learn to feel guilty.
If children live with encouragement,
They learn confidence.
If children live with tolerance,
They learn patience.
If children live with praise,
They learn appreciation.
If children live with acceptance,
They learn to love.
If children live with approval,
They learn to like themselves.
If children live with recognition,
They learn it is good to have a goal.
If children live with sharing,
They learn generosity.
If children live with honesty,
They learn truthfulness.
If children live with fairness,
They learn justice.
If children live with kindness and consideration,
They learn respect.
If children live with security,
They learn to have faith in themselves and in those about them.
If children live with friendliness,
They learn the world is a nice place in which to live.

7. Freedom and Individuality

Human beings are inherently free- (see Note 107) and possess individuality. While we naturally seek social coexistence, we also desire to maintain our personal boundaries, autonomy, and uniqueness. The notion of captivity and subjugation is deeply unsettling to us. Though modern civilization may appear to have built a cage around individuals, seemingly restricting their personal freedom, it is important to recognize that an existence outside of civilization—akin to our primitive ancestors—offers no true alternative. The romanticized idea of absolute freedom in the absence of societal structure collapses at the first sign of conflict, leaving individuals defenseless against chaos. Within the family unit, respecting a child's individuality and freedom is fundamental to their healthy development. While there is no universal formula for parents, one of the most effective approaches is spending quality time with children and encouraging their connection with nature—an age-old companion that nurtures the spirit of freedom. Trust is also vital; parents should avoid being overly intrusive. Genuine trust fosters self-esteem and independence. When children feel trusted and respected, they will naturally regard home as a safe, welcoming space rather than a place they long to escape. In religion and spirituality, extreme systems that misinterpret teachings to demand absolute obedience and submission deprive individuals of their free will, stunting their spiritual and moral development. Societies governed by rigid doctrines become fragile and stagnant, suppressing the natural evolution of human thought and ethics.

In the long run, it is highly unlikely that such restrictive religious approaches can persist without encountering resistance from the innate human drive for freedom.

Note 107*-Your God created you free, be free.*
— Abū Saʿīd Abū'l-Khayr

8. Justice and Fairness

Human beings are inherently seekers of justice, both for themselves and others. In terms of well-being and livelihood, people recognize that everyone has the right to earn an honest living and provide for their families. This understanding forms the foundation for the expectation that society should offer equal opportunities for education, employment, and social mobility. A just society ensures that all individuals, regardless of background, have access to these opportunities and can fully participate in political and economic structures without discrimination. The presence of social and medical insurance systems, which provide a fair distribution of resources, aligns with humanity's sense of justice. These systems foster hope, satisfaction, and a sense of security, while reducing the risk of unrest and social upheaval. From a legal perspective, individuals expect a fair judicial system that ensures their rights are protected and that justice is served impartially. This system should act as a safeguard against injustice, defending the people against aggression and maintaining balance across all levels of governance. To maintain these principles of justice, civil society depends on taxes—sometimes burdensome taxes—that citizens pay. These contributions are essential for sustaining the societal structures that uphold justice and equality. In this way, the system of taxation is not in conflict with the foundations of justice but is rather a necessary mechanism for ensuring continued welfare and fairness.

9. Curiosity, Exploration, and Creativity

Human beings are inherently curious and driven to explore, innovate, and express creativity in all domains—intellectual, artistic, and practical. This inherent drive for discovery is a cornerstone of mental health and personal fulfillment. The freedom to explore and choose within this context is vital for well-being. However, extreme religious interpretations and rigid ideologies that restrict this natural curiosity can severely hinder human happiness and growth. History consistently shows that no limiting belief system can withstand humanity's pursuit of knowledge and truth. It is essential for families and societies to cultivate an environment free from dogmatic prejudices and superstitions, allowing individuals—particularly children—the freedom to explore their interests without constraint. This freedom should extend across all fields: crafts, technology, science, art, philosophy, theology, and mysticism. By fostering an open and supportive environment for exploration, societies create conditions for personal and collective prosperity, contributing to individual fulfillment and societal advancement.

10. Channeled Aggression, Play, and Sports

Aggression, like the instinct for hunting, is an inherent part of human nature. - (see Note 108) If our evolutionary roots trace back to species closely related to bonobos and chimpanzees, then we must acknowledge that, alongside the instinct for kindness, peace, and social bonding—traits particularly prominent in bonobos—we also share traits with chimpanzees, such as strength, hierarchical structuring, and aggression. However, this aspect of human nature is not inherently destructive. When properly understood and channeled, aggression manifests as courage, resilience, and determination—qualities that have historically played a crucial role in human survival and progress.

Note 108 *- Konrad Lorenz, the renowned Austrian ethologist, explores the issue of aggression in his book On Aggression. He categorizes human motivations into four primary instinctual drives: hunger (feeding), love (reproduction), flight (fear), and aggression. According to Lorenz, these instincts operate within an interconnected system, and for a person's psychological and social well-being, each instinct must find an appropriate outlet. Lorenz's central concern is that in modern urban life—far removed from the environmental conditions in which human instincts evolved—there are often insufficient opportunities for the natural release of aggression. Unlike earlier times, when aggression had adaptive outlets in survival activities such as hunting, territorial defense, and structured social competition, contemporary society offers fewer such channels. As a result, unexpressed aggression may accumulate, leading to psychological distress and behavioral abnormalities. Lorenz warns that without proper outlets, this pent-up energy may manifest destructively, emphasizing the importance of structured physical activities, sports, and socially accepted forms of competition to maintain a healthy balance.*

Modern scientific understanding suggests that when an individual's physiological and psychological needs are met in a balanced way—and when society does not provoke aggression through systemic frustration or oppression—- (see Note 109)-this innate energy can be directed toward constructive and beneficial paths. Society has a responsibility to provide structured outlets for this energy, particularly for young people, by investing in gymnasiums, sports facilities, and accessible recreational programs.

Note 109 *- Norman L. Mann, a renowned American psychologist, in his book "The Fundamentals of Human Adjustment", describes aggression as a motivation that typically arises when an individual is deprived of fulfilling either physiological or acquired needs. Mann emphasizes that social norms play a crucial role in either suppressing or encouraging this motivation. For instance, the Arapesh tribe of New Guinea is known for its peaceful nature, discouraging anger and aggression. In contrast, the Mundugumor tribe actively instills aggression in their children from an early age. A striking example of this is seen in their feeding practices: infants are given only brief opportunities to nurse, and if they fail to latch quickly, they lose their chance to eat. A researcher studying this phenomenon observed that this practice fosters a strong inclination toward struggle. The infant learns to grip the mother's breast forcefully and suckle with maximum speed and intensity, often to the point of choking. The mother's reaction—one of frustration and punitive behavior—further fuels the child's aggression. As these children grow, especially the boys, they are explicitly trained to fight for survival. Similar contrasts can be observed among the Zuni and Comanche tribes of North America. The Zuni people, known for their pastoral lifestyle, prioritize peace, which is symbolized in their long, flowing robes that reflect their non-confrontational nature. In contrast, the Comanche, renowned warriors, raise their men to embrace combat. Their aggressive disposition is evident even in their dress and weaponry, which are designed to project strength and dominance. Mann challenges earlier psychological theories that framed aggression as a fundamental, universal, and indestructible drive. Some psychologists once believed that war was an unavoidable consequence of human nature. However, Mann argues that modern research disproves this notion. He suggests that human behavior is largely shaped by environmental conditions. If individuals can satisfy their basic needs without significant obstacles, and if their social groups do not encourage conflict, they are more likely to exhibit peaceful behavior.*

Many professions, such as firefighting, emergency medical services, law enforcement, military service, and competitive sports, also serve as natural channels for individuals with high energy levels and a strong drive for challenge and competition, allowing them to utilize their abilities meaningfully.

At the family level, any initiative—no matter how small—that provides a healthy outlet for children's aggression is a wise investment. Parents who engage in playful wrestling, physical activities, or simple games like tearing up scrap paper, popping balloons, or building makeshift hills from blankets and cushions instinctively understand this. These seemingly trivial activities serve an important function: they allow children to release their pent-up energy in a positive way, reducing the need for constant restrictions such as "Don't do this," "Don't run," or "Stop that"—commands that either turn children into passive conformists or rebellious resistors. - (see Note 110)

Note 110 *- It has been said that in order to raise healthy, balanced, and prosperous children, three fundamental principles should be considered:*

• ***Love them*** *– Shower children with affection, kindness, and emotional support, creating a nurturing environment where they feel valued and secure.*

• ***Lift their potential*** *– Provide opportunities and guidance that encourage growth, independence, creativity, and the development of their unique talents.*

• ***Lead them wisely*** *– Offer guidance to help them recognize challenges, make informed decisions, and navigate life's complexities with confidence and resilience.*

These three essential principles form what we call ***The Three L's (LLL)****—a comprehensive parenting approach that fosters love, growth, and wisdom, preparing children to flourish in both their personal and social lives.*

The result of such joyful, creative play is that children see their parents not as rigid enforcers but as trusted allies and companions in their journey of growth. Through these interactions, aggression transforms from a potential source of conflict into a force for happiness, friendship, and mutual trust. Wise parents recognize that their children are not possessions to be controlled but honored guests in their lives. During the fleeting time they share together, they uphold the sacred duty of hospitality—nurturing, guiding, and ultimately preparing their children to step into the world with confidence and strength.

11. Art

Art is an essential and inseparable part of human existence. When not constrained by cultural, religious, or ideological restrictions, it naturally flourishes as a powerful form of expression. Art brings tenderness, balance, and richness to the human spirit, grounding individuals in the present moment and offering a respite from suffering. It is profoundly connected to our spiritual and emotional well-being, allowing people to express their innermost thoughts and feelings. A society that suppresses art weakens its cultural fabric, depriving its members of the deep, transformative experiences that art can offer. Conversely, when the marketplace distorts art by promoting inhumane morals and values, using it as a tool against human nature, art loses its true power. Instead of uplifting, it becomes a source of harm and detracts from human happiness. In the embrace of the family, children are naturally drawn to music, one of the most beautiful and harmonious forms of art. Music embodies joy, rhythm, and movement, helping children connect with their own emotions and those of others. Families must nurture a love for the arts from an early age, offering children opportunities to explore and develop their artistic abilities. This nurturing awakens one of the most noble aspects of their holistic nature and contributes to their personal growth and happiness.

12. Restlessness and the Drive for Progress

Restlessness, coupled with a sense of innocent adventurism, is an intrinsic part of human nature. This drive stems from the complexity of the human brain and its unique ability to navigate the boundary between reality and imagination. Its evolutionary roots may lie in preparing humans to face challenges that our ancestors never encountered. Regardless of its origin, this innate emotion needs an outlet for expression. In modern society, avenues for safe adventure and excitement abound. From sports like boating, skiing, rock climbing, mountaineering, and scuba diving to activities such as parachuting, festivals, carnivals, concerts, and religious celebrations, people find ways to channel this restlessness. Even the global tourism industry serves as an outlet, allowing individuals to explore new horizons and relieve the tensions of everyday life. On a larger scale, this restlessness sometimes manifests in social and political contexts, particularly when masses of people feel discontent with the status quo. Driven by a desire for change, they may seek to alter their circumstances—often without fully understanding the consequences. In well-established democratic systems, where peaceful transitions of power are possible, this desire for change is addressed through structured processes. The result is typically a return to stability as the consequences of such transitions become apparent over time. However, in societies where power remains centralized and change is restricted, this unfulfilled drive for progress often leads to widespread social unrest. Revolution, rebellion, and chaos may follow, resulting in significant upheaval. The lack of a legitimate mechanism for the circulation of power can turn a natural human impulse for progress into a force for societal destabilization.

13. Connection with Nature and Survival Skills

Humans are inherently drawn to nature, seeking to return to the embrace of their true mother—the natural world. Time spent in nature restores balance and peace, nurturing both body and mind. In suburban areas, this connection is more easily maintained, but in urban environments, barriers such as concrete walls, dense populations, apartment living, and a lack of green spaces can create man-made cages, severing individuals from nature's soothing presence. – (see Note 111). Addressing this disconnection is essential for a healthy society. Governments and local municipalities carry the responsibility to create accessible parks, green spaces, and natural sanctuaries throughout the year. While national parks and outdoor excursions like camping provide valuable opportunities to reconnect with nature, their distance and cost can limit access. Instead, communities and municipalities should prioritize the establishment of nearby, year-round, and welcoming natural spaces, ensuring everyone can experience nature's benefits regardless of season or circumstance. These spaces do far more than beautify the environment. They foster physical and mental well-being, ease social tensions, and contribute to the prevention of numerous societal problems. Beyond access to nature itself, there is immense value in rediscovering the fundamental skills of surviving in nature. For countless generations, human life depended on understanding and harmonizing with the natural world—knowing how to find food, water, and shelter; how to navigate unfamiliar terrain; how to remain patient, observant, adaptable, and resourceful.

Note 111- Desmond Morris and the Human Zoo-*Desmond Morris, the renowned English zoologist, argues in his influential book* The Human Zoo *that by embracing urbanization, humans have essentially placed themselves in a situation similar to that of animals in a zoo. He points out that while cities are undeniably impressive in their splendor and complexity, they also share the same characteristics and limitations found in traditional zoos. The high population density and various cultural pressures in urban environments create both visible and invisible cages for individuals. These social constraints confine people in ways that hinder their natural and biological behaviors, sometimes leading to maladaptive or problematic actions. In this sense, the behavior of humans in the modern city can be compared to that of captive animals in zoos. Morris does not reject urban life. On the contrary, he recognizes it as a remarkable achievement of human ingenuity and creativity. However, he raises a critical question: How can we ensure that the biological needs of humans are addressed within the confines of modern urban environments to prevent maladaptive behaviors from emerging? This, he suggests, is a challenge that requires careful attention.*

Hunting was historically one essential aspect of survival, particularly in rural and indigenous communities where it remains part of daily life. – (see Note 112) Yet, even in societies where hunting is no longer a necessity, the qualities it cultivated—tact, foresight, courage, cooperation, patience, and planning—remain part of our human heritage. These attributes continue to guide us through the complexities of modern life.

While the NUSAWVO school of thought does not encourage hunting for sport and considers unnecessary harm to animals as ethically questionable, it acknowledges that certain aspects of hunting—such as the art of careful observation, understanding animal behavior, learning to move silently and mindfully—can still provide valuable lessons. These skills foster awareness, presence, and respect for the delicate balance of life.

Ultimately, it is the responsibility of families and communities to help the younger generation rediscover this connection to the natural world—not only as an escape but as a fundamental part of human growth. Whether through hiking, camping, gardening, learning survival techniques, or simply spending time outdoors, this return to the nurturing arms of our ancient mother will greatly contribute to the flourishing of individuals and the well-being of society.

Note 112- *It comes as no surprise that, especially with the advent of digital cameras, countless people have embraced nature photography, referring to it as "hunting the moment."*

14. Honesty, Integrity, and Promise-Keeping

Human beings are inherently drawn to truth and honesty. While trickery is a natural part of human nature—an inheritance from the need to hunt—it stands in conflict with the transcendent nature of the human spirit. Deception, falsehood, and dishonesty demand great energy, leading to anxiety and unrest. Thus, although humans possess the ability to deceive, it does not inspire true passion; their lasting peace is found in truth. The wisdom of ancient civilizations reflects this understanding. Over 2500 years ago, Darius the Great, one of Iran's most renowned Kings from Achaemenid dynasty, inscribed a powerful invocation: "May God protect this land from drought, lies, and enemies." His words recognize falsehood as a fundamental threat, akin to natural disasters and external foes—capable of eroding the foundations of a nation.

A society that succumbs to a culture of deception—where falsehood governs its beliefs, people, and rulers—inevitably faces profound cultural, economic, social, and political crises. In such an environment, the well-being and happiness of its people are always at risk. Within the family, children learn honesty not from words but from the actions of their parents. When parents uphold their promises, they instill loyalty and integrity in their children, fostering deep trust among family members. This trust, in turn, forms a strong foundation for personal growth and societal harmony.

15. Awareness of Moral Defilements (Alterations-The Abysses)

Throughout human history, certain universal and enduring taboos have shaped our evolution. As discussed earlier, the taboo against incest and the necessity of exogamy are not only embedded in our cultural memory but also rooted in our genetic reservoir. Other taboos—though not all—share a similar foundation, even if they remain less apparent in cultural consciousness. In this sense, widely accepted taboos can be seen as the product of cultural-biological evolution, refined through generations of trial and error. Taboos have historically emerged in contrast to defilements—the abysses that threaten human growth and well-being. While the specific definition of defilement may vary across societies, its essence remains remarkably consistent. No culture has ever sanctified or encouraged vices such as murder, theft, deceit, adultery, reckless indulgence, or substance abuse. Societies that disregard these moral defilements inevitably decline into depletion and decay.

However, the number of moral defilements—and the corresponding ethical boundaries—should remain limited yet fundamental. If excessive moral restrictions are imposed, they may provoke rebellion and psychological unrest, hindering human freedom, choice, and growth. Instead, moral principles should serve as a core framework, while laws and social rules function merely as extensions of these foundational truths. Importantly, every individual—especially children—has the right to be aware of these defilements and their dangers. Just as a novice driver must learn road signs to navigate safely, a child must recognize the moral abysses that threaten their well-being. However, moral awareness is not instilled through sermons or lectures. Children do not learn the "untainted state"—the avoidance of defilement—through words alone, but rather through the real and objective behavior of their parents and environment.

Thus, if families wish to prevent their children from straying into moral peril, they must first distance themselves from defilement. There is no doubt that a child's happiness is endangered when they lack awareness of these moral abysses—especially when they grow up without strong, healthy ethical models in their family and society. That said, history has shown that even individuals from immoral and unstable families can rise to great moral and social heights. Having experienced the consequences of corruption firsthand, some develop an unshakable commitment to virtue. Ultimately, every person determines their own path through the power of choice.

16. Tools and Technology

Human beings are both tool makers and tool users—they derive satisfaction from working with tools, and more accurately, they need tools to maintain mental balance and fulfill the complex motivations of the brain. Tools, in all their forms, serve as extensions of the human body, rooted deeply in our socio-biological evolution. Whether industrial, artistic, commercial, or domestic—such as in sewing or cooking—tools are an inseparable part of human existence.

Wise families, organizations, and governments recognize the importance of tools and technology. They provide appropriate resources, facilities, and education to children, the general public, and especially to students and job seekers. This support extends to all sectors, ensuring that even prisons, sanatoriums, and homes for the elderly are equipped with the tools necessary for personal development, work, and well-being.

17. Peace-Seeking, Love, Compassion, and Voluntary Service-(see Note 113)

The innate desire for peace, love, and compassion is deeply ingrained in human beings. At our evolutionary core, we share common roots with bonobos and chimpanzees. Bonobos, in particular, are known to be sentimental and peace-seeking creatures. Remarkably, no bonobo has ever been observed to kill another. Similarly, chimpanzees are highly emotional animals, with strong feelings of love, care, and friendship.

Note 113- *"The only ones among you who will find true happiness are those who are devoted to serving others and have discovered their path to do so."*

— Albert Schweitzer

Looking back at our evolutionary history, we can identify numerous factors that have shaped human development over the last 5 million years. Drives such as "organization," "assistance and solidarity," "brotherhood and union," and the "nuclear family" have played vital roles. Fortunately, the development of these drives has promoted peace, friendship, love, and compassion—principles that have allowed over 8 billion people to coexist and work together in the first quarter of the 21st century.

Beyond familial love and compassion, the highest expression of these emotions is seen in charity work and voluntary service. Many individuals dedicate their time to serving others without expecting anything in return, demonstrating the essence of love and compassion. People who choose to spend part of their lives—or even their entire lives—in such selfless acts embody the highest form of humanity.

A society and family that actively engages in universal philanthropy, voluntary service, and charitable work ensures the happiness and well-being of its members, particularly its children.

18. Wishing Happiness and Benevolence for All, Without Exception and Without Expectation

The principles of peace, friendship, love, compassion, and voluntary service extend to the desire for the happiness of all, without exception. The pursuit of joy is one of humanity's most fundamental drives. No one seeks suffering for themselves or others. Turmoil and conflict arise from imbalances within our transcendent human nature and from ignorance about our integral existence in a shared essence called consciousness. As children grow, it is crucial for parents to emphasize this fundamental truth: the beliefs, words, and actions of a person are distinct from their transcendent essence. When someone causes harm or discomfort, it often stems from misguided teachings or, at worst, from an internal imbalance. A child should not learn to hate individuals but instead be taught to separate people from their actions.

19. Skepticism and the Avoidance of Dogma

Some cultures and belief systems tend to revolve around three fundamental attitudes:

1. We are right.
2. Everyone else is entirely wrong.
3. We are superior to all others.

One of life's greatest fortunes is being raised free from these rigid beliefs during childhood. However, no individual or belief system is entirely immune to these tendencies. The moment we express an idea, we do so because we see some truth or value in it—at least to some degree. This is not problematic in itself; in fact, it drives intellectual progress and cultural openness. The real issue arises when we assume that we hold the absolute truth, failing to recognize that every idea or phenomenon has countless dimensions. To fully grasp even a single aspect of reality requires immense effort and collective inquiry. Ultimately, we come to understand that all knowledge is relative, always subject to refinement and deeper understanding.

The complexity of existence—whether in ideas, phenomena, or fundamental truths—is so vast that even the simplest issue cannot be fully dissected without encountering endless chains of causes and effects. – (see Note 114) If such is the case with everyday matters, how much more intricate are philosophical inquiries into subjects like the first and second realms or the paradoxes of non-locality and non-causality in the second realm?

Note 114- *In this context, please refer to the discussion on pragmatism in Appendix No. 2.*

Open-minded parents will ensure that rigid beliefs do not take root in a dogmatic or biased manner. More importantly, they will give their children the freedom to explore and form their own beliefs. Likewise, intellectuals, elders, and religious or cultural leaders must promote critical thinking, skepticism, and an awareness of the dangers of dogma. They must strive to balance the three factors that often lead to ideological rigidity. The Czech thinker Ivan Klíma wisely noted: "No idea in the world is so good that it justifies a fanatical effort to impose it upon others." Given these complexities, it is essential for parents to foster open-mindedness in both themselves and their children. They must ensure that rigid beliefs do not take root in a dogmatic or biased manner. More importantly, children must be given the freedom to explore and form their own beliefs. Fortunately, with the increasing free flow of information in today's world, dogma and bias are gradually diminishing—whether by intention or not.

20 - Spirituality and Transcendence

Beyond physical, emotional, and intellectual needs, human beings also have spiritual needs. Therefore, families and society must provide enough freedom for individuals to explore and fulfill these aspects of their existence. Suppressing or restricting religious and spiritual activities—as was the case in some communist regimes—inevitably leads to the sterility of human nature in its entirety. At the same time, rigid insistence on religious teachings, particularly the promotion of altered or misinterpreted doctrines that contradict reason, humanity, and compassion, fosters violence and division. It is likely that such distortions have been amplified over time, diverging from the original essence of these traditions. Consequently, indoctrinating children with dogmatic or manipulated religious ideas not only harms their well-being but also destabilizes society, leading to religious conflicts.

Forcing adherence to a specific religion or ritual while degrading other faiths and spiritual perspectives as false or deviant is equally damaging. Spirituality, by its nature, is an instinctive pursuit that does not require coercion or forceful intermediaries. When imposed, it often breeds resentment and aversion. Moreover, those who make extreme and irrational spiritual claims frequently become symbols of misguidance and destruction themselves. Additionally, the emergence of risky mystical practices—especially when intertwined with narcotics—poses serious dangers and leads to further confusion.

If it can be introduced without appearing self-centered or arrogant, we might consider presenting the concept of 'meaning' to children in a way aligned with the NUSAWVO school of thought—if, of course, we find value in this perspective. This approach fosters a gentle and inclusive spirituality, offering a broad and insightful foundation that resonates with the teachings of true spiritual guides while remaining compatible with the later spiritual journeys of religious individuals, regardless of their faith. This approach presents consciousness as the essence of the universe and imparts to children the understanding that we are all one—that others, all beings, and all things are extensions of ourselves. In doing so, we establish the natural foundations of universal spirituality and conscience, ensuring their happiness more than ever before.

Continuation of the Discussion

Through the 20 factors we have explored, we sought to approach the "common natural essence"—the "natural gem of humanity"—within all human beings. In other words, we examined the meeting points of external and internal influences that form the foundation of our happiness, viewing this "human gem" from 20 different perspectives. Some of these aspects may appear redundant, while others may have been unintentionally overlooked. We make no absolute claims in this regard. However, if we wish to delve deeper into the subject, appreciate its subtleties, and advance our understanding, we must continue our exploration through these 20 elements. To move forward, we assume that every individual benefits, to some extent, from some of these foundational elements. If people experience varying degrees of happiness, perhaps this difference stems from how much they derive from these gifts—though this remains only an assumption.

We may even hypothesize a direct correlation between happiness and the degree of access to these factors—suggesting, for example, that the complete fulfillment of all 20 foundations equates to absolute happiness, while complete deprivation leads to total despair. There may be some truth in this idea, yet it does not capture the full complexity of happiness. In reality, happiness holds many mysteries, and to unravel them, we must examine different perspectives.

Daphne du Maurier insightfully stated:
"Happiness is not a possession to be prized; it is a quality of thought, a state of mind." Carl Gustav Jung similarly observed:
"It all depends on how we look at things, and not how they are in themselves."
Masiha Barzegar added:
"Comfort, leisure, and happiness lie in the way we perceive life, not in life itself. It is our perspective that shapes the world we see."
Francis Bacon reinforced this notion:
"To live happily, we must create favorable circumstances rather than wait for them."
And finally, Randy Cage affirmed:
"Your situation changes when you change your actions, and your actions change when you transform your beliefs."

Each of these statements offers a profound perspective, but the key takeaway from all of them is this: happiness is not something external to us—it is shaped by our own actions and perceptions. Ultimately, all roads lead back to us. If this conclusion is valid, the next crucial question arises: What specific actions must we take? And how do they relate to the 20 foundations of happiness? To deepen our inquiry, let us consider a thought-provoking statement that may initially seem unsettling, yet aligns with our exploration and ultimately leads to clarity: Antoine de Saint-Exupéry, the renowned French writer, once said: "Whatever a man clings to with the most passion is taken away from him." This statement holds a certain truth, but it raises a vital question: Is happiness something beyond human grasp? Do we need to "seize" happiness, or should we seek it? Is the act of pursuit itself what leads to its eventual loss? In reality, no—happiness is not something to be seized, nor is it an object to be sought. Happiness is like breathing. We do not need to chase it, nor do we need to grasp it desperately. Just as all beings are born into existence, they are also born into happiness. Life itself is happiness. Yet, within this realization lies an essential truth: while we do not need to pursue happiness as if it were a distant prize, we do need to place ourselves on what we call "the path of happiness." Happiness, after all, is not a fixed goal—it is a journey. Importantly, this path is not separate from our being. It is entirely possible that we are already on this path, yet unaware of it. Or perhaps we are not actively engaged in it.

This leads us to the final, most crucial question:

How can we become actively and consciously present on the path of happiness?

NUSAWVO's Approach to Happiness

To answer this question, the NUSAWVO school of thought offers a deeper and more comprehensive understanding of human happiness. As we've seen, the NUSAWVO school embraces a singular definition worthy of humanity: "the happy human." In this context, the school focuses on the "common natural essence of humanity," examining it from 20 perspectives—referred to as the "20 foundations of human happiness."

However, as the discussion progresses, NUSAWVO does not rely solely on these 20 foundations for its definition of humanity and happiness. The school moves beyond the role of a neutral observer, stepping into the position of a humble and unpretentious consultant, offering its unique approach to the subject of happiness. At this stage, the school sets aside questions of whether we are benefiting from the 20 foundations and shifts its focus to a broader and more holistic understanding of happiness. This exploration aims to uncover the "path of happiness," independent of our access to the 20 foundations, seeking to reveal the deeper secrets of happiness itself.

In this regard, NUSAWVO introduces a fundamental, timeless, and profoundly important concept: "Unified human nature." It proposes that true happiness can only be fully realized when an individual's nature is activated in all aspects of their being.

According to NUSAWVO, human nature consists of four essential aspects: physical, emotional, mental, and spiritual. When these aspects are activated in a complete and balanced way, the entirety of human nature comes into full expression. But the question remains: How do we achieve this activation in the real world?

NUSAWVO offers a simple and natural solution. This solution involves a specific arrangement of key natural elements that define our humanity. Central to this arrangement are the "six natural human characteristics," encompassing the four aspects of human nature and, in effect, the 20 foundations of happiness. These six characteristics not only represent happiness but, in a way, recreate it. We call these six characteristics the "six agent gems of human happiness."

The NUSAWVO school believes that these six gems hold the potential to activate each of the four aspects of human nature (and, correspondingly, the twenty foundations of happiness) in a balanced and harmonious way.

The six active gems of human happiness are:

- **Naturalness (N)**
- **Unaltered-Untainted (U)**
- **Seeking the Truth (S)**
- **Abiding in Truth (A)**
- **Wishing Happiness for Everyone (W)**
- **Volunteering to Serve Everyone (VO)**

As observed, the first letters of these gems form the acronym "NUSAWVO."

According to NUSAWVO, the six gems hold the power to shift the responsibility for our happiness from external factors—such as society and others—and place it firmly within ourselves. In doing so, they empower us to actively pursue our own path of happiness. When these six gems operate in harmony, they activate our Full-Scale Human Nature, or what we refer to as our sublime nature.

The next natural question is: How do we activate these six characteristics?

To activate any natural human characteristic, the first step is awareness: we must become conscious of the nature and function of the characteristic in question. The next step involves identifying the means and power necessary to activate it. These solutions and powers may range from discipline and control to force and fear. While these methods may have utility in certain contexts, the NUSAWVO school does not rely on such forces in its teachings. Instead, NUSAWVO assigns the task of activating the six gems (NUSAWVO) to the power of love and compassion, which align with the law of attraction. In a world full of countless influences and motivations, the NUSAWVO school believes that the only true guiding force for activating and organizing a person's wholeness is the power of love and compassion—the force that draws us toward harmony and fulfillment.

What is this "Love," really?

Understanding love is not a difficult task, for it is woven into the very fabric of our being, nourishing every aspect of our existence. This is the love that brought our ancestors together, the love that sustained us in our mother's womb and granted us life. It is the love that empowered our fathers to endure life's hardships, granting them the unwavering will to protect and support their families. And it is the love that continues to accompany us, embracing every moment of our presence in this world. From before we existed to the moment we are here—and beyond—love governs both our being and non-being, shaping the world itself.

It would not be wrong to regard love as an aspect of spirituality, nor would it be incorrect to see love and consciousness as two sides of the same coin—two expressions of a singular reality that transcends time and space, indeed, that creates time, space, and all things. Thus, the human being is, in essence, a lover—one who knows that they are in love. And what could be more glorious than this? Love—and, more precisely, love and compassion—has another dimension, known as the law of attraction. As we shall see, love is the force that activates Unified Human Nature (Full-Scale Human Nature) and all the sublime characteristics that define our highest potential. Compassion and the law of attraction further illuminate the subtle dimensions of this profound truth. In future explorations of this school, we will delve deeper into these insights.

Who Should Apply This Force?

Who should apply this force of love and compassion? At first, it falls upon parents, caregivers, society, and culture to nurture it. However, ultimately, it is the individual who must take responsibility—who must seize control of their own happiness and destiny and, through this divine force, shape their own path.

And what is the final outcome of all this?

When the six agent-gems of NUSAWVO are activated through love, compassion, and the law of attraction, Full-Scale Human Nature—or Transcendent Nature—is realized. As a result, a person takes direct control over their fate and happiness, stepping consciously and intentionally onto what we call "the path of happiness."

Let us state it once more:

"When the six agent-gems of NUSAWVO are activated with the power of love and compassion (and the law of attraction), Full-Scale Human Nature is fully realized. As a result, the individual embarks on the journey of happiness in a conscious and self-determined way." From "Happy Human" to "Active-Agent Happy Human"

Through its teachings, the NUSAWVO school elevates the concept of a "happy human" to that of an "active-agent happy human." - (see Note 115) However, this insight does not stop at a mere definition. It takes the final step in completing the process of happiness by inviting happiness itself into the realm of human consciousness and self-awareness. Bringing happiness into human awareness is not a complex matter. On the contrary, it requires only a few simple reminders. For anyone who takes agency over their destiny (or is already walking the path of happiness), happiness already exists within their subconscious. However, like many profound truths, it must be emphasized and brought into full consciousness. This irreplaceable gift—self-aware happiness—manifests through five exalted qualities. Simply knowing and living by them, whether consciously or not, will elevate our happiness into self-awareness. These five characteristics-traits are:

- **Thanksgiving (Appreciation, Gratitude)** – The Thanksgiver (T)
- **Hope** – The Hopeful (H)
- **Astonishment and Elation** – The Astonished and Elated (A)
- **Opportunity Finding and Opportunity Making** – The Opportunity Finder and Maker (O)
- **Seeking Meaning** – The Seeker of Meaning (S)

Note 115- The famous English artist ***John Lennon*** *once said:* ***"You don't need anyone to tell you who you are or what you are. Whatever you are, you are who you are!"*** *Similarly,* ***George Bernard Shaw*** *remarked:* ***"Life isn't about finding yourself. Life is about creating yourself."*** *At first glance, these perspectives may seem distinct—one emphasizing self-acceptance, the other self-creation. However, within the framework of the* ***NUSAWVO school****, we seek to* ***harmonize*** *these two profound insights. We recognize that a person's* ***essence*** *is already present within them—unaltered, untainted. Yet, this essence is not merely something to be passively discovered; it must be actively realized, shaped, and fully* ***activated****. In this way, NUSAWVO integrates the wisdom of both views:* ***you are who you are, and at the same time, you must become who you are meant to be.***

As observed, by taking the first letters of these five traits, they can be arranged as **THAOS**—or in any other preferred order.

Through this dynamic interpretation and the unveiling of happiness's secret, the NUSAWVO school refines its concept of an "agent happy human," further elevating it to a **"self-aware agent happy human."**

The Two-Step Path to Happiness

In summary, the entire proposed solution of this school regarding human happiness is encapsulated in the following two fundamental steps:

1. **Activation of Full-Scale Human Nature** – Through the power of love and compassion and the law of attraction, the six characteristics known as NUSAWVO awaken the unified human nature, enabling a person to actively shape their own happiness.
2. **Integration of Happiness into Self-Awareness** – Once happiness manifests through the five exalted qualities-traits, it is fully invited into self-awareness, transforming into what can be called **"self-aware agent happiness."**

At this point, once this gift is fully realized, the name we give ourselves becomes secondary. Whether we call ourselves a **"self-aware agent happy human,"** an **"agent self-aware human,"** a **"happy human,"** or simply **"human,"** the task is accomplished—the secret of happiness is revealed! - (see Note 116)

Note 116- *It is clear that, beyond the cases mentioned, many others could be considered but have been left unaddressed. We entrust the continuation of this exploration to* ***seekers of truth****—including you.*

The topics of **"Agent Happiness"** and its key components—including **"Full-Scale Human Nature," "The Six Agent Gems of Human Happiness (NUSAWVO),"** and **"The Power of Love and Compassion"**—will be explored in depth from the fourth to the eleventh insights. Additionally, in the **twelfth insight**, we will delve into the concept of **"Self-Aware Happiness"** and its five exalted traits in detail.

Summarizing the Third Insight

With this, the third insight reaches its conclusion. Throughout our exploration, we traced the natural evolution of humanity and uncovered the foundation of **happiness-bliss**. We familiarized ourselves with the beliefs of the NUSAWVO school, which envisions a transition from **"unconscious happiness"** to **"self-aware agent happiness."**

Ultimately, we arrived at the image of a truly happy human—one with a **full-scale (unified) nature**, embodying love, care, gratitude, hope, astonishment, elation, the ability to find and create opportunities, and the pursuit of meaning. In other words, a **self-aware, agentic, happy human being.** Yet, despite these insights, one may still feel that the question of happiness remains unanswered. We make no absolute claims in this regard. Instead, we invite you to continue with us through the upcoming insights, where further dimensions of our response unfold. However, a full understanding of happiness, as seen through the NUSAWVO school, emerges only when all **13 insights** are taken together.

That being said, undoubtedly, any claim to have discovered the "secret" of happiness—or worse, that adhering to a particular teaching could eliminate all unhappiness—is not only naive but absurd. **Sorrow, too, has an honorable place in our nature.** It is not a flaw in our existence but an integral part of it. When understood correctly, even sadness can be a force for growth, transformation, and excellence.

In the next insight, we will delve into **unified(full-scale) human nature**—humanity's transcendent essence—and its many facets. The symbol of the third insight is a **circle,** representing the completeness of the happy human. This circle, as we will see, extends to encompass the related insights that follow, from the **fourth to the twelfth.**

Happy Human Being

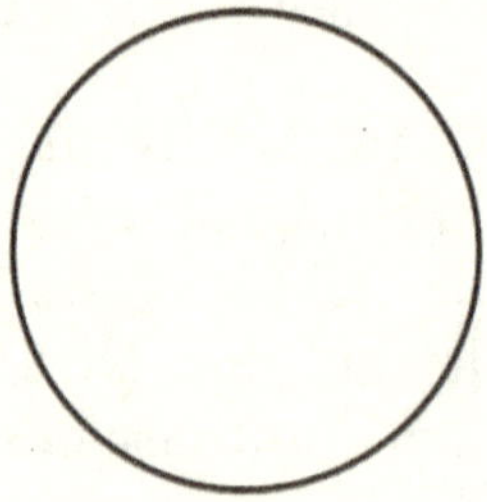

Below, the interconnection of these insights has been depicted in the figure:

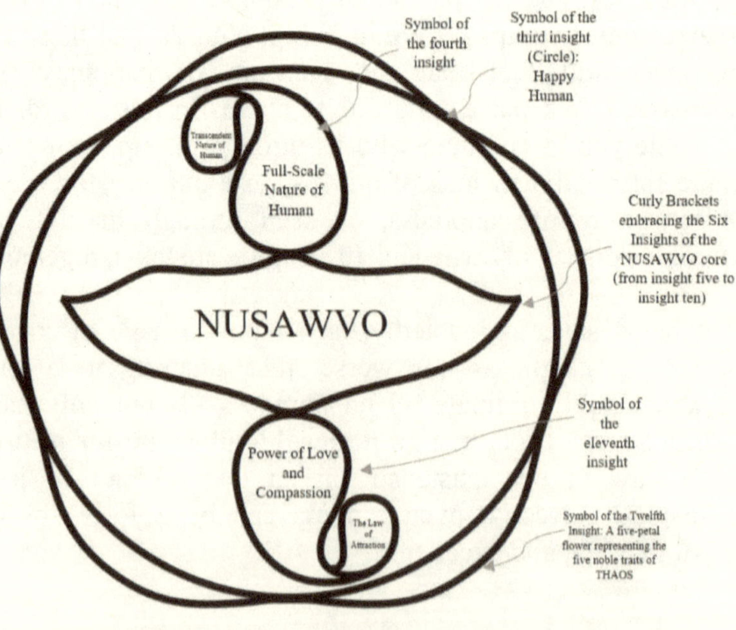

The symbolic sequence of the first three insights in this school is as follows.

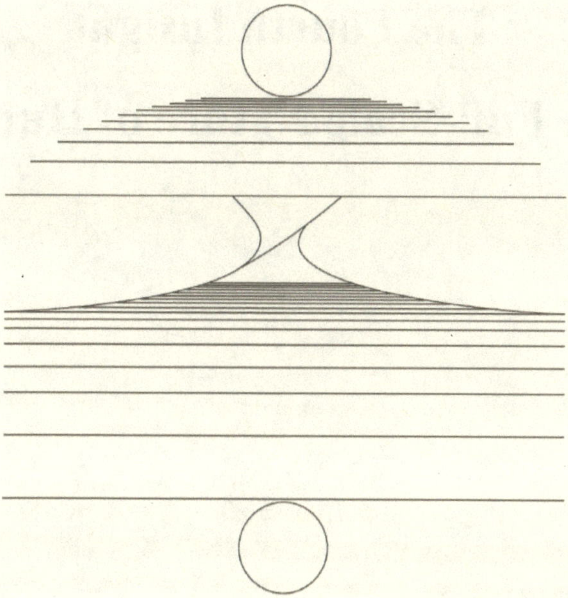

The Fourth Insight
"The Full-Scale Nature of Human"

Humanity is lost in dust and stone,
A fleeting shadow, yet unknown.
Another world must take its flight,
To birth a human bathed in light.
-Hafiz

Not in fine robes does humanity shine,
But in the soul, in acts divine.
Look beyond the silk and gold,
To where true human worth unfolds.
-Saadi

The Fourth Insight: The Full-Scale (Transcendent) Nature of Human Beings

The fourth insight of the **NUSAWVO school** is dedicated to the concept of **Full-Scale Human Nature**, also referred to as the **Transcendent Nature** of human beings. As mentioned in the third insight, this school views the transition from **"unconscious happiness"** to **"self-aware (agentic) happiness"** as dependent on the activation of this **unified** nature.

The **Full-Scale Nature** of human beings consists of four fundamental aspects:

1. **Physical** – The biological dimension of human existence, encompassing everything related to the body.
2. **Emotional** – The domain of feelings and experiences, including both positive and negative emotions.
3. **Mental** – The realm of cognition, encompassing perception, thinking, beliefs, values, interpretations, desires, and goals.
4. **Spiritual** – The dimension where diversity and plurality dissolve—where "I" and "others" lose their distinctness. Here, we are not just parts of a whole, not mere waves in an ocean—we **are** the ocean itself. This mysterious ocean is called **Consciousness**. Moreover, this aspect represents our innate ability to **connect with meaning (God).**

These four aspects are not rigidly separate; they interweave seamlessly, like colors blending through a dome of stained glass, each hue offering a different interpretation of the whole.

While these aspects were referenced in previous insights—particularly in the **20 Foundations of Human Happiness**—they remained **scattered and abstract**. To transform them into **practical and structured principles**, they must be embodied as **fundamental traits** that shape our thinking and behavior.

The Six Gems: Embodying the Full-Scale Nature

The **NUSAWVO school** proposes **six essential characteristics**—referred to as **the Six Gems**—as the embodiment of **all four aspects** of unified human nature. These traits form an **organic, interwoven system**, refining and complementing one another to cultivate a **balanced, dynamic, and self-aware personality**. Ultimately, they transform happiness from a passive state into an **agentic and conscious force** in life.

The **Six Gems** are as follows:

- **N** – Naturalness-The Natural state
- **U** – Unaltered-Untainted-Undefiled
- **S** – Seeking the Truth
- **A** – Abiding in Truth
- **W** – Wishing Happiness for Everyone
- **VO** – Volunteering to Serve Everyone

These initials form the name **NUSAWVO**.

The Role of Love: Activating the Six Gems

These **six characteristics** exist as **potentials**, but their **activation** depends on a fundamental force: **Love** (and the Law of Attraction).

Among the countless possible guiding forces, the **NUSAWVO school** recognizes **Love and Compassion** as the **primary** forces that awaken, refine, and harmonize these traits. Love is not merely an emotion—it is a **force woven into our very nature**, emerging from the essence of existence itself.

Thus, it is no surprise that **Love** can be seen as a **translation of Consciousness**. Nor is it surprising that **Love serves as the foundation of our connection with meaning**. - (see Note 117) Love is what **sets the Six Gems into motion**, transforming **potential into action** and harmonizing them into a **unified and balanced composition**—what we call the **Full-Scale (Transcendent) Nature of Human Beings.**

On the other side of **Love and Compassion** lies **the Law of Attraction**, which will be explored further in the **11th Insight**.

Summary of the Fourth Insight

In this insight, we explored the concept of **Full-Scale Human Nature**, also known as the **Transcendent Nature** of human beings, along with its four fundamental aspects: **physical, emotional, mental, and spiritual**.

We also introduced **the six natural characteristics**, referred to as **the Six Agents of Happiness**, which embody and integrate these four aspects:

- **N** – Naturalness
- **U** – Undefiled
- **S** – Seeking the Truth
- **A** – Abiding in Truth
- **W** – Wishing Happiness for Everyone
- **VO** – Volunteering to Serve Everyone

Note 117- *In eternity's hush, before time took flight,*
A ray of your beauty longed for the light.
Desire awakened, a spark took form,
And Love was born in the heart of the storm.
Hafiz

Together, these characteristics form the acronym **NUSAWVO** and serve as the **executive and operational core** of the **NUSAWVO school**.

Additionally, we emphasized the role of **Love and Compassion**, along with the **Law of Attraction**, as the **fundamental force** that **activates, harmonizes, and coordinates** these six traits.

The **Six Gems of NUSAWVO** will be explored in detail in **Insights Five through Ten**, while the **power of Love and Compassion** will be examined in the **Eleventh Insight**.

The symbol representing the **Full-Scale (Transcendent) Nature** of human beings, as introduced in this insight, is as follows:

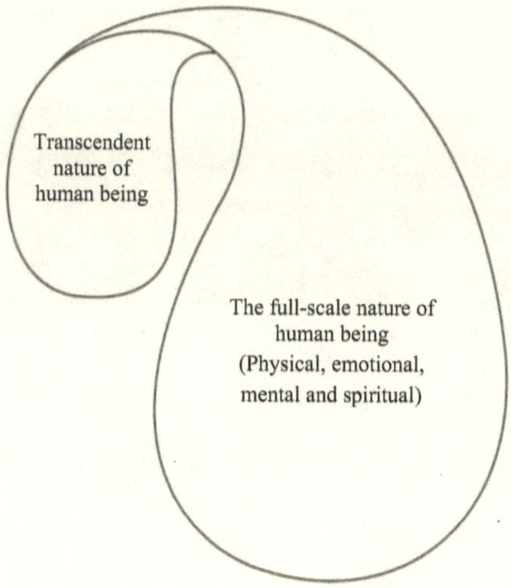

The Fifth Insight

"Naturalness"
(The Natural State)

"Don't get me wrong, I admire elegance and have an appreciation of the finer things in life. But to me, beauty lies in simplicity."

-Mark Hyman

"The unlimited human being, in his continuous flow and insightful growth, instinctively turns to himself. At times, he is creative; at times, he is childlike. He surrenders only to what truly makes sense to him—what he believes in, desires, and chooses to live by."

-Wayne Dyer

Through the streets, our sheikh did roam,
A lamp in hand, its light his own.
"I tire of demons, I tire of beasts—
A human soul is what I seek."

"None exist," the people cried,
"We have searched both far and wide!"
He smiled and said, "Then all the more,
I long for what is yet denied."

-Rumi

"By allowing things to follow their own nature, they will eventually find their own way to rest and fulfillment."

-Laozi, Tao Te Chin

The Fifth Insight: Naturalness

The fifth insight of the NUSAWVO school centers on the concept of the "natural." "Natural" is the first of the six agentic gems of human happiness that form the core of the NUSAWVO worldview.

When we speak of the "natural human" in this context—without any prefixes or suffixes—we seek to present human nature in its simplest and most complete form. This approach avoids unnecessary complexity or abstraction, allowing us to grasp the essence of our existence as it truly is. Once we achieve this clear and unaltered understanding, we can begin to extract the other agentic gems of happiness from its core and expand upon them.

The message of this insight is simple: we must trust our nature. We believe that deep within us, there are no inherent tendencies toward hard-heartedness or harm toward others. Such behaviors are rare, even among our close evolutionary relatives, such as chimpanzees and bonobos. From this perspective, it becomes clear that the root of significant moral and social deviations does not lie in our nature but in external factors—namely, "cultural shortcomings" and "cultural deviations." **Cultural shortcomings** refer to the failure to nurture the constructive aspects of our nature, while **cultural deviations** represent the imposition of values, beliefs, and actions that stray from these constructive elements. Both, in essence, are forms of **ignorance**—a disregard for the natural foundations of our happiness.

We will explore these points further. As we discussed in the Third Insight, after diverging from our evolutionary relatives, humans underwent extensive evolutionary changes over millions of years. These changes affected all dimensions of our being—biological, cultural, and social. Fortunately, this trajectory has moved us toward greater connection and harmony, allowing billions of people to coexist today. Among these changes, **culture** has played a pivotal role, emerging as a supra-genetic force that operates somewhat independently of our biological nature. This cultural development has shaped much of human behavior, bridging the vast gap between our prehistoric and modern states.

However, as an abstract category, culture is not immune to manipulation. It can be distorted, leading us toward unnatural values, beliefs, and actions. Edgar Morin eloquently captures this idea when he states:

"In the absence of culture, man is a helpless and backward creature who will not be able to continue his life."

Yet, he also reminds us:

"Becoming human is not only biological evolution, nor solely social-cultural evolution, nor only spiritual evolution. It is a complex and multifaceted process involving genetic, biological, cultural, social, and spiritual factors."

Here, we encounter a mystery. If Morin's words hold true, and human nature is indeed a complex, multifaceted blend of biological, social, cultural, and spiritual elements, then we face a paradox. Our initial understanding of the "natural human" seems to stand at odds with this complexity. When we focus on the natural human, we are not dealing with a simple subject but with a category that integrates various dimensions.

There is no doubt that a profound truth lies within these interpretations. Yet, we believe that beneath this intricate, multifaceted structure, there is an **inner core**—a substance yet to be fully uncovered. Our aim is to examine it independently, free from the constraints of various cultural lenses. After all, human cultures differ from one society to another and even from one individual to another. Despite these differences, a common thread unites us all: **simplicity, sincerity, and love**. This is the **natural gem** of humanity.

Earlier, in the twenty titles of the Third Insight under the theme *Foundations of Human Happiness*, we sought to approach this

inner essence—the shared nature of all humans, beyond external differences. Perhaps this is the **natural human being** we have lost touch with. It can be viewed from twenty different windows, which we call the **constructive elements of our nature**, or more simply, the **natural foundations of human happiness**. We will now briefly revisit these twenty titles.

1. Balanced Environment:

Human beings are inseparable from their environment. We are all part of one interwoven fabric, and our environment—which encompasses the entire planet—is fragile. Its delicate balance is threatened by the rapid development of our civilization. It is crucial that we urgently address this imbalance, identifying and halting the harmful aspects of progress. Undoubtedly, human happiness lies within a balanced environment.

2. Security:

Security is the fundamental priority for every living being. For humans, whose ability to think about the future has evolved, the need for security has grown even more critical. Without security, there can be no trust in tomorrow. This is why a person prioritizes the safety of themselves, their family, and others.

3. Livelihood and Purposeful Work:

Adequate nourishment, shelter, and work are vital to human well-being. Without proper nutrition, both physical and mental health suffer. Without work, mental equilibrium is lost. Purposeful engagement is essential for human happiness.

4. Health and Vitality:

Health plays a crucial role in maintaining the natural balance of the human being and ensures survival. Vitality is a cornerstone of happiness.

5. Close-knit Family:

Humans need a family for love, shelter, and support in growth. Without a family, we would not have reached the advanced biological, cultural, and spiritual development we have today.

6. Self-Dignity and Respect:

A human being possesses inherent dignity, and it is essential to be respected for this unique gem. Respect fosters decency and honor, keeping our nature from falling into danger and guiding our growth.

7. Freedom and Individuality:

Human beings are inherently free and possess individuality. An ideal social system promotes solidarity while respecting the freedom and individuality of each person, allowing for their full participation in political and social spheres.

8. Justice and Fairness:

Justice resides within the human institution, although its realization requires constant effort. Despite its challenges, the current state of human civilization, with all its flaws, is far more just compared to the conditions of our ancestors.

9. Curiosity, Exploration, and Creativity:

Human beings, like all living creatures, are inherently curious and searching. However, due to the advanced capabilities of our brain and culture, humans possess an extraordinary capacity for exploration and creativity, encompassing science, art, philosophy, and spirituality. This irreplaceable gem must not be suppressed or limited by ideological or extreme religious doctrines.

10. Channeled Aggression, Play, and Sports:

While aggression may seem at odds with play and sports, these elements may actually share a common origin in the human drive for survival and the need to coexist and interact. The instinct of aggression should not be suppressed but rather guided toward constructive and uplifting paths through activities such as games, sports, group work, and nature walks. In high-energy social ranks—such as the military, police, firefighters, or Red Cross—aggression can be channeled into responsible roles that promote safety and well-being. Healthy aggression is vital for ensuring the security of humanity, and sports complexes are crucial for both the physical and mental health of society.

11. Art:

Art is inseparable from human nature. It helps to elevate humanity and develop our capacities, serving as a form of meditation through the silence it brings to the mind. This fosters spiritual growth, making art an essential component of human well-being.

12. Restlessness and the Drive for Progress:

Restlessness is inherent in human nature, stemming from the highly developed brain that operates between reality and imagination. If society and families create conditions to nurture this drive, it can be channeled into creativity and personal growth. However, if suppressed, it can lead to emotional explosions that harm both the individual and society.

13. Connection with Nature and the Fundamentals of Surviving in Nature:

Humans are children of nature and rely on it for balance and well-being. Reconnecting with the natural world is essential for relaxation, clarity, and inner rejuvenation. To support this connection, cities should prioritize the creation of parks, promenades, and year-round greenhouses—spaces where people can experience the presence of nature even amid urban life. Throughout history, survival in nature has shaped essential human qualities such as cleverness, foresight, cooperation, and resilience. Hunting was once a fundamental part of this process, contributing to the development of our intelligence and adaptability. Today, these traits can be nurtured through various activities such as survival training, camping, hiking, or even nature photography—ways of harmonizing with the natural world without causing harm, while fostering awareness, skill, and appreciation.

14. Honesty, Integrity, and Promise-Keeping:

Though humans have evolved the ability to deceive (a necessary skill for hunting), at their core, they are deeply honest and honorable. Children who observe honesty, integrity, and promise-keeping in their parents and communities will naturally embody these qualities as they grow.

15. Awareness of Moral Defilements (Abysses):

Awareness of moral defilements—ancient taboos and ethical boundaries—is part of our biological and cultural evolution. This awareness is passed down to children through the behaviors, beliefs, and teachings of parents, religious groups, teachers, and community elders, guiding them away from moral abysses.

16. Tools and Technology:

Tools are integral to human evolution, serving as extensions of our natural capabilities. They should be readily available throughout human development, aiding in both survival and progress.

17. Peace-Seeking, Love, Compassion, and Voluntary Services:

Human beings are inherently peace-loving and possess a profound capacity for love and compassion. These natural, noble qualities should be nurtured through group work, healthy emotional relationships, a sense of responsibility, and voluntary services. They are essential to creating a harmonious and supportive society.

18. Wishing Happiness or Benevolence for Everyone Without Exception and without expectation:

The instinct for peace, love, and compassion is deeply connected to the desire for the happiness of all. In its natural state, humanity wishes not only for their own happiness but for the happiness of others. When this is not the case, it is due to harmful beliefs or ideas that have been imposed. These false beliefs must be identified and replaced with the understanding that true happiness lies in wishing well for everyone, without exception-universal and without expectation.

19. Skepticism and Avoidance of Dogma:

The human capacity for cooperation, intellectual thought, and spiritual growth fosters a natural inclination toward skepticism and intellectual flexibility. A society that values critical thinking, skepticism, and avoids rigid dogma will cultivate open-minded, optimistic, and transcendent individuals, contributing to a flourishing and open society.

20. Spirituality:

Spirituality—the belief in meaning and faith in the unity of existence—is intrinsic to human nature. Even without traditional teachings passed down through generations, it naturally seeks to flourish. It is crucial that spirituality is not suppressed by ideological deviations or restricted by extreme religious concepts. Additionally, it should not be diminished by market-driven, unethical behaviors that seek to undermine its essence. Spirituality must remain untainted, preserving its deep-rooted significance within the human experience.

By reviewing the 20 factors that embody the essence of human nature—what we refer to as the foundations of our happiness—we aimed to present a comprehensive picture of humanity's true nature. Through this, we sought to clarify the core message of this insight: "Natural." Before concluding, we must return to the fundamental question raised at the beginning: If our nature is inherently noble, peaceful, and honorable, where do major moral and social deviations-perversions- originate?

As explained earlier, the root of these perversions lies in ignorance—specifically, ignorance of the foundations of our happiness. This ignorance manifests in two forms: cultural shortcomings and cultural deviations (or perversions). Although these two concepts are fundamentally connected—often reflecting different aspects of the same issue—we can, with some leniency, distinguish them as follows:

- **Cultural shortcomings** refer to a lack of awareness about the 20 foundational aspects of our happiness, leading to their neglect.
- **Cultural deviations or perversions** arise from a misunderstanding of these foundations, resulting in the imposition of values, beliefs, and behaviors that contradict them.

The solution to cultural shortcomings lies in increasing awareness and ensuring that these foundational principles gradually become an integral part of human culture. However, cultural deviations or perversions give rise to a range of common distortions and anomalies. Below are some of the most prevalent cultural perversions and deviations (which, in a broader sense, can also be considered cultural shortcomings):

Perversion Caused by Greed and Excessiveness

Greed and excessiveness lead to offenses against the interests and rights of others. In extreme cases, they manifest as invasions, wars, colonization, and other actions that foster hostility among people. Prehistoric tendencies of greed and ignorance, whether among individuals or groups, dominate in this context.

Perversion Caused by Fear of Others and Alienation (Xenophobia)

The division of people into rigid categories—friend/enemy, good/bad, self/non-self, believer/non-believer, high class/low class—reduces individuals to mere objects. In an attempt to justify this ignorance, people become tools to exploit, harass, or abuse. This fear and panic, stemming from a lack of awareness of our common roots, is a deficiency in real spirituality. Hidden within this fear is also a form of greed and excessiveness.

Perversion Caused by Market-driven Anti-Ethical Behaviors

In the pursuit of profit, certain systems turn a blind eye to unsound methods such as sex trade, prostitution, human trafficking (modern slavery), and the narcotics trade. Worse still, under misleading labels such as "market production," industries like weapon manufacturing fuel warmongering. Here, greed, excessiveness, and ignorance prevail.

Perversion Caused by Extremist Ideological Teachings

Extremist ideologies, through totalitarian systems, divide people into factions (partisan/non-partisan, etc.), leading to political suppression, oppression, and even war. In this context, greed, excessiveness, and ignorance dominate.

Perversion Caused by Extremist Religious Doctrines

Manipulated religious doctrines and self-serving interpretations create extremist religious systems that divide societies, leading to suffocation in both social and political spheres. Within this context, greed, excessiveness (evident in practices such as slavery and the spoils of war), and profound ideological ignorance are apparent.

Why Do We Consider These Cultural Deviations Unnatural?

Even among our closest relatives in nature, behaviors such as war-mongering, torture, genocide, slavery, the spread of narcotics, and ideological massacres do not exist. These actions are unnatural; they are products of cultural perversion and ignorance.

Fortunately, all of these issues can be addressed. The solution lies in trusting the noble and honorable nature of humanity. Through the balanced growth of awareness and culture—both personally and socially—we can overcome these challenges, eliminating all forms of cultural deviations and perversions. This requires:

- Recognizing and integrating the 20 foundations of human happiness into public awareness and collective conscience.
- Healing greed and excessiveness, which trace back to prehistoric offenses against others, including warfare.
- Cultivating an awareness of existence, rooted in consciousness, to nurture natural spirituality and heal the conflicts and abuses caused by alienation.
- Reducing market-driven anti-ethical behaviors by exposing their harm and raising awareness of their consequences.
- Stopping ideological brainwashing through gradual reforms of oppressive ideological systems.
- Reforming extremist religious interpretations that exploit selective, aggressive doctrines for personal gain while ignoring the comprehensiveness and mercy of their teachings.

In the upcoming insights, we will delve deeper into these subjects.

Before concluding this discussion, let us explore an example of profound cultural perversion.

To illustrate this, we turn to the reflections of a school principal who witnessed unspeakable tragedies. It is difficult to comprehend how ideological (or cultural) perversion can, in a very short period, deeply corrupt the minds of a select few individuals from an otherwise honorable, hardworking, and artistic nation. This perversion can strip them of their humanity, leading to the destruction of millions of innocent lives—simply due to religious, cultural, or so-called racial differences. –(see Note 118)

Here is an account from a survivor of forced labor camps, recounting what he witnessed:

"I am one of the survivors of forced labor camps. My eyes have witnessed tragedies that no one should ever see. I saw engineers build gas chambers, doctors poison children, nurses kill children, and university-educated individuals executing children and women. This experience made me question the effectiveness of education. My request is this: Your efforts should not focus solely on producing university-educated humanoids. Reading, writing, history, and geometry are important, but they should serve to humanize students, not dehumanize them."
(Excerpt from Haim Ginott)

Note 118- *Axel Kahn, a geneticist and head of the European Commission's high-ranking group of life sciences experts, asserts that, from a biological perspective, there are no human races. The genetic differences within individuals of a so-called "race" are far greater than the differences between individuals from different "races." However, he quickly emphasizes that this issue is not the core of the matter. If we address "racism" merely by arguing that humans do not have biologically distinct races, we inadvertently imply that "racism" would be acceptable if such races did exist.*

Summary of the Fifth Insight

The core message of this insight is simple: **"Trust in your true nature, which is noble, honorable, and benevolent. Strive to fulfill the 20 foundations of your happiness."** Its central concern is equally straightforward: **"If values, beliefs, or actions are imposed upon you that contradict the 20 foundations of human happiness, recognize that something is working against your 'sublime nature' and 'full-scale nature.'"** Some of these distortions include **greed and excessiveness, immorality, substance abuse, dogma, deception, violence, superstition, xenophobia, war, laziness,** and other harmful tendencies.

To fully grasp this insight, one must recognize a fundamental truth: despite individual differences and cultural backgrounds, all human beings share a **'common natural gem'**—the 20 foundations of our true nature. This insight aims to raise awareness of that shared essence. Acknowledging these core attributes is the first and most crucial step toward evolving both individual and collective culture.

While we do not expect immediate transformation simply through awareness of these principles, we understand that consciously recognizing the natural foundations of our own happiness—and that of future generations—is the first and most essential step in realizing our **full-scale nature.**

In Summary:

This insight serves to **reveal and restore** the 20 foundations of human happiness, at least within the realms of thought and conscience. Over the next five insights—the remaining core insights of NUSAWVO—these principles will be gradually **actualized,** ultimately leading to the full realization of our nature. As with all sextuplet insights of NUSAWVO, the key driving force in achieving this realization is the **power of love and compassion (and the law of attraction).** No other force is recognized in this context.

A Crucial Reminder:
Trust in your true nature, but do not allow it to be eroded in the abysses.
Do not recklessly test its resilience by exposing it to the abyss, nor play with its simplicity and innocence under the guise of optimism.

What Is the Abyss?
The abyss is not merely a place; it is a state of being—a condition that distances us from the exalted and noble essence of humanity. It takes many forms:
• The consumption of alcohol and opioids, which endanger health and balance, leading to addiction.
• The seduction of nightclubs, which strip away dignity.
• The abyss of pornography and prostitution.
• The recklessness of debauchery, which erodes the gem of our existence.
• The seemingly harmless small lies that cascade into avalanches of deceit.
• The 'innocent' petty theft that numbs the conscience.
• The casual flirtation with a colleague, which betrays trust and integrity."

The abyss is always closer than it appears!

Emphasizing the noble, pure, and honorable nature of humanity—or explicitly identifying these harmful tendencies—does not mean advocating asceticism or rigid piety. Rather, it is a call to **remove the rust** from the balanced gem of human nature, so that we may witness its true, harmonious, and mysterious beauty. That is all.

In the following insight, we will explore the abyss further.

The symbol of this insight is a part of NUSAWVO's **smile,** which will be fully introduced at the conclusion of the twelfth insight.

The Sixth Insight

"Untainted – Undefiled – Unaltered – Untampered"

*"When you see a greater joy, let go of the lesser joy.
The wise abandon smaller pleasures when they see a greater happiness."
— The Buddha*

*"Hasten toward righteousness and avoid mistakes.
He who is lazy in doing good, his mind delights in wrongdoing."
— The Buddha*

*"Whatever you wish for others, wish for yourself as well.
This is the way of the prophets."
— Jesus Christ*

*"One who covers his mistake with a good deed
illuminates the world like the moon emerging from behind the clouds."
— The Buddha*

*"A truly good deed is one that leaves no regret,
and whose outcome is received with enthusiasm and vision."
— The Buddha*

*"If a person makes a mistake,
let him be cautious not to repeat it,
for the accumulation of mistakes leads to suffering."
— The Buddha*

*"One who commits adultery faces four consequences:
punishment, loss of peace, public disgrace, and ultimately, the torment of the afterlife."
— The Buddha*

*"Whoever considers the consequences of their actions
will not weep at the end."
— Unknown*

"Life has reached its final moment.
You stand at the gates of death.
There is no refuge left for you, and you have not gathered what was needed.
Light your path. Strive earnestly. Hurry and return to yourself.
Cleanse yourself of defilements and free yourself from the cycle of aging and rebirth."
— The Buddha

"Like withering leaves on a tree,
the messengers of death draw near.
You are about to depart, yet you have not secured what is essential.
Light your path. Strive earnestly. Hurry and return to yourself.
Wash away the stains of impurity.
And go to paradise, to the land of the noble."
— The Buddha

"O My servants who have transgressed against themselves,
do not despair of the mercy of God,
for indeed, God forgives all sins."
— The Quran (39:53)

"Steer clear of situations that invite slander"
—Ali Ibn Abi Talib

The Sixth Insight: Untainted, Undefiled, Unaltered, Untampered

All living beings instinctively avoid suffering and seek pleasure. However, only human beings, when recognizing greater pleasures, willingly abandon smaller ones, prioritizing what is truly fulfilling over what is fleeting. This insight, the second pillar of the NUSAWVO Core, is named "Undefiled." Its purpose is to discern the difference between profound and superficial pleasures, thereby forming a natural and logical understanding of defilements. But what does "undefiled" truly mean? To be undefiled is to recognize and avoid defilements. And what exactly is a defilement? It is any belief, speech, or action that harms oneself or others. While different cultures may define defilements in slightly varied ways, their fundamental principles remain largely universal. No society openly endorses acts such as murder, theft, deceit, adultery, or substance abuse. If a culture were to normalize such vices, it would inevitably collapse, descending into moral and existential depletion. Some universal moral prohibitions—such as the avoidance of incest and the preference for exogamy—are not merely cultural constructs but are embedded in our very genetic and evolutionary history. - (see Note 119) As explored in the Third Insight, these taboos have been preserved not only in our collective cultural memory but also within our biological inheritance.

Note 119- *Alfred Russel Wallace, a contemporary and fellow thinker of Charles Darwin—the pioneer of the theory of evolution—did not conceal his astonishment at the profound differences between various moral values. He observed that certain moral standards, along with individual and collective ethical ideals, provide their adherents with stability and a higher likelihood of success in life. Wallace argued that while* ***natural selection*** *operates in both the human and animal realms, its mechanisms and levels of influence differ. In humans, natural selection has, in the current stage of natural history, transformed into* ***social selection****, manifesting itself through societal beliefs, institutions, and values.*

(Quoted from "Ethical Science and Progress" by Howard Selsam.)

Many other taboos—though not necessarily all—are similarly rooted in the historical trials and spiritual-biological evolution of humanity. Taboos emerge in direct opposition to defilements. A defilement is an abyss—a deviation from the full-scale nature of human beings. It not only inflicts harm upon individuals and society but also hinders personal and collective growth, obstructing the path to true happiness and fulfillment. However, for a moral system to be effective, it cannot be burdened with excessive prohibitions; otherwise, it would provoke rebellion, confusion, and mental dissonance. True morality is based on a few fundamental ethical principles, while other social norms and formalities serve as extensions of these core values.

Awareness of defilements is a fundamental right of every individual—especially children. Just as a novice driver must first understand road signs to navigate safely, a child must be educated about moral abysses to lead a balanced and meaningful life. Yet, the key to this education is not verbal preaching but the lived example of parents and role models. Children do not internalize the concept of being untainted through words alone; they learn it through the genuine, consistent, and ethical behavior of those around them. If parents truly wish to protect their children from straying into defilements, they must first embody the principles they wish to instill. Without awareness of defilements—or in the absence of strong and virtuous role models—a child's happiness becomes fragile and vulnerable. However, history has also shown that some of the most morally resilient individuals have emerged from troubled and dysfunctional backgrounds. Having directly experienced the consequences of defilements, they have consciously chosen a higher path, proving that in the end, our choices define our future.

In this insight, we introduce 14 defilements (Abysses). Some of these defilements primarily affect individuals, some impact society as a whole, and some are deeply interwoven between the two. The truth remains that personal and social morality are inseparable: seemingly private choices have profound societal consequences, and collective social conditions, in turn, shape individual destinies.

The Fourteen Defilements

1. **Murder** – The intentional taking of another person's life. This is a grave violation of human rights, causing irreversible harm and deep societal trauma.
2. **Theft** – Taking another's property without consent. Theft erodes trust, creates insecurity, and disrupts both social and economic systems.
3. **Lies and Dishonesty** – The deliberate falsification of information or misleading others. Deception corrodes trust, sows confusion, and leads to broader societal harm.
4. **Adultery** – Engaging in romantic or sexual relations outside of a committed partnership. Adultery undermines trust, inflicts emotional harm, and destabilizes families and relationships.
5. **Alcohol and Drug Abuse** – The consumption of substances that impair judgment, health, and behavior. Substance abuse often leads to addiction, violence, neglect, and severe physical and psychological harm.
6. **Violence** – The use of physical force to harm or intimidate others. Violence inflicts suffering, fosters fear, and has long-term psychological and social consequences.
7. **Dogmatism and the Imposition of Beliefs** – Rigidly adhering to a belief system while forcing it upon others. This stifles intellectual freedom, breeds intolerance, and can result in oppression and conflict.
8. **Malice, Slander, and Abusive Speech** – Intentionally harming others through speech, spreading falsehoods, or using demeaning language. These behaviors degrade individuals, fuel division, and destroy reputations.
9. **Laziness and Neglect** – The avoidance of responsibility and the failure to fulfill duties. This leads to stagnation, inefficiency, and the deterioration of both personal and societal progress.
10. **Greed and Excessiveness** – An insatiable desire for wealth, power, or pleasure. Greed fosters exploitation, deepens inequality, and damages both individuals and the environment.

11. **Slavery** – The forced labor or ownership of individuals as property, depriving them of freedom and dignity. Slavery constitutes a severe violation of human rights and perpetuates systemic injustice.
12. **Discrimination** – The unjust treatment of individuals based on race, ethnicity, religion, gender, sexuality, or perceived superiority. Discrimination fosters division, sustains inequality, and obstructs social harmony.
13. **Totalitarianism and Authoritarianism** – The establishment of a political system where power is concentrated in the hands of a single ruler or elite group, suppressing individual freedoms and dissent. Such systems breed oppression and human rights violations.
14. **Militarism, Warmongering, and Aggression** – The glorification of military power and the pursuit of war for territorial expansion or dominance. Militarism fuels conflict, causes immense suffering, and diverts resources from peaceful development.

1. Murder (Murder, Suicide, and Beyond)

The act of taking a life—whether through murder, execution, or any other means—is condemned by countless mystical and spiritual traditions. The sanctity of human life transcends individual actions and beliefs, marking the difference between the sacred and the profane in human existence. It is important to recognize that even the gravest of human mistakes, such as taking another life, often stems from distorted teachings or deep psychological struggles. In either case, the wrongful act is distinct from the person's divine and noble essence. This perspective does not negate the necessity of justice or penalties for harmful actions but categorically condemns the act of taking life. No individual, least of all the state, holds the right to extinguish this inherent right—the right to life. Fortunately, many nations have moved away from the death penalty, reflecting an evolving moral consciousness that honors human dignity.

1.1 Suicide: A Form of Self-Destruction

Suicide is a tragic and complex act. Whether arising from heartbreak, professional failure, or overwhelming sadness, it is, ultimately, an act of self-inflicted harm. However, it is important to differentiate this from the decisions made by individuals facing terminal illnesses. In cases where a person suffers from unbearable physical pain and repeatedly requests assistance to end their suffering, the situation is vastly different from suicide driven by emotional or mental struggles.

Understanding the complexity of suicide is crucial to providing compassionate support and addressing the root causes of suffering.

1.2 Extending the Discussion to Animals

Our ethical concern should also extend to animals. As civilization progresses, so should our relationship with other living beings. The rise of plant-based alternatives to animal products could represent a significant step forward, not only in reducing animal suffering but also in promoting human health and environmental sustainability. Additionally, inhumane practices such as animal testing must be reevaluated. With technological advancements, there are more ethical and effective ways to conduct research, and we should urgently move toward methods that do not involve cruelty to animals.

2. Theft (Theft, Bribery, and Gambling)

Theft—whether stealing from a neighbor, the government, or an employer—violates the fundamental trust that holds society together. It disrupts social stability, fosters insecurity, and perpetuates corruption.

Bribery operates on a similar principle. By distorting legal and ethical processes for personal gain, it undermines justice, creating inequality and sowing corruption. A society in which bribery is normalized cannot thrive; it loses the meritocracy that sustains fair governance. Gambling, while often viewed as a personal choice, carries its own ethical dilemma. It encourages the redistribution of wealth through chance, where the winners are often the establishments themselves, not the individuals. This creates an unfair system that ultimately harms its participants, even when individual victories occur.

3. Lies and Dishonesty

Lying is a natural, albeit destructive, human instinct. Initially used for survival, lying becomes an energy-draining cycle that spirals into greater falsehoods, eroding personal integrity and societal trust.
Falsehoods corrode relationships, harm reputations, and dismantle social cohesion. It is no coincidence that in ancient Persia, people prayed for protection from lies as a threat to their land. Promise-breaking has a similar effect—it weakens the trust between individuals and can have long-lasting emotional consequences. Particularly when promises are made to children, breaking them erodes the foundational trust in relationships and sets a harmful precedent for the future.

4. Adultery (Adultery, Incestuous Adultery, Prostitution, Sexual Novelty-Seeking, Pornography, and Nudity)

Adultery is a clear violation of trust and commitment. It inflicts emotional damage and disrupts the core unit of society—the family. Engaging in relationships outside of a committed partnership betrays the mutual understanding that should define these bonds. Prostitution, sadly, has been a product of societal inequalities, often tied to economic necessity. However, in societies where basic living standards are met, prostitution is an outdated and damaging practice, undermining human dignity.

The pursuit of constant novelty in relationships—often termed libertinism—might seem enticing in the short term but is ultimately self-destructive. Much like drinking seawater, it provides temporary relief at the cost of greater harm to one's emotional and spiritual well-being. Pornography, in all its forms, creates an illusion of intimacy while deeply harming participants. It distorts natural sexual inclinations and fosters isolation instead of meaningful connection. As for nudism—the public exposure of intimate body parts—its implications should not be underestimated. This issue is often discussed with a focus on women, but it is important to include men in the conversation as well. Regardless of how it is presented, whether subtly or overtly, nudism can erode both personal dignity and the collective well-being of society. Often promoted by market-driven forces with little regard for ethics, nudism is increasingly being portrayed as a liberating subculture. However, it is crucial to recognize the potential harm it can cause and to address its rise thoughtfully in order to protect human dignity and promote genuine happiness.

5. Alcoholic Beverages and Narcotic Substances (Alcohol, Narcotics, Psychotropic Substances, and Tobacco)

The consumption of alcohol, while accepted in many cultures in moderation, has undeniable harmful effects. Regardless of the quantity consumed, scientific research continually uncovers the extensive damage alcohol inflicts on individuals and society. The primary issue with alcohol is that it induces a temporary, false sense of euphoria, often used to numb daily worries or fill the existential void caused by a lack of meaning in life. However, engaging with our full-scale human nature and finding genuine meaning in life—without resorting to alcohol—allows us to avoid the risks of addiction, organ damage, impaired judgment, and the harm caused to others while intoxicated. Alcohol abuse is often observed in broken families, diseases, violence, and regretful actions.

The same truth applies to drugs and psychotropic substances: true meaning cannot be found through shortcuts. As Dr. Viktor Frankl explains in *Man's Search for Meaning*, individuals who turn to drugs may end up like laboratory rats in self-stimulation experiments. These rats, when given the ability to induce pleasurable sensations at will, ultimately ignored real mates and food, choosing instead the false pleasure of the electrical current. In the same way, those who seek meaning in drug-induced euphoria bypass the real, deeper meaning available to them in life.

Smoking, while not as immediately harmful as narcotics, also remains dangerous. Beyond the well-known health risks, smoking serves as a gateway addiction, increasing vulnerability to other substance dependencies.

6. Violence (Violence and Abuse of Others, Child Abuse, Animal Abuse, Sexual Misconduct with Spouse, and Sexual Assault)

Violence and aggression are innate human capacities, necessary for the survival of the species. However, in the NUSAWVO school of thought, we advocate for understanding these impulses and channeling them constructively. When balanced with love, compassion, empathy, and cooperation, aggression can be directed toward positive outcomes, expressing the highest potential of human nature. It's crucial to raise awareness about the many forms of violence. Often, individuals may not realize they are inflicting harm, mistaking their actions as normal or even justified. In some societies, a parent or teacher may punish a child, sometimes physically or verbally in public, unaware of the lasting harm caused. Similarly, someone who harms an animal might not recognize its suffering. These actions reflect a lack of empathy, conscience, and mutual respect, which urgently needs to be addressed. Sexual misconduct, especially rape, represents one of the most extreme forms of violence. It not only devastates the victim physically and mentally but also leaves profound emotional scars. This violent act can be seen as a kind of spiritual murder, damaging the victim's dignity and emotional well-being.

Sexual misconduct within marriage, too, is a significant form of violence. It undermines the core of mutual respect and trust that should exist between partners. Such actions, whether physical, emotional, or coercive, violate the sanctity of the relationship and the inherent dignity of both individuals. It is essential to recognize that a marriage is not a license to dominate or mistreat; it is a partnership rooted in love, respect, and equality. The impact of such violations can destroy the emotional connection between spouses, leaving long-lasting psychological scars.

7. Dogma and the Imposition of Beliefs (Dogmatism, Coercion of Belief, Religion, Ideology, and Inquisition)

The seventh insight (seeking the truth) and the eighth insight (abiding in truth) in this school, also as discussed in Appendix 2 of this book, directly address the dangers of dogmatism and provide ways to counter its influence. Dogmatism has historically been the root cause of some of the deadliest wars, resulting in millions of lives lost. It manifests in the imposition of beliefs, whether religious or ideological, often with devastating consequences. The misinterpretation and misuse of religious teachings have exacerbated this problem. - (see Note 120) Any form of belief imposition contradicts humanity's inherent drive for free thought and the search for truth. It harms individuals and society by stifling genuine inquiry and growth. The Inquisition is a prime example of dogmatism's harmful consequences. How many innocent lives were destroyed by religious institutions in the Middle Ages, and how many still suffer today from similar extremism in ideological or religious communities? These practices reveal the dangers of coercing belief upon others.

Note 120- *A wise man once said, "God did not compel people to accept His existence, which is central to monotheistic religions. While He could have created everyone on the right path—unified and guided—He chose not to. If this is true, how can anyone justify forcing others to adopt a specific religion?" Moreover, it is important to recognize that a significant portion of the world's population does not believe in God's existence at all.*

Carl Gustav Jung, the Swiss psychologist, argued that without freedom, true morality cannot exist. In the same spirit, the French philosopher Voltaire famously stated: "I may not agree with what you say, but I will defend to the death your right to say it."

8. Malice, Slander (Malice, slander, foul language, talking behind one's back- (see Note 121...)

Benevolence and the genuine wish for happiness—both for oneself and others—are inherent qualities in every human being. However, several factors can distort and damage this natural disposition:

- Deficiencies such as ignorance, excessiveness, and greed lead us to prioritize our own interests and rights above all else.
- Our ignorance of the world's universal awareness-consciousness causes us to perceive others through external distinctions, often viewing them as threats rather than fellow human beings.
- Market-driven perversions—rooted in ignorance and greed—nurture a sense of malice within us.
- Certain ideological doctrines foster fear, division, and artificial groupings, such as white vs. colored skin, bourgeois vs. communist, and so on.
- Religious extremism promotes hatred, superstition, and xenophobia, further distorting our innate sense of unity.
- Daily anxieties and psychological disturbances arising from interactions, and sometimes conflicts, with others are left unaddressed, festering without proper resolution.

Note 121- *As often said (though its attribution is debated), 'Great minds discuss ideas; average minds discuss events; small minds discuss people.'*

Clearly, these issues must be understood at their core and properly addressed. We will explore these matters further in the Ninth Insight: Wishing Happiness for Everyone.

9. Laziness and Neglect

(Laziness and indifference toward security, livelihood, safety and health, education, knowledge, art, truth-seeking, love and compassion, family, and voluntary service.)

Throughout the 20 Foundations of Human Nature, we have seen that key factors such as security, livelihood, health, the pursuit of knowledge, art, truth-seeking, love and compassion, family, and voluntary service play a crucial role in human survival and growth. - (see Note 122)

Indifference toward these vital areas—such as neglecting work and activity, disregarding personal health and hygiene, and being apathetic about personal and family safety—leads to countless harms. These impacts are felt first by the individual, then by the family, and ultimately by society at large. Clearly, this issue requires serious attention, and its underlying causes must be carefully examined. From a social perspective, it is the responsibility of conscientious individuals and compassionate governments to prioritize education and awareness in these areas, ensuring that efforts to improve society remain central to their agendas.

Note 122- *Tolstoy emphasized that work and wisdom are virtues, while idleness and belief in superstitions are among the greatest vices.*

10. Greed and Excessiveness-Extravagance

Greed and excessiveness stand in direct opposition to piety, contentment, and a simple life. Acknowledging this issue is not meant to condemn hard work, entrepreneurship, or economic activity. The concern is the culture of extravagant lifestyles and environmentally destructive excess.

The root problem lies in the Earth's limited capacity. We are pushing our planet to its breaking point, straining its ability to regenerate, maintain stability, and preserve balance. Across many parts of the world, our consumption habits—whether in energy, greenhouse gas emissions, food, meat, water, or other resources—are entangled with wastefulness, extravagance, and insatiable greed. It is as though we've emerged from famine, unable to stop consuming. Meanwhile, the number of people suffering from diseases like diabetes, resulting from excess, continues to rise.

We are not satisfied with one or two cars—we are always chasing the next model, the latest upgrade. Our computers, phones, clothes, home accessories—nothing is exempt. More, bigger, newer. The farther, the better. The more, the better.

Even in reproduction, we impose no limits, as if the world exists solely to fulfill our endless desires, as though this fragile planet owes us and our countless children eternal abundance.

It is time for humanity to urgently reflect on its greed and excesses. This insatiable thirst undoubtedly stems from millennia of hardship, hunger, and survival struggles—instincts we share with all predatory creatures. A wolf attacking a herd doesn't stop at one kill; a fox in a henhouse slaughter indiscriminately, beyond its needs.

But the world has changed. Whether we accept it or not, the circumstances are no longer the same. There is no justification for abusing ourselves, others, or this fragile planet.

The solution must begin with us—each individual. We must put an end to extravagance, waste, and unnecessary consumption. Not tomorrow, not later—right now, from this very moment! - (see Note 123)

11. Slavery

Slavery stands as one of the most painful and tragic blights in human history. It inflicts irreparable harm on the enslaved, exposing them to destruction while embodying violence and aggression. Yet, it also degrades the enslaver, reducing them to a state of moral corruption that has no parallel in the animal world.

No human being is deserving of being owned or possessed; such a notion is fundamentally false. The inhumane allure of slavery, rooted in these profound defilements, has driven countless historical campaigns and actions, causing untold suffering and violence.

Note 123- *If we could change ourselves, the tendencies in the world would also change. Commonly attributed to Gandhi.*

12. Discrimination (Based on race, ethnicity, religion, sexuality, or perceived superiority)

Since all of us—human beings, living creatures, and the entire universe—are intrinsically interconnected, there can be no true hierarchy or distinction in our inherent worth. Thus, racial, - (see Note 124) ethnic, religious, and sexual discrimination, as well as any belief in human superiority over other species, are utterly meaningless.

If any special status exists, it is found only in love, compassion, and voluntary service to others. The only form of ranking that holds value is the capacity to serve others selflessly. In fact, such service is not a rank but a form of spirituality—a true elevation of the human spirit.

13. Creation of a Totalitarian Political System (Authoritarian, Autocratic)

A totalitarian political system is one in which power remains unchallenged and does not rotate. In such systems, a particular governing body—typically the central authority in national administration—retains control for extended periods, often for life, through hereditary succession or predetermined structures. There is no reciprocal relationship between this entrenched power and the people; the public is powerless to judge, criticize, or replace the ruling entity, even when dissatisfied.

Note 124 *- Interestingly, recent biological and genetic research reveals that the concept of distinct human races, as commonly understood, doesn't actually exist.*

The form this power takes can vary, it may be embodied by the central committee of a communist party, a Qutb, a Murad, a monarch in an absolute regime, or a president or prime minister with lifetime tenure. Regardless of its specific structure, centralized power in a totalitarian system remains immune to criticism, reform, and structural change. Even when elections occur, they often serve as a superficial exercise, affecting only a minor segment of governance while leaving the true core of power untouched.

The defining feature of such systems is the absence of freedom—freedom of thought, association, the press, political parties, and, most critically, freedom of elections at the highest levels. History has shown that in the absence of freedom, even the promise of social justice—the very ideal all political systems claim to pursue—inevitably collapses. As we will explore, freedom and social justice are inseparable; one cannot survive without the other.

When governance is controlled by an unchangeable minority and public opinion becomes symbolic, those in power, even if initially well-intentioned, gradually detach from the concerns of the people. Shielded from judgment, criticism, and electoral accountability, they inevitably distance themselves from society, creating a wall of mistrust and fear between the rulers and the public. Over time, a single individual often consolidates control, dominating others and ultimately assuming the title of dictator.

In such a system, where criticism and course correction are stifled, inefficiencies and flaws compound, leading to stagnation and corruption. Instead of achieving its original goals of justice, the government becomes trapped in cycles of inefficiency and corruption. To preserve its power, it resorts to suppression, silencing dissent rather than addressing the root causes of unrest.

The Root Cause: Lack of Awareness

The persistence of totalitarian systems stems from the low political awareness and social experience of the general public, particularly intellectuals and societal elites. When literacy, awareness, and political experience are insufficient, society leans toward totalitarianism. However, as public consciousness rises, the demand for accountability grows, leading to gradual reforms—whether peacefully or, tragically, through violent upheaval.

Revolutionary overthrows, however, often repeat the same cycle. A new minority emerges, believing itself more enlightened than the previous one, only to replicate the same patterns of unchecked power, suppression, and disillusionment.

Why Freedom Cannot Survive Without Social Justice

Conversely, freedom itself is fragile without social justice. A society burdened by poverty and insecurity is easily manipulated. Desperate individuals may trade their freedom for empty promises—whether from unqualified populists in democratic societies or revolutionaries aiming to overthrow the existing order. In both cases, personal autonomy is sacrificed, and true freedom is lost.

Yet, history's trajectory moves toward greater literacy, awareness, and ultimately the transformation of totalitarian systems. As people and elites recognize the fundamental interdependence of freedom and social justice, new systems will emerge—systems that uphold both principles simultaneously.

The Path Forward

A just and sustainable political system requires two key pillars:

1. **A fully democratic system with genuine power rotation at all levels**
   No individual or political program should be shielded from public judgment, rejection, or approval.
2. **A robust social safety net supported by progressive taxation**
   - Ensuring access to education, healthcare, and essential welfare services, including livelihood

protection, unemployment benefits, and retirement security.
- Minimizing extreme wealth disparities through incremental taxation.
- Establishing institutions such as labor unions to facilitate dialogue and negotiation between different social groups, preventing unrest and upheaval.

Examples of such systems can be observed in Scandinavian countries and beyond, where political and economic structures promote both freedom and social justice. However, cultural values also play a crucial role. Without a foundation of truth-seeking, integrity, and diligence, no system, no matter how well-designed, can thrive. A society built on laziness, falsehoods, hypocrisy, and superstition cannot sustain meaningful progress.

Why This Awareness Must Be Instilled from Childhood

The dangers of totalitarianism must be ingrained in every child's conscience and intellect, ensuring that a fundamental, lasting solution can take root. After all, today's child is tomorrow's adult, intellectual, elite, and politician.

14. Militarism, Warmongering, Invasion, and Colonization

The insatiable greed of mankind, coupled with ignorance of our fundamental unity and the resulting fear of others, stands as one of the primary causes of warmongering and invasion. These tendencies often originate from totalitarian (authoritarian, despotic) systems, where the core of power is aggressive, immune to criticism, driven by excess, and fueled by xenophobia.

Beyond territorial and material conquests, ideological and religious invasions must also be acknowledged—those waged under the pretext of spreading belief systems. However, these too can be traced back to the same root causes of greed, ignorance, and fear. It would not be an exaggeration to say that warmongering and expansionism are the ultimate manifestations of humanity's accumulated misjudgments, misguided behaviors, and collective perversions. Time and again, they have brought nothing but destruction, suffering, and setbacks to human civilization at various stages of history.

Moreover, war and invasion often disguise themselves under the deceptive banner of colonization, falsely claiming to bring civilization, cultural advancement, or religious enlightenment to native populations. In reality, such justifications only add layers of hypocrisy and deceit to the violence. Far from benefiting the indigenous people, colonization has historically led to their displacement, suffering, and even extinction—whether through direct conflict, forced assimilation, or the spread of foreign diseases. What is presented as a mission of progress is, in truth, a cycle of oppression and loss.

Summary of the Sixth Insight

The essence of human existence is inherently noble and untainted, with no natural inclination toward defilement—especially when the 20 fundamental aspects of human nature, or the foundations of human happiness, are fully realized. This means that the full-scale nature of humanity must be active and fulfilled in all its dimensions.

In this school of thought, the focus is not primarily on defilement and purity. These concepts arise only after addressing the fourth insight (full-scale human nature) and the fifth insight (natural fulfillment). This intentional order reflects the belief that activating the full-scale nature of humanity naturally leads to profound joys—such as knowledge, art, awareness, love, and spirituality. The depth and intensity of these higher pleasures are so vast that they leave no room for fleeting, harmful indulgences.

Cultivating our full-scale nature requires patience and effort, as explored within the framework of the six insights forming the NUSAWVO core. As Zoroaster wisely suggested, we are not here to fight the so-called demons of defilement—we are simply lighting a candle. That is all.

This does not mean disregarding the achievements of those who came before us. On the contrary, we honor their wisdom by raising awareness of defilements, for this awareness enhances our ability to navigate life safely and fulfill our potential. Understanding defilements can be likened to knowing traffic rules. Just as road signs and regulations guide us toward safe driving, recognizing defilements offers essential guidance for a fulfilling life journey.

To clarify further, consider the law of attraction. This book teaches that to apply it effectively, one must focus on its positive aspect. For example, if we aim for safe driving, we should not dwell on fear or accidents. Instead, we focus on mindful, smooth driving, aligning our thoughts with a positive outcome.

One might ask: Does learning about road signs like "Do Not Enter" or "Stop" mean we are negatively engaging with the law of attraction? Not at all! In reality, understanding traffic laws is essential for safe driving. Signs like speed limits or restricted entry points do not limit our freedom; they enable us to exercise freedom in harmony with others in a civilized world. These signs complement the law of attraction, guiding our interaction with the environment—considering factors such as roads, vehicles, pedestrians, and weather conditions.

Similarly, when it comes to defilements, the purpose of this insight is not to dwell on them but to recognize them as signposts along life's journey. Just as road signs prevent accidents, understanding defilements helps us avoid pitfalls that may harm our well-being. In this way, this insight serves as a compass or plumb bob, guiding us to distinguish between the straight and the crooked.

This compass, this plumb bob, is the invaluable heritage of our ancestors—not belonging to any one group but to all of humanity. By holding onto it, we can discern right from wrong, avoiding deception and unnecessary suffering. Of course, no one is a saint, and everyone experiences defilements in some form, learning from them while inevitably losing something—often time, health, or vitality. What truly matters is using the intelligence inherited from millions of years of human evolution, ensuring that we do not fall into the same traps repeatedly.

Feel free to revisit these descriptions of defilements as often as needed, reflect on them, and if they resonate with you, apply them to your life. Ultimately, you understand these truths better than anyone else, so do not take this as a directive—rather, it is a simple reminder.

Above all, if you recognize the truth in these principles and virtues, we hope you will not trade the precious gems of life, health, and dignity for anything less than their true worth. That is all.

In the next insight, we will explore the search for knowledge, art, and truth. The symbol of this insight forms part of NUSAWVO's smile and will be introduced at the end of the twelfth insight.

As with all parts of the NUSAWVO core, the force that activates this insight is love, compassion, and the law of attraction.

The Seventh Insight

"The Seeker of Truth"

"If you suffocate the truth and bury it underground, it will rise again with such explosive force that when it erupts, it will shatter everything in its path."
— Émile Zola

"Truth is the proper name of God."
— Mahatma Gandhi

"Can anything be more extraordinary to me than the truth?"
— Fyodor Dostoevsky

"The nature of the human mind is to constantly strive for greater growth and awareness."
— Shakti Gawain

"Seek the truth with absolute neutrality and a mind free from all prejudice."
— René Descartes

"If you wish to be a true seeker of truth, at least once in your life, you must doubt everything as much as possible."
— René Descartes

"Music is the voice of God."
— Avicenna (Abu Ali Sina)

"Delving deeper into the mystery is the most beautiful experience I can have. This sense of wonder and excitement forms the foundation of humanity. Whoever loses this capacity—to be moved, to be astonished—might as well be dead, his eyes forever closed to the unknown."
— Albert Einstein

"The problem is not that truth does not exist—there is truth, and there may even be only one truth. The problem is that no one can be absolutely certain they have attained it."
— Viktor Frankl

The Seventh Insight: (The Truth Seeker)

The Seventh Insight: The Seeker of Knowledge, Art, and Truth

The seventh insight of the NUSAWVO school is dedicated to the pursuit of truth. As the third pillar of the *NUSAWVO* core, this insight emphasizes the profound value of truth in human existence. According to this school, truth is the most cherished virtue—because, as Socrates said, everything becomes vulgar where there is no truth. - (see Note 125)

Among the thirteen insights of this school, two are adorned with the name *truth*, signifying its central place in this worldview. This emphasizes that free thinking, inquiry, the rejection of dogma, honesty, and everything associated with the sacred name of *Truth* hold a sublime position within this school.

The full title of this insight is *Seeking Knowledge, Art, and Truth*. It emerges in response to humanity's instinctive and natural drive for curiosity and exploration. Human beings are inherently inquisitive, drawn to discovery, trial and error, and creative expression—intellectually, artistically, and practically.

Note 125- *Anatole France regarded the question* "What is truth?" *as the most profound question in human history, because no other question is independent of it. (Will Durant,* The Pleasures of Philosophy)

This boundless pursuit across all domains is vital, and the freedom to explore is essential for mental well-being. In the cultural-biological evolution of humankind, hunting and its related processes played a crucial role in the development of intelligence and mental faculties, including curiosity and the search for knowledge. These evolutionary changes were built upon an ancient and universal trait—one shared by all living beings. Hunting merely amplified it. Curiosity is not unique to humans; any child with a pet hamster is aware of the intense curiosity and exploratory nature of even small creatures. Indeed, no form of life exists without curiosity and a drive to seek—it is simply expressed in varying forms and intensities.

However, as discussed in the *Third Insight*, humanity underwent a qualitative transformation in the evolutionary process. With increasing intelligence and awareness, two distinct traits emerged: *foresight* and *self-awareness*. - (see Note 126) Humans became conscious of their own reactions and, as a result, extended their curiosity beyond mere survival—to the vastness of the universe, the past and future, and the fundamental nature of existence itself.

Note 126- *Albert Jacquard, a Swiss geneticist, posits that human self-awareness doesn't stem from a specific "self-awareness gene." Instead, he suggests that our capacity for thought, socialization, and societal development—facilitated by various genes—culminates in the creation of culture, which in turn fosters human self-awareness. This perspective is discussed in his book "They Have Not Written the Future." Jacquard emphasizes that while our genetic makeup provides the foundation for cognitive abilities and social interactions, it's the cultural environment that shapes and enhances our self-awareness. This view challenges the notion of a singular genetic determinant for complex traits like self-awareness, highlighting the intricate interplay between our biology and cultural experiences. In essence, Jacquard's argument underscores the importance of culture as a dynamic force that, built upon our genetic predispositions, leads to the emergence and development of human self-awareness.*

This unique ability has allowed humans to evolve socially and culturally, refining their way of life, behaviors, and guiding principles.

Erich Fromm, the renowned psychoanalyst, highlights this in *The Anatomy of Human Destructiveness*:
"There is abundant psychological and neurological evidence indicating an extraordinary drive for creativity within the structure of the human brain."

It is precisely because of this intrinsic drive that the NUSAWVO school regards the search for truth—closely linked to creativity—as one of humanity's most defining characteristics. The pursuit of truth plays a crucial role in the fulfillment of our *full-scale human nature* and, ultimately, our happiness.

In today's civilization, the conditions necessary for such searches—such as the free flow of information and unrestricted access to educational institutions and knowledge resources—must be recognized as fundamental obligations of daily life. The absence of these essential freedoms inevitably leads to the stagnation of the creative capacities of the human mind.

The Search for Truth: A Never-Ending Journey

Seeking the truth is as natural, enjoyable, and effortless as breathing—yet it also holds its mysteries. This ability, born from our intellectual, cultural, and neurological evolution, constantly challenges us with the endless sea of the unknown. However, the process demands immense energy, and at times, even the most dedicated seekers of truth experience fatigue and exhaustion.

This is normal; such weariness is usually fleeting, soon replaced by renewed curiosity, enthusiasm, - (see Note 127) and creative energy. It is through creativity that the search for truth becomes a living, dynamic process, one that continually reinvents itself and expands in scope.

But beyond the subtleties of truth-seeking and the occasional fatigue it brings, a greater concern arises: sometimes, we completely suspend this extraordinary and superior human ability—our drive to search and create—and instead, we resign ourselves to passively accepting the conclusions and declarations of others. There is nothing inherently wrong with learning from others; indeed, this is often a conscious choice. However, we must be mindful of the risks:

1. If this essential aspect of our nature is suppressed, our full-scale human nature—or, in other words, our transcendental nature—remains unfulfilled. This leads to an incomplete sense of happiness and a loss of true agency over our intellectual and spiritual growth.

Note 127*- Gandhi, in the context of seeking the truth, said:*

"What one perceives as truth is often not the same in another's eyes. However, the seeker should not be troubled by this. If the effort is sincere, it will become clear that these seemingly different truths are like the countless leaves of a tree—distinct in appearance but part of the same whole. The truth is the true name of God. Therefore, not only is it natural for each person to search for the truth through their own perspective, but it is also their duty. In this pursuit, if one errs, they will correct their mistake and be guided to the right path. For truth-seeking demands the willingness to endure suffering—sometimes even unto death—and in this journey, there is no room for selfishness. In such a sincere search, any momentary confusion or hesitation will not last long."

2. This passive approach not only harms the individual but also has far-reaching consequences for society. When we abandon our search for truth, we expose ourselves to dogmatism—a state where beliefs become rigid, unquestionable, and immune to criticism. This is a dangerous state of mind that requires deeper examination. - (see Note 128)

From a broad perspective, there are only two fundamental approaches to engaging with ideas:

- **The Skeptical Approach**: This method recognizes that all beliefs, thoughts, and conclusions—without exception—are subject to error. Therefore, skepticism requires continuous self-criticism, open inquiry, and an active search for counterarguments and alternative perspectives. It values the process of seeking over any single conclusion. It is through this constant reevaluation that creativity thrives, as it fosters innovation and new ideas. Creativity becomes a necessary tool in this process, allowing us to generate new ways of understanding and challenging existing frameworks.
- **The Dogmatic Approach**: This mindset, by contrast, treats its beliefs as absolute, beyond question or debate. Whether consciously or unconsciously, it leads to the dangerous triad: "We are right, others are completely wrong, and we are superior to all." History has shown that this mindset is at the root of the greatest human conflicts and tragedies.

Every individual, whether consciously or not, chooses one of these two paths in how they engage with their beliefs and worldview. There are no other alternatives. This means that every person follows one of two courses:

Note 128*- We have explored this issue in depth in Appendix Two.*

1. **The Seeker's Path**: A person who honors their highest human faculty—the relentless search for truth—avoids fanaticism and ideological rigidity. Through constant study, research, and free thinking, they engage with knowledge, art, and truth in all their dimensions. In doing so, they not only cultivate their own transcendental nature and experience deep, authentic happiness, but they also contribute to the intellectual and cultural enrichment of society. Creativity is essential to this journey; it is through creativity that we bring new perspectives and solutions into the world, continually shaping our understanding of truth.
2. **The Path of Stagnation**: A person who neglects or suppresses their innate drive for truth—whether due to ignorance, ideological dogma, religious extremism, or any other rigid belief system—falls into intellectual inertia. Their mind becomes silent, deprived of creativity and the necessary skepticism required for genuine insight. The immediate consequence is personal stagnation, an unfulfilled life devoid of intellectual and spiritual progress. But beyond this personal loss, there is a far greater danger: when intellectual passivity creates a vacuum, dogmatic ideologies inevitably fill the void. A mind that has abandoned the pursuit of truth becomes highly susceptible to rigid doctrines that demand blind faith and discourage critical thought.

This is why, historically, authoritarian ideologies—whether political, religious, or cultural—thrive in environments where critical thinking is weak. Such ideologies are appealing because they absolve individuals of the responsibility to think for themselves. However, when the seeker's mind falls silent and dogmatic ideas take root, the consequences extend beyond the individual: intellectual stagnation spreads, diverse perspectives are silenced, and society becomes vulnerable to superstition, fanaticism, and, in extreme cases, war-mongering ideologies that disrupt peace and progress.

The Imperative of Eternal Seeking

For these reasons, the pursuit of truth is enshrined as one of the thirteen insights of the NUSAWVO school. It is not merely a suggestion but an imperative—a nonstop, permanent duty for every human being. The dignity of a person lies in never bringing an end to their search.

This insight, as part of the path toward realizing the full-scale human nature, encompasses three fundamental aspects:

- Seeking truth in all fields—including knowledge, art, philosophy, mysticism, and spirituality.
- Abiding in truth—adhering to honesty, integrity, and correctness (a theme that will be emphasized further in the next insight, the Eighth Insight).
- Rejecting dogmatism and embracing skepticism—not only in intellectual pursuits but also in personal beliefs, societal structures, and ethical considerations. (This topic will also be expanded upon in the Eighth Insight.) - (see Note 129)

***Note 129-** The contemporary philosopher Karl Popper, in his study of scientific development, emphasized that for a theory to be genuinely scientific, it must be falsifiable—it must have the potential to be proven wrong. Scientific progress, he argued, does not occur by accumulating examples that confirm a theory but by identifying instances where it fails. The refutation of a scientific theory paves the way for the emergence of new and more refined theories.*

This principle also applies to philosophy: in a truly philosophical approach, claims should only be regarded as valid if they can be tested and critically examined.

Above all, this insight is built upon a fundamental acknowledgment: No truth is entirely free from error and approximation. In the realms of science, philosophy, and thought, it is highly unlikely—if not impossible—for anyone to claim possession of absolute truth. Historically, those who have made such claims have often been proven wrong in due time.

In other words, everything is relative, and it is on this basis that seeking must be eternal.

The Search for Knowledge: Safeguarding the Pursuit of Truth

If truth-loving and the pursuit of knowledge, art, and truth are fundamental instincts and rights of every human being, then the means to pursue and fulfill them must be ensured. This vital matter is safeguarded through formal education in schools and universities, freedom of the press and political parties, and broad social, political, cultural, artistic, and religious liberties. Furthermore, access to libraries, the internet, philosophical and mystical societies, art galleries, museums—especially those dedicated to natural history and anthropology—plays an equally crucial role in fostering this pursuit.

If you, the reader, have made it to this point, you should recognize that your journey itself is a perfect example of seeking knowledge and truth. You have had the courage to suspend, even temporarily, your assumptions about the world and yourself and to listen to new ideas. Perhaps much of what you have read has reinforced your existing knowledge, added new insights, or even introduced viewpoints that challenge your current understanding. That is wonderful! The purpose of this book is not to impose conclusions upon you but to support your ongoing research and reflection. As you have likely noticed, this book presents ideas alongside their contradictions, inviting you to engage in intellectual challenge rather than passive acceptance. The NUSAWVO school does not seek to think on your behalf but rather to provide material that encourages further thought and exploration. At this point, it may be valuable to highlight a scientific finding related to the physiological roots of dogmatism—insights that might grant a degree of innocence to our mental laziness. As we know, the human brain is the most energy-consuming organ in the body. Neuroscientific research has shown that rejecting a claim requires activating multiple brain regions, demanding significantly more energy than simply accepting a claim at face value. This is why children often believe everything they hear—it is cognitively easier. Doubt, in contrast, requires not only more mental effort but also a set of intellectual skills.

This physiological reality underscores the fact that "gullibility is natural." If a culture does not foster skepticism from an early age, beliefs transferred to children as certainties take root deeply in their minds. As a result, individuals raised in such an environment may grow into adults who rarely question their inherited beliefs, regardless of the superstitions, hostilities, or irrational elements embedded within them. This, in turn, can compromise their happiness by binding them to unexamined convictions.

The Boundless Capacity for Knowledge

Finally, it is essential to emphasize a crucial point within the framework of this insight: there is no inherent limitation to human ability in knowing the universe and truth. Any concept that enters the human mind or any experience, no matter how subtle, is within the realm of human comprehension.

Moreover, in the pursuit of knowledge, humans employ multiple methods of understanding. As will be further discussed in Appendix No. 2, the English philosopher Walter Terence Stace, in his renowned book *Hegel's Philosophy*, strongly argues from both a philosophical and epistemological standpoint that human cognition has the capacity to grasp the entirety of existence and truth, thereby rejecting the notion of an "unknowable thing" once and for all. Human perception and knowledge operate across two seemingly distinct but deeply interconnected domains: "logical-intellectual knowledge" and "intuitive knowledge." As seekers, human beings appear to be fully equipped for exploration in both realms. In the second insight and Appendices 1 to 3 of this book, you will encounter further discussions on these dimensions of knowledge.

Summary of the Seventh Insight

The pursuit of knowledge, art, and truth in all its forms is an inseparable aspect of human completeness—what we may call full-scale human nature or transcendent nature. Likewise, a love for truth is deeply interwoven with the noble and honorable essence of human beings.

Throughout the various insights of this school, we have examined the concept of truth from multiple perspectives, seeking answers while simultaneously subjecting those very answers to scrutiny and challenge. In this way, skepticism and uncertainty have been established as the foundation of all genuine inquiry. Ultimately, the search for truth has manifested as two of the thirteen core insights of this school, underscoring its central role in the NUSAWVO worldview.

Given the undeniable significance of this noble human instinct, it is the responsibility of both family and society to recognize its value and cultivate an environment that nurtures the pursuit of knowledge, art, and truth—free from baseless prejudices, rigid beliefs, and superstitions. Only by doing so can we unleash the gifts of creativity and the boundless potential of free and fertile minds, allowing society to flourish and take pride in its intellectual and cultural achievements.

The force that activates this profound human capacity—alongside the other foundational elements of the NUSAWVO school—is love, compassion, and the law of attraction.

The symbol representing this insight is also a part of NUSAWVO's smile and will be introduced at the conclusion of the twelfth insight.

The Eighth Insight

"Abiding in Truth"

"Where there is no truth, everything is vile, ugly, and vulgar."
— Socrates

"Truth is a flower that grows even in the wasteland."
— Thomas Mann

"The only virtue I seek is truth—never the use of force."
— Gandhi

"No idea in this world is so noble that it warrants fanatical efforts to impose it by force."
— Ivan Klíma

"Doubt is the first step toward truth. It must be applied everywhere, for it is the true touchstone."
— Denis Diderot

The Eighth Insight: Abiding in Truth

(*Adherence to Truth* and *Conditional Persistence in Interests, Rights, and Truth*)

The eighth insight represents the fourth pillar of the NUSAWVO core. It is titled *"Abiding in Truth,"* with its full designation being *"Adherence to Truth"* and *"Conditional Persistence in Interests, Rights, and Truth."*

The first part, *Adherence to Truth*, emphasizes a transparent and unwavering commitment to honesty, truthfulness, and integrity. This distinguishes NUSAWVO from doctrines that justify achieving goals by any means necessary. In this worldview, truthfulness and honesty are paramount—even when they conflict with personal interests. The second part, *Conditional Persistence in Interests, Rights, and Truth*, highlights a crucial issue that underlies many human conflicts: the tendency toward excessiveness and one-sided perspectives regarding benefits, rights, and rigid, dogmatic thinking. These tendencies, when unchecked, lead to both individual and societal crises.

The following sections elaborate on these key aspects.

Adherence to Truth

Human beings are inherently drawn to truth and honesty. While deception may be part of our evolutionary inheritance—born out of survival strategies like hunting—on a social and personal level, dishonesty is energy-draining and anxiety-inducing. It remains in constant conflict with the higher faculties of human nature. Though people are capable of deception, it does not lead to deep fulfillment. True and lasting peace is only found in truthfulness.

From the perspective of NUSAWVO, *Adherence to Truth* means upholding honesty, integrity, and correctness in our thoughts, actions, social interactions, and professions. It is highly unlikely that someone who engages in deceit, dishonesty, or theft can attain true happiness. Even if such individuals seem outwardly successful, they ultimately fail to achieve inner peace and familial harmony. Keeping one's promises is also an essential expression of truthfulness. Through this insight, we learn that no goal, no matter how noble, justifies sacrificing the truth. This directly opposes the belief that *"the end justifies the means."* As Gandhi wisely put it, *"Pure goals can only be achieved through pure means."* He further emphasized: *"The relationship between the goal and the means is like that between a tree and its seed."*

Conditional Persistence in Interests, Rights, and Truth

In this part of the insight, we acknowledge that human nature is susceptible to distortions that have historically been the root of major crises. Therefore, it is essential to recognize and actively address these distortions.

These distortions include:

- **Excessiveness and one-sided perspectives** on personal and collective interests and rights—whether individual, societal, ethnic, or territorial.
- **Dogmatic, rigid, and inflexible thinking** about beliefs—leading to intellectual stagnation and division.

Conditional Persistence in Interests

Everyone naturally seeks to secure their own interests. However, both instinctively and culturally, we recognize that individual well-being can only be achieved when collective interests are also safeguarded. Since there is no clear boundary between personal and collective interests, we are all, to some extent, vulnerable to violating the rights and interests of others. To truly assess where our interests and rights stand in relation to others, we must take their perspectives into account. To do this effectively, we need to step into their shoes and embody their role. For example, if we are dissatisfied with our employer or employee, we must strive to understand their concerns. This requires expanding our knowledge of their responsibilities and challenges. We can also approach the situation more intelligently by imagining ourselves in the roles of our spouse, child, parent, sibling, or even a third party, considering how we would feel if the situation were reversed. If one of our loved ones were in the position of our employer or employee, would the third party have the right to feel as offended as we do?

Additionally, we should reflect on the benefits others provide us. Are we showing gratitude for these contributions? Often, we overlook the positive impact others have on our lives and focus only on perceived conflicts. In fact, we may find ourselves caught in excessive demands or unhealthy competition, perhaps believing we deserve more than we do. We might also realize that we are not fully suited to our roles—whether as an employer, manager, spouse, or in any other capacity. When we recognize these truths, we must address our shortcomings, stop unhealthy competition, and offer understanding and comfort to others. This helps heal any negative emotions we may be harboring.

Conditional Persistence in Rights

The principles we discussed regarding benefits also apply to rights. In many cases, as spouses, employees, workers, employers, children, parents, citizens, neighbors, landlords, and tenants, we claim rights that are often rooted in ignorance, narrow-mindedness, or greed. For instance, when we break traffic laws, avoid paying taxes, cut in line, berate employees in front of others, or expect unreasonable service from our spouse, we are often unaware of the consequences of our actions. These behaviors stem from ignorance and can be corrected in several ways.

One method is to put ourselves in the shoes of others or even take it a step further by imagining a loved one in their position and a third party in our own. Would we be comfortable with the rights we're claiming if the roles were reversed? For instance, if we treat our spouse poorly, how would we feel if our child or sibling were in their place? Alternatively, we can imagine that we are no longer present—either having passed away or moved far away. What would we say about our loved ones in our absence? Would we better understand their vulnerability and limitations? In cases like tax evasion or traffic violations, it's important to consider the larger consequences. If everyone behaved this way—avoiding taxes or disregarding driving regulations—what kind of world would we leave for future generations? Would society collapse? These reflections help us understand the broader impact of our actions and the importance of adhering to shared societal responsibilities.

There are countless ways to curb greed and excessiveness while respecting the rights of others. The key is to recognize the oneness of all people, as discussed in the second insight of this book. Once we realize our interconnectedness, we must nurture the power of love, compassion, and justice within ourselves. These two qualities will guide us toward the right path. Additionally, reading enlightening literature and considering issues from other people's perspectives can further broaden our understanding. We can activate our full-scale nature through voluntary acts of service, offering our time and resources without expectation of anything in return. If we still feel slighted by small infringements on our rights or benefits, we can choose to view these disruptions as a form of giving—perhaps even considering them as a "tithe," a small, regular donation of our time or resources for the greater good.

The same principles also apply to territorial and international issues. It is vital to recognize the excessive or illusory rights we may claim for ourselves and protect the gem of peace. Over the past century, numerous countries have experienced significant territorial changes due to war, independence movements, and other factors. Many of these nations have accepted these changes with foresight, focusing on reconstruction and development. However, conflicts still arise due to ignorance about the value of peace, and these disputes continue to this day. In all of these cases, we can take action within the framework of conditional persistence in interests and rights.

Even if we don't fully believe in it, we can sincerely apologize to others. Apologizing is not an admission of fault but an expression of honesty, humility, and a commitment to growth.

Conditional Persistence in Truth

No truth, no matter how self-evident it may seem, can be regarded as absolute. Even seemingly indisputable truths, such as those in mathematics and physics, are relative, holding validity only under specific conditions. For instance, Newtonian physics, once considered definitive, revealed its limitations only with advancements in observational and measurement tools. After centuries of dominance, it ceded ground to relativistic physics in fields dealing with cosmic dimensions or extremely high velocities. Similarly, Euclidean geometry, once universally accepted, was found in the 19th century to be limited, prompting the development of both Euclidean and non-Euclidean branches of geometry.

We need not look far to find other examples. Until five centuries ago, the flatness of the Earth was considered an unquestionable fact. Likewise, anyone who dared claim that the Earth orbits the Sun rather than vice versa risked being condemned as a heretic by the religious authorities of the Middle Ages, often facing dire consequences.

Thus, it is essential to recognize that no truth is absolute. Every proclaimed truth belongs to specific conditions and is valid within those conditions. Following the principle that "everyone knows something," we must acknowledge that every perspective, no matter how opposed to our own, holds at least a fragment of truth. Even if we strongly believe that a particular religious idea contradicts freedom and human rights, there will always be individuals who see things differently—who not only find meaning in that idea but might even perceive our beliefs as a form of modern barbarism, materialist reductionism, or, at best, a colonialist tool. As Niels Bohr, the brilliant Danish physicist, once said, opposites do not contradict but rather complement one another. Therefore, we should respect the truths held by others while remaining open to questioning our own certainties. In doing so, we may uncover a convergence point where knowledge and truths interconnect—though this point will always be fluid, evolving as culture and understanding advance.

Needless to say, the driving force behind this outlook is love, compassion, and the law of attraction. This is crucial because persistence in interests, rights, and truth must also be conditional. We must continuously examine our own biases and excesses in what we claim as our rights and beliefs. At the same time, our pursuit of truth must be guided by love and compassion, rejecting force and violence. This principle holds particular significance in social and political activism.

The Role of Love and Nonviolence in Truth

Mahatma Gandhi's Satyagraha—"persistence in truth through nonviolence"—exemplifies this approach. Despite facing immense violence in his struggle for independence, he firmly rejected armed resistance. He believed that in Satyagraha, there are no true losers—both sides become partners in resolving conflicts. The departure of the British Empire from India in 1947, amid cheers and mutual expressions of goodwill, is a testament to this philosophy. As historian Arnold Toynbee noted, "Gandhi did not only free India; he also freed Great Britain." The legacies of Nelson Mandela in South Africa, Dr. Martin Luther King Jr. in the United States, and countless other activists who wielded love against hatred affirm this human potential. The Buddha articulated this immutable law: "Hatred can never be eradicated by hatred; it is only eradicated by love." Likewise, Rumi declared, "Blood cannot be washed away with blood."

An Example from Satyagraha

To illustrate the power of love and nonviolence in the pursuit of truth, consider a striking example from the Satyagraha movement. During a massive strike for Indian independence, a few wealthy Persian businessmen sided with the British, breaking the strike. In response, some strikers resorted to violence against them. Upon hearing this, Gandhi immediately called off the strike nationwide. He declared that unless the Indian people attained spiritual maturity—internalizing nonviolence and love as their guiding principles—continuing the struggle would be futile. Many at the time viewed this as political suicide. However, history reveals that this very decision helped elevate the spirit of the Indian nation, paving the way for its remarkable achievements in the years to come.

Summary of the Eighth Insight

The eighth insight naturally extends from the previous one. There, "truth" was presented as a "sublime gem" for the "sublime human being"—a treasure to be sought and cherished. In this insight, the focus shifts to how we safeguard this noble gem. It highlights two profound concepts: "adherence to truth" and "conditional persistence in truth." Above all, it emphasizes the indispensable role of love and compassion, which not only accompany truth but allow it to flourish and bear fruit.

The next insight, titled "*Wishing Happiness for Everyone*", will delve deeper into these themes.

The symbol of this insight is a part of NUSAWVO's smile, which will be fully introduced alongside the symbol of the twelfth insight.

The Ninth Insight

"Wishing Happiness for Everyone"

"Nothing in the world is as good as a good intention."
— Immanuel Kant

"The opposites are not contradictory but complement each other."
— Niels Bohr

"Be kind, for everyone you meet is fighting a hard battle."
—Plato

"There is nothing simpler than judging what is ugly and disgusting, and nothing more difficult than understanding it."
— Dostoevsky

"The one who lives in himself sees himself in everyone and everyone in himself. Love for the Lord of love erases all selfish hopes—which are the torture of the soul—in Him."
— Bhagavad Gita

"If you are not happy with the happiness of others, know that you will never be happy."
— Arthur Schopenhauer

"The people you love and those you have not yet found a reason to love are yourself, merely in another place."

—...

The Ninth Insight: "Wishing Happiness for Everyone – Universal Benevolence" (W)

The pursuit of happiness is one of humanity's most profound individual and collective aspirations. No one actively seeks suffering or sorrow, neither for themselves nor for others. Even a child instinctively recoils from pain, and the suffering of others—whether human or animal—resonates deeply within us. -(see Note 130) Psychologists often highlight two primary forces that propel the world forward: the pursuit of pleasure and the avoidance of suffering.

Note 130- *Developmental psychology studies reveal that goodness is innate in humans. Even children as young as six months old demonstrate a clear understanding of friendship and cooperation. One such study, conducted at Yale University under the title* Children: Their Amazing World, *sought to determine whether children have a preference for good behavior, and at what age this preference emerges.*

To explore this, researchers presented a puppet show to infants under one year old, featuring characters in distinct shapes and colors that represented good or bad moral roles. In the show, a red circle struggles to climb a hill, while a "bad" blue square attempts to push it down. Simultaneously, a "good" yellow triangle assists the red circle by helping it up the hill. After the show, the children were given the choice of which character to play with—the "bad" blue square or the "good" yellow triangle. All the children chose the yellow triangle, the one that helped and demonstrated friendship. Notably, even babies as young as seven months made this selection.

Further confirmation of these findings came from a study conducted at Kyoto University, which showed that children as young as six months old could distinguish between a victim, a bully, and a helper. In similar tests, most of the infants chose the helper, reinforcing the notion that children are naturally inclined toward empathy and kindness. A similar study at Harvard University produced identical results, further supporting the idea that children have an inherent inclination toward friendship and affection.

The ninth insight of the NUSAWVO school (as the fifth core pillar of NUSAWVO) aligns with this fundamental theme—the quest for collective, universal happiness. We refer to this as "benevolence"—a pursuit of joy and well-being that transcends individual boundaries, embracing both personal and collective fulfillment. Within the NUSAWVO school, benevolence is recognized as an essential element in realizing our full-scale human nature—or, in other words, our "agentic happiness."

However, it is evident that various obstacles hinder the realization of this noble human virtue. These obstacles, which manifest as sources of conflict and distress among individuals and societies, must first be identified before they can be overcome. This insight begins by illuminating these barriers and then offers pathways for their resolution.

Through the cultivation of awareness, it is hoped that insightful and perceptive individuals will come to appreciate the inherent beauty of benevolence, embrace it fully, and experience the profound blessings it bestows.

The Fundamental Question

To begin, we must address a crucial question:

If every human being naturally seeks to avoid suffering, desires happiness, and wishes well for others, why do turbulence, friction, and conflict persist among us? This question has long been central to all humanitarian schools of thought.

Throughout history, various perspectives have sought to explain the roots of human discord. Some, like Buddhism, attribute these struggles to selfish desires, greed, and excessive attachment to worldly matters, viewing them as the primary sources of suffering and enmity. Others have pointed to private property and class divisions as fundamental causes, advocating for the abolition of private ownership and the establishment of a labor-based government as a solution. - (see Note 131).

The NUSAWVO School's Perspective

The NUSAWVO school shares a fundamental belief with other humanist traditions: human nature is inherently noble, honorable, and inclined toward peace. However, it also acknowledges that human nature comprises multiple aspects, and when the balance between these aspects is disturbed—due to specific reasons we will soon explore—conflicts and abnormalities arise.

According to NUSAWVO, the root causes of discord lie in a lack of awareness regarding the natural foundations of happiness. This manifests in **cultural shortcomings**—such as the failure to fulfill natural human needs—and **cultural deviations**, which involve the imposition of values and actions that contradict the foundations of our twenty virtues. Both can be considered forms of **perversion**. In this context, the NUSAWVO school identifies five major anomalies that contribute to human friction, resentment, and hostility. While these issues have been previously discussed, it is necessary to revisit them from a fresh perspective, particularly in relation to benevolence. To further illuminate the factors that hinder and disrupt goodwill, we introduce a sixth category, expanding our analysis into six key sources of conflict:

Note 131*- The general aspects of these perspectives, along with their historical and empirical outcomes, have been examined and analyzed in the second appendix under the title* "Communism - Marxism," *within the discussion on the philosophy of* pragmatism.

Perversions Hindering Benevolence and Universal Goodwill:

1. Perversions Caused by Greed and Excessiveness-Extravagance

One fundamental source of conflict stems from humanity's insatiable greed, excessive and extravagant desires. The resources of our planet and human societies are finite; they cannot accommodate boundless demands. However, mankind's deep-seated greed—predating recorded history—continuously fuels competition and discord.

From the earliest times, humans have sought to take what belongs to others:

- A hunter sees more game in neighboring forests.
- A shepherd covets greener pastures beyond his own.
- A farmer desires the fertile lands of others.
- An industrialist seeks to dominate markets and outcompete rivals—often through conquest, expansion, and colonization.

This cycle of greed and overreach remains **untreatable** so long as humanity fails to reform its worldview—so long as it does not recognize the unity of all existence, does not see others as extensions of itself, and remains disconnected from spirituality. Unless human beings restore balance among the elements that define their true nature and the foundations of their happiness, this pattern will persist.

As the Native Americans wisely say in this context: *"We do not inherit the earth from our ancestors; we borrow it from our children."*

This profound truth reminds us that the world is not ours to exploit without regard for future generations. Greed and excessiveness arise from the illusion of ownership, as if the earth and its resources are ours to take without responsibility. Yet, once we recognize that we are merely caretakers, our approach to life, nature, and one another begins to change.

Yet, there is hope. As collective awareness grows, worldviews inevitably shift, whether one actively seeks it or not. As understanding deepens, the different aspects of human nature gradually return to harmony—bringing healing to these longstanding anomalies.

2. Perversion Caused by Xenophobia

When a person has not realized their unity with existence and does not perceive the world as an extension of themselves, they suffer from a fundamental alienation—both intellectually and emotionally. Regardless of greed or ambition, such a person lives in anxiety and isolation, fearing even those closest to them: relatives, friends, and neighbors. This fear breeds hostility, leading them to see others not as fellow human beings but as mere "things"—threatening, disposable, or exploitable objects.

Worse still, this dehumanizing perception manifests in different forms:

- **Objects of exploitation:** People become mere tools for personal gain—things to enslave, steal from, or use for pleasure.
- **Objects of hierarchy:** Individuals are sorted into rigid categories—weak vs. strong, rich vs. poor, believer vs. nonbeliever, privileged vs. oppressed, and so on.
- **Objects of division:** People are judged based on race, class, gender, religion, or nationality, fostering prejudice, malice, and conflict.

This fundamental error in perception—this inability to recognize the unity of all existence—serves as the foundation of countless human clashes. When we correct our worldview, cultivating an awareness of unity and fulfilling the **twenty foundational virtues of happiness** (particularly its spiritual dimension), these anxieties, prejudices, and hostilities naturally dissolve.

3. Perversion Arising from Market-Driven Unethical Behaviors

For centuries, a subculture has emerged that prioritizes the production of wealth above all else, stripping human and moral values from economic activity. This perversion manifests in various destructive ways:

- The commodification of human dignity (e.g., prostitution, human trafficking, and slavery).
- The exploitation of addiction and suffering (e.g., the drug trade).
- The profit-driven orchestration of violence (e.g., wars fueled by the arms trade).

At its core, this deviation is fueled by **unchecked greed and ignorance**, both at the individual and collective levels. However, as public awareness grows and humanity attains a more balanced understanding of its higher nature, these deviations will become increasingly evident, and the power of such exploitative markets will gradually diminish.

4. Perversions Caused by Narrow-Minded Ideological Views

At times, authoritarian ideological systems emerge, distorting or exaggerating certain principles to justify dividing people into rigid categories. Over time, these categorizations evolve into hostilities, fueling widespread conflicts and divisions within society.

Similarly, extremist religious interpretations contribute to this problem:

5. Perversions Caused by Extremist Religious Beliefs

Misinterpretations of religious teachings—often exaggerated or manipulated for control—can rapidly instill fear, justify discrimination, and escalate into violence. History provides numerous examples of how extreme religious ideologies have led to wars and mass persecution.

The root of these two perversions (ideological and religious extremism) can ultimately be traced to the first two:

1. **Excessive ambition and greed.**
2. **Intellectual and ideological ignorance.**

Though some individuals may initially act with good intentions, the hollow and extreme nature of these beliefs inevitably leads to corruption and destruction. The cure lies in **a transformed worldview**—one that embraces the unity of all existence—and the integration of the **twenty foundational virtues of happiness**, ensuring a balanced and enlightened approach to human life.

6. Anomalies Caused by Unresolved Personal Resentments and Disorders

At times, personal misunderstandings, unresolved grievances, and psychological struggles give rise to ill will. If left unaddressed, these conflicts can escalate, fostering animosities that spread through families, communities, and entire societies.

Recognizing and addressing these issues is essential for personal, familial, and societal well-being. While modern psychology and various guiding principles have made progress, the work remains unfinished. Humanity, undergoing a profound metamorphosis toward a higher level of awareness, now stands at a crucial turning point.

If we are to overcome these perversions and embrace **true benevolence**, we must cultivate awareness, correct our worldview, and strive toward the realization of our **full-scale human nature**.

In this context, a circular diagram called the **"Attitude Cycle"** has been introduced within this insight, which will be addressed soon.

The Approach of the NUSAWVO School

To address the root causes of human conflicts—essentially reviving the theme of "benevolence for all"—the NUSAWVO School has outlined nine general goals, structured around thirteen insights. These insights serve as foundational pillars for the school's philosophy:

1. **The Concept of "Meaning" (God) as a Unifying Principle**

   The first and final insights present "meaning" (God) in the most innocent and unbiased manner—free from anthropomorphic attributes, adjectives, and verbs. This approach seeks to counter one of the most significant causes of human conflict: self-constructed and exclusionary interpretations of "meaning." Furthermore, based on this principle, the NUSAWVO School introduces a universal resolution to seemingly insoluble conflicts: "Suspending disputes and entrusting the generality of contentious matters to 'meaning.'"

2. **Consciousness as the Singular Basis of Existence**

   The second insight highlights that existence, as understood by the NUSAWVO School, is fundamentally rooted in consciousness. By shifting people's perspectives and attitudes, this approach aims to dissolve conflicts stemming from isolation, xenophobia, greed, and excessiveness—elements that have long fueled human discord.

3. A Comprehensive Perspective on Truth

The seventh and eighth insights redefine "truth" across three fundamental axes, distinguishing this school from other approaches that do not prioritize truth. These key elements include:

- **Commitment to the continuous pursuit of knowledge, art, and truth**: This approach minimizes superstition and dogma, mitigating conflicts arising from them while fostering intellectual and emotional maturity.
- **Emphasis on truthfulness, integrity, and moral correctness**: This framework opposes ideologies that justify unethical means for an end, rejecting the "end justifies the means" mentality, which underlies many of humanity's deepest problems.
- **Advocacy for a balanced pursuit of interests, rights, and truth**: This perspective makes individuals aware of their own ingratitude, ignorance, greed, and excessiveness while discouraging dogmatic thinking—alleviating one of the primary sources of conflict. Additionally, because love and compassion serve as the activating forces of this school's philosophy, even when addressing issues such as oppression and ideological disputes, non-violent methods remain paramount.

4. The Role of the Law of Attraction

The NUSAWVO School underscores the pivotal role of human agency in attracting both positive and negative experiences. Awareness of this principle is crucial, as it suggests that individuals largely shape their own circumstances. In this view, enmity and conflict often originate within ourselves, and the path to resolution lies in our own hands.

5. **A Positive Definition of Human Nature**

The third insight rejects corrosive interpretations of human nature—such as the belief that humans are inherently sinful. Instead, it defines humanity as "happy and aware," ensuring that all philosophical and practical teachings rest on a constructive foundation.

6. **Understanding Full-Scale Human Nature**

Through the third and subsequent insights, the school explores the four fundamental aspects of human nature (physical, emotional, intellectual, and spiritual)—not merely in general terms, but with specific guidance. This includes the "20 Foundations of Human Happiness," the concept of "human agency in happiness," and its practical mechanism (NUSAWVO Core), as well as teachings on "self-awareness in happiness." These principles aim to balance human nature and counter destructive tendencies such as greed and excessiveness.

7. **Addressing Defilements with Transparency and Modesty**

The sixth insight openly but humbly identifies the sources of human perversion, speaking directly to the conscience of both individuals and society. By increasing awareness of these defilements, it promotes inner and collective transformation.

8. **Love, Compassion, and the Law of Attraction as Activating Forces**

The eleventh insight introduces "love, compassion, and the law of attraction" as forces that activate full-scale human nature. By shifting the responsibility for personal balance from external factors to an individual's internal emotional and rational faculties, this approach invalidates any justification for violence. This principle removes one of the most persistent roots of human conflict: the use of force.

9. Universal Benevolence as the Key to Agentic Happiness

This ninth insight emphasizes the unconditional role of universal benevolence in achieving "agentic happiness." Rather than relying on vague moral imperatives, the school aims to uncover the root causes of conflict and offer practical methods for healing grievances. This is illustrated through the "Attitude Cycle"—a circular model designed to facilitate reconciliation and inner peace.

The Foundation of Flourishing and Growth

The NUSAWVO School asserts that human flourishing and growth depend on internal and external peace, conflict resolution, and the revival of collective goodwill. These goals can only be achieved through individual and societal efforts to deepen awareness, insight, and commitment to balancing the four aspects of human nature. While external factors may influence this process, the school does not validate any external force as a determining factor in human progress. Since the essence of this school is rooted in "meaning" (God), the relationship between belief in "meaning" and its effect on personal destiny, happiness, and world peace has already been addressed in the first insight. Nothing further can be added in this context.

The Role of Welfare and Social Justice

One might wonder why the NUSAWVO School does not explicitly address welfare and social justice, despite their prominence as key drivers of human conflict. The answer lies in the school's core teachings: welfare and justice are intrinsic to human happiness, already embedded within the "20 Foundations of Happiness" and other insights. Overemphasizing these concepts separately would, therefore, be unnecessary.

While welfare and social justice are vital for individual and collective well-being, the NUSAWVO School posits that societies are built by individuals. Consequently, efforts to achieve social justice must begin with transforming people's thoughts and beliefs. This transformation is facilitated through the eight principles outlined above, which are essentially encapsulated within the "20 Foundations of Happiness."

Once this foundational work is achieved, larger systemic changes—including welfare and social justice—will naturally emerge within a logical and sustainable framework. The relationship between social justice and democracy, as well as the mutual expectations between citizens and a just government, is explored throughout various insights. In summary, the NUSAWVO School asserts that social justice and democracy are inseparable; one cannot exist sustainably without the other. Also, in terms of the relationship between social justice and democratic governance, we can add that for a democracy to function effectively,. certain structural elements must be in place, including:

- Free elections, ensuring full power rotation from the top down
- A free media, independent political parties, and NGOs
- Equal opportunities for all

Similarly, expectations from a just democratic government include:

- Support for entrepreneurship
- Minimum livelihood assistance for vulnerable populations
- Access to free healthcare, education, and social insurance (e.g., unemployment and old-age insurance)
- Equitable taxation, including inflation-adjusted taxes on the wealthy to fund public services

Importantly, this model rejects approaches based on asset confiscation and forced redistribution. Instead, it emphasizes a culture of hard work, honesty, and ethical living. If societal problems such as laziness, ignorance, dogmatism, and hostility persist, addressing them through cultural and educational reforms must be the first step. Through these principles, the NUSAWVO

School aims to create a sustainable framework for human peace, prosperity, and collective well-being.

Addressing Grievances and Restoring Benevolence

As promised, we now move beyond generalities to explore practical ways—at least on a personal level—of overcoming grievances and offenses while restoring the virtue of benevolence. These approaches are not confined to personal matters; in reality, they aim to address all forms of human conflict, whether individual or societal, by transforming our attitudes toward them. We hope that humanity, in alignment with its lofty and unparalleled nature, will rediscover both its singularity and its deep connection to others. Just as one seeks to sustain and enrich their own happiness, so too should they genuinely wish for the happiness of all—without exception, without discrimination. Regardless of the circumstances, nothing should interfere with this sacred intention.

To facilitate this, a circular diagram known as the **Attitude Cycle** is presented as a guide for resolving grievances, transforming attitudes, and overcoming emotional disturbances. To use this diagram, we must first identify the person or situation that has caused us pain. Then, we begin at step one and progress through the cycle, repeating the process as needed until a shift occurs. If, after completing the cycle, we still cannot sincerely wish for the happiness of the person who wronged us, it indicates that something remains unresolved, and we must go through the cycle again. The process may be difficult, even painful, as it requires us to confront our grievances rather than avoid them.

However, the effort will be worthwhile, for the outcome will be healing. Alternatively, we may find that repeatedly revisiting these grievances is so uncomfortable that we choose to let go of the offense entirely, simply to avoid the process. In that case, too, the goal has been achieved—a release of the burden.

Attitude cycle:

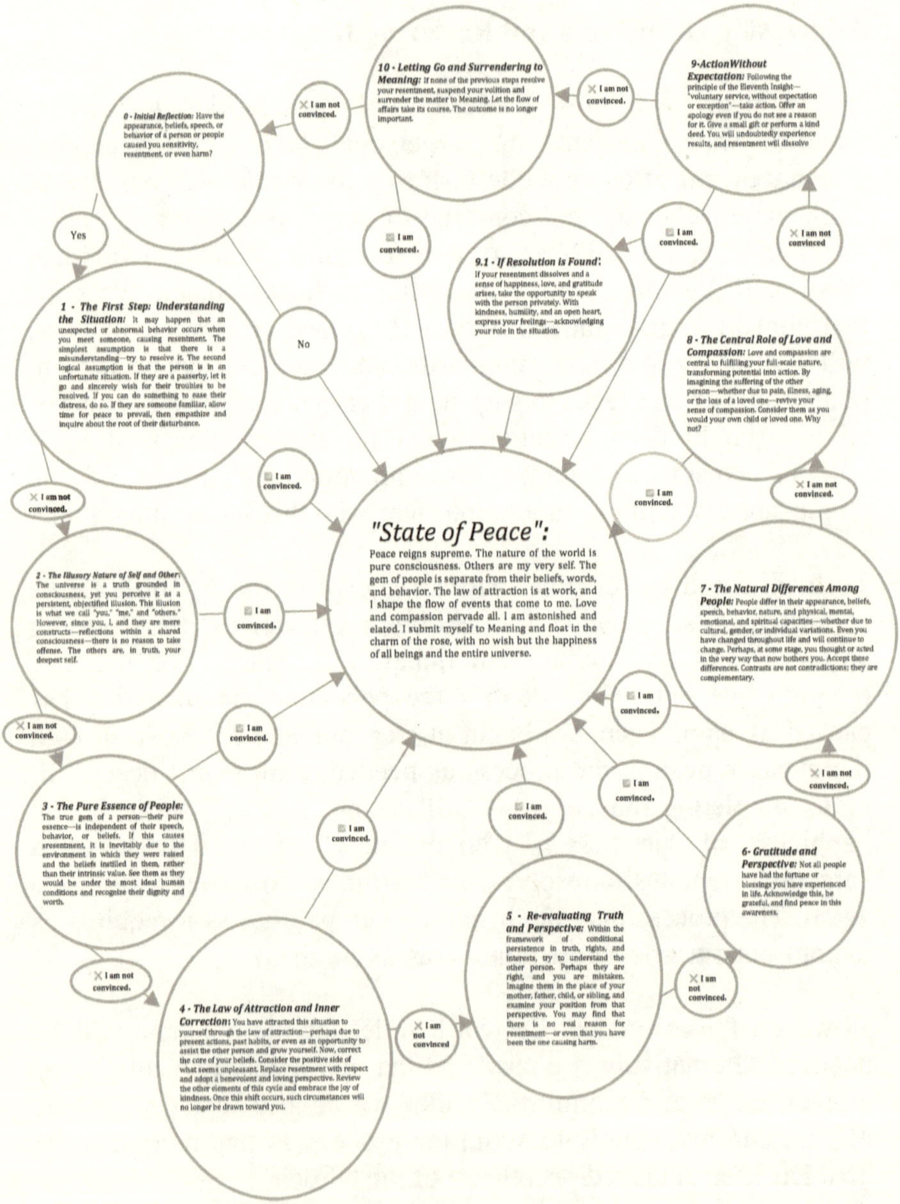

Progressing in the Attitude Cycle

The Attitude Cycle serves as a structured guide for transforming grievances into opportunities for personal and collective growth. To progress through this cycle effectively, we must approach each step with sincerity, patience, and a commitment to inner peace.

1. **Identifying the Source of Pain**

   Begin by acknowledging the specific grievance or emotional disturbance caused by a person, situation, or event. This step requires honest reflection and the willingness to confront discomfort head-on. Avoid bypassing this stage, as identifying the true source of pain is essential for meaningful progress.

2. **Reevaluating the Situation**

   Once the source of pain is identified, take a moment to reassess it from a broader perspective. Consider the possible reasons behind the behavior or event that caused the grievance. This process involves empathy, as we attempt to understand the situation through the lens of the other person's or circumstance's experience.

3. **Shifting Focus Toward Benevolence**

   In this stage, the goal is to consciously shift your focus from resentment or anger to benevolence. This is where true transformation occurs—by consciously choosing to wish for the well-being of the person or situation that caused you pain, regardless of how difficult or unjust it may seem. Benevolence does not mean condoning harmful actions but rather recognizing that healing begins with a compassionate heart.

4. **Releasing Emotional Attachments**

Release any attachment to the pain caused by the grievance. Emotional attachments often bind us to negative experiences, preventing healing and reconciliation. Letting go is not about forgetting or excusing the offense but freeing yourself from the emotional weight that prevents growth. This step facilitates mental and emotional clarity.

5. **Reaffirming Positive Intentions**

As you progress, reaffirm your commitment to universal benevolence and peace. In this stage, reinforce the intention to act with goodwill and to allow these intentions to guide your future interactions. This is not just a mental affirmation but an emotional and spiritual commitment to transformation.

6. **Reflecting on the Shift**

After completing the cycle, pause to reflect on any shifts in perspective or emotional release. You may notice that your feelings toward the situation or person have changed, and you can now approach them with more understanding or neutrality. If this shift hasn't fully occurred, repeat the cycle as many times as necessary, giving yourself permission to revisit and rework any unresolved emotional responses.

7. **Integrating the Shift into Daily Life**

The final stage involves integrating the shifts achieved through the Attitude Cycle into your daily life. This is an ongoing process of embodying the new, benevolent mindset in your actions, interactions, and choices. The cycle should become a regular practice that enhances emotional resilience and fosters long-term peace within yourself and your community.

By following this process, we cultivate an inner environment of peace, compassion, and resilience. The Attitude Cycle is not a one-time fix but a continuous practice that transforms not just individual grievances but societal conflicts as well.

Now, let us embark on this transformative journey.

0. Initial Reflection
Has someone's appearance, beliefs, words, or actions caused you discomfort, annoyance, or emotional pain? Take a moment to recognize and acknowledge these feelings without judgment.

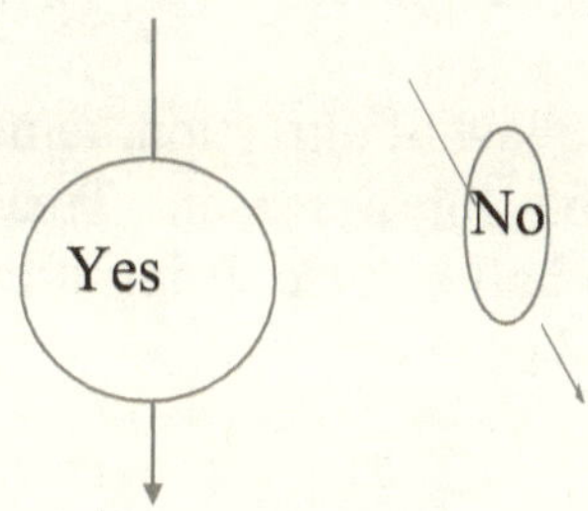

1. The First Step: Understanding the Situation

When you encounter someone's behavior that triggers discomfort, resentment, or even hurt, it's often helpful to first consider that a misunderstanding may have occurred, or that the person might be experiencing a difficult moment—perhaps they've received bad news or are dealing with adversity. In either case, they might not be in their usual state of mind.

If possible, let go of resentment, and genuinely wish for their well-being and the resolution of their distress. If a misunderstanding seems likely, take steps to resolve it. Should the situation go beyond your tolerance, approach it with kindness and sincerity.

With the same warmth and smile as you would normally offer, express that their behavior has caused you discomfort and that you are open to helping them find peace. If the person is someone you know, offer them a chance to clarify or work through the issue. Then, in a respectful and understanding manner, try to understand the underlying cause of their behavior.

Often, in such conversations—whether intended by them or not—you may uncover the root of their actions. However, if their situation is too painful to discuss, respect their space and offer patience. If the issue remains unresolved, you can refer to other steps in the Attitude Cycle.

Reflecting on the law of attraction, remember that this situation has come before you for a reason. Treat it as an opportunity for personal growth and service, using it to deepen understanding and compassion.

Thus, when encountering someone's unusual or unpleasant behavior, always assume that it stems from a misunderstanding or a personal struggle. If they are a stranger, release resentment and, if possible, offer help. If they are someone you know, offer the chance for them to explain and reach out privately to foster a resolution.

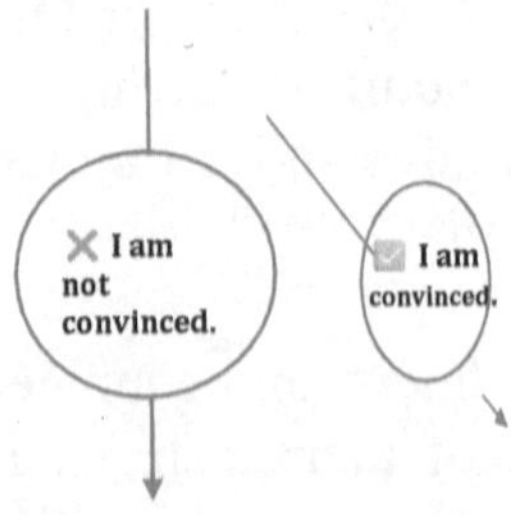

2. The Illusory Nature of Self and Other: "The others are my truest, deepest self."

As explored in the second insight of the NUSAWVO School, the structure of the world is grounded in consciousness, which manifests as a persistent illusion. In reality, both we and everything that exists in the universe are one; we are singular. There is no true separation between "me," "you," "others," or "things."

Therefore, the other—whether in appearance, beliefs, speech, or actions—is, in essence, my truest, deepest self. When we recognize this, we realize there is no reason for suffering.

Remember: **"You," "I," and "others" are merely interpretations of a singular reality. The other is not separate from you; they are your truest, deepest self.**

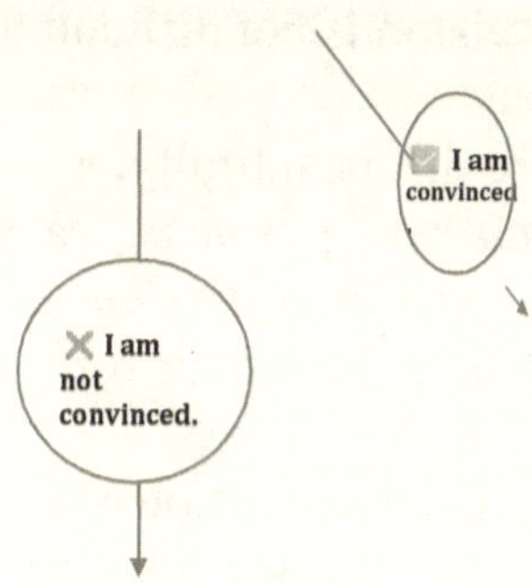

3. The Pure Essence of People: The Distinction Between Essence and Appearance, Beliefs, Speech, and Behavior

Our true nature is grounded in consciousness, which has taken on an objective form as a persistent illusion—manifesting as "me," "you," and "others." It's important to recognize that within this persistent illusion, a person's appearance, beliefs, speech, and behavior are separate from their essence and intrinsic nature. If someone's actions cause pain or harm, this is often a result of their environment, upbringing, or conditioning. Whether it's due to insufficient education, intellectual immaturity, or personal suffering, such behaviors can usually be traced back to external factors. In these cases, the problem lies not within their essence, but in the education, culture, or time required for maturity to emerge.

Reflect on your own life: you've likely experienced a transformation in your behaviors and beliefs over time. This process of change is maturity, and maturity requires patience. To separate a person's essence from their actions, imagine them in their highest and noblest state—as a saint, a spiritual guide, or an honorable being. This vision is not difficult to create.

Goethe's words capture this beautifully:
"When we treat a human being as he is, we reduce him to less than what he is. When we treat him as if he has already reached his full potential, we elevate him to what he is truly meant to be."

Similarly, Osho, the Indian sage, emphasized the unity of all beings in *The Secret*, explaining that we must separate people's actions from their sacred essence:

"You, those you love, and those you hate—you are all manifestations of God. This simple sentence can change your entire life. It can transform your entire perspective. The moment

one realizes that everything is one, love arises naturally." Even the teachings, cultures, or ideologies you may find inadequate, or undesirable should not be sources of offense. In our ever-changing world, every progressive idea and culture influences others, constantly absorbing and exchanging valuable elements. If significant misunderstandings or disagreements persist between cultures, it suggests that both perspectives are incomplete. There may not yet be a universally agreed-upon culture—or perhaps such a model is unnecessary. Instead, cultural diversity is essential to human civilization and presents a great opportunity. In truth, there is no fundamental battle between good and evil.

Even if a particular ideology or culture is flawed, Zoroaster's wisdom shines through:
"Do not draw a sword to fight the darkness—light a candle instead."

Therefore, always separate a person's appearance, beliefs, speech, and behavior from their sacred essence—because in truth, their essence is your truest, deepest self. These matters are unrelated to their singular being, which, ultimately, is you.

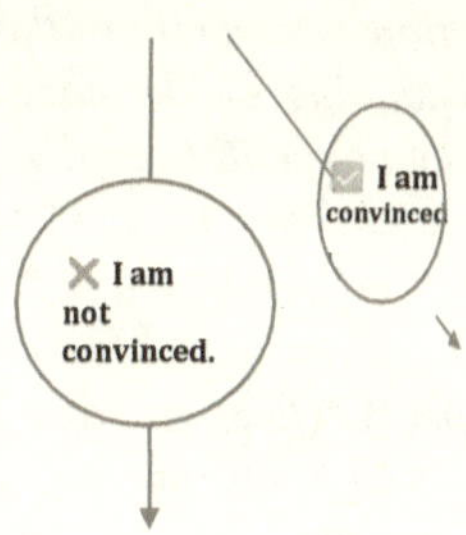

4. The Law of Attraction and Inner Correction: The Role of the Law of Attraction in Full-Scale Human Nature

From our previous insights, we've established that the driving force behind activating our full-scale nature is love and compassion. On the other side of this force lies the law of attraction in its positive aspect. - (see Note 132) Love represents the emotional dimension, while the law of attraction embodies the intellectual and logical aspect of a singular, underlying truth. - (see Note 133) Known as *Asha* (to some extent) in Zoroastrianism and *Karma* in Hinduism, the law of attraction suggests that everything we encounter in life is generally drawn to us through this law, and it will continue to operate as such. - (see Note 134) The law of attraction can be understood as a newly conceptualized force, akin to a combination of gravity and magnetism, with a distinct characteristic of polarity. Just like gravity, which pulls objects toward a central point, the law of attraction pulls experiences, situations, and conditions aligned with the core of our thoughts, emotions, and actions.

Note 132 – *The law of attraction is essentially the law of action and reaction, manifesting in two aspects depending on the nature of our actions. Its positive aspect emerges when we act benevolently, in alignment with our noble nature. Conversely, its negative aspect takes shape when we stray from our higher nature and engage in harmful or malicious actions.*

Note 133 – *In reality, both aspects reflect the same underlying truth that shapes our world—***consciousness***. This truth affirms that we are all inherently united and inseparable. It is not just our minds that are interconnected, but our very essence, which, though appearing as separate individuals, remains fundamentally one.*

Note 134 – *At the very least, in human interactions, there is no doubt that we exist within a highly sensitive and intelligent web of relationships—a web governed unmistakably by the law of action and reaction.*

Yet, unlike magnetism, where opposite charges attract, the law of attraction operates in a different way: positive energies attract positive experiences, and negative energies attract negative experiences, creating a dynamic flow that mirrors our inner world.

Much like gravity's bending of space-time, the law of attraction bends the fabric of our reality, shaping the events and conditions around us in accordance with the dominant forces of positivity or negativity within us. This force operates within the framework of the core of an idea, meaning that the framework itself—positive or negative—determines what is drawn into our lives. From another perspective, the law of attraction acts like a mirror, reflecting our thoughts and actions back to us. In essence, our thoughts, speech, and behavior set in motion equivalent responses in our lives. For example, if a particular issue disturbs us, we can assume (with certain exceptions, which we will discuss) that our past or present words and actions have, in some way, attracted or invited these conditions. This doesn't mean we deserve suffering, but rather that our consciousness has either created, invited, or at least interpreted the situation in a way that causes us distress. Understanding this law empowers us to take control of our lives and reshape our reality.

How Does the Law of Attraction Work?

As the name suggests, the law of attraction operates by attracting whatever occupies the core of our thoughts and influences our behavior. For instance, imagine someone telling you, "Don't think about lemons!"—your mind immediately conjures an image of a lemon, and you might even feel your mouth salivate. The mind does not register negation in such a context; it focuses on the core idea. Similarly, if someone warns you, "Control your anxiety and be careful not to have an accident," your focus shifts to anxiety and accidents, increasing the likelihood of a mishap. On the other hand, if you are told, "Drive calmly and focus on safety," your mind aligns with calmness and safety, making it the central focus of your thoughts, thus making your driving more secure.

Now, let's return to a troubling issue. If you wish to avoid disrespect (where disrespect represents the negative counterpart to respect), to transform your reality, you must focus on the positive aspect—respect—rather than the absence of disrespect. Trusting the law of attraction, cultivate joy in the very idea of respect and the favorable conditions it brings. This positive feeling will become the foundation of your subsequent actions, and as a result, respect will naturally be drawn toward you.

Transforming Reality Through Awareness

When faced with hurtful words or actions, and you seek resolution, simply shift your perception of the person involved. By applying the ten transformative factors—including the law of attraction—you can reframe the situation, find inner peace, and experience the happiness that comes with a new, more harmonious state (just as you would feel confident and at ease when driving safely).

Over time, your behavior toward that person will change, and their behavior toward you will also shift. Eventually, whether sooner or later, circumstances will realign, setting both of you on a path toward mutual understanding and human happiness.

However, there is an important point to remember: The law of attraction is not solely about personal gain—it operates universally. Whatever thoughts or behaviors you cultivate must promote happiness and well-being not just for yourself, but for others and the entire web of existence. True alignment with this law is unconditional—it does not discriminate, nor does it permit exceptions. If you focus solely on personal benefit without considering the well-being of others (who are, in reality, extensions of your deepest self), the law of attraction will still function—but in time, you will inevitably experience its negative repercussions.

The law of attraction is a neutral teacher. The situations you encounter are reflections of your inner state—your thoughts, emotions, and behaviors. If something in your life troubles you, simply redirect your focus toward its positive counterpart, believe in its realization, and allow yourself to experience the joy and peace that accompany this shift. From there, the law of attraction will adjust the conditions accordingly.

The Skeptic's Perspective: Three Scenarios

Before we conclude, let's examine this concept through the lens of skepticism. Everything we've discussed about the law of attraction and its impact on reality can be viewed as Scenario One. Now, let's contrast it with two alternative viewpoints.

- **Scenario Two: A World Without Consciousness**

Let's assume that consciousness is not the foundation of reality and that we are not fundamentally connected to the world or to each other. In this view, people's behavior is indistinguishable from their existence—they are exactly what they appear to be, nothing more. Additionally, there is no law of attraction.

If we dismiss attraction, Asha, Karma, and the cause-and-effect dynamics in human interactions, we might feel justified in pursuing our interests by any means necessary—since, in this worldview, moral and spiritual principles are mere subjective preferences—as long as we stay within legal and civil regulations.

At first, this approach may seem effective. However, over time, we may find ourselves inhabiting a world that feels cold, empty, and devoid of meaning. Even if we achieve our personal goals, we are left standing in a lifeless reality, disconnected from any deeper purpose. Relationships with those we once cherished may deteriorate, and we face an unfeeling world of stone and steel—a world we have conquered, yet one that ultimately fails to fulfill us.

- **Scenario Three: The Fortress of Fear**

Let's consider a middle ground. We accept that human interactions involve action and reaction, but we reject the idea that thoughts and words carry energy or shape destiny. We believe safety lies in vigilance—shielding ourselves with suspicion, caution, and strong defenses. In this view, we preoccupy ourselves with avoiding negativity. Our thoughts are consumed with fear of coercion, aggression, accidents, disease, crime, and misfortune. If someone disrespects us, we retaliate with force. If confronted by hostility, we strike back with full power, ensuring that our adversaries are punished.

The result? A life consumed by anxiety, where we live perpetually on guard, fearing every encounter and bracing for conflict.

Your Choice

Now, you are free to choose among these three perspectives. However, if you align with the first scenario—if you accept that consciousness is the foundation of existence and the law of attraction governs reality—**then**:

When faced with a troubling issue, reflect on the ten transformative factors, shift your focus toward the positive counterpart of what disturbs you (e.g., respect instead of disrespect), trust in the law of attraction, and immerse yourself in the joy and fulfillment that arise from this alignment. The law of attraction will then bring into your life exactly what corresponds to this new state of being.

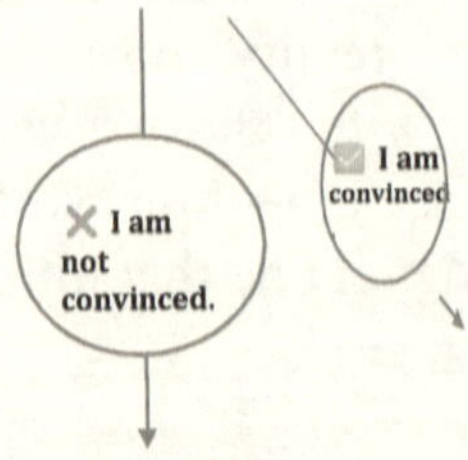

5. Re-evaluating Truth and Perspective: Conditional Persistence in Interests, Rights, and Truth

Within the framework of the eighth insight—conditional persistence in interests, rights, and truth—we recognize that we are always at risk of exaggeration, encroaching on others' boundaries, and falling into dogmatism or bigotry. Unconsciously, we may engage in these behaviors, harming others under various justifications, only to find ourselves hurt in return.

Thus, we must learn to accept that our persistence in defending our interests, asserting our rights, and upholding what we consider to be the truth must always be conditional. This means we should remain open and actively seek out contradictions to our own positions.

For example, when it comes to our interests, we must consider the possibility that we have—consciously or unconsciously—compromised the interests of others. In claiming our rights, we must reflect on whether our gains have come at the expense of someone else's rights or whether our very presence in a situation may unintentionally cause harm. Similarly, in the realm of truth, we must make a conscious effort to challenge our own beliefs, as no truth is entirely free from falsehood.

Therefore, within this framework of conditional persistence, as a progressive human being, you must acknowledge that you are always susceptible to overestimating your interests and rights. Moreover, within any "truth" you hold, there are inevitably unfounded or false elements waiting to be uncovered. It is your responsibility to find them, refine your thinking, and evolve.

With this understanding, there is no reason to be offended by anyone.

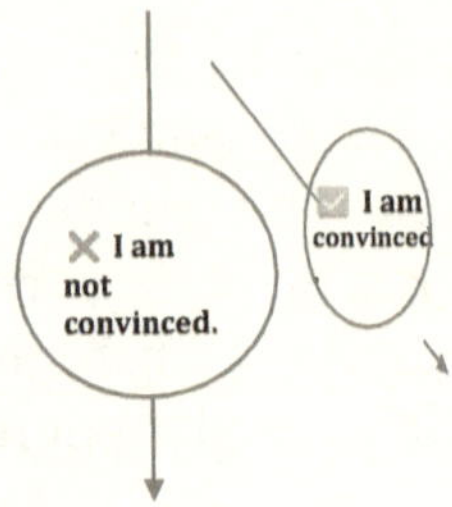

6. Gratitude and Perspective: Not Everyone is as Lucky as You

If you take pride in your qualities, thoughts, speech, and behavior (hopefully without any delusions in this regard!), remember that not everyone in the world has been as fortunate as you. Many have not had access to the same opportunities—whether in family, country, work environment, culture, friends, education, or other advantages. Even if they have, factors beyond their control may have hindered their growth.

In light of this, whatever harm or offense you perceive from others (if there is any truth to it) is often not a matter of their deliberate choice or will.

Therefore: Be humble and grateful for the opportunities you've had. Share your happiness with others, and with this understanding, you will find no reason to feel offended by anyone.

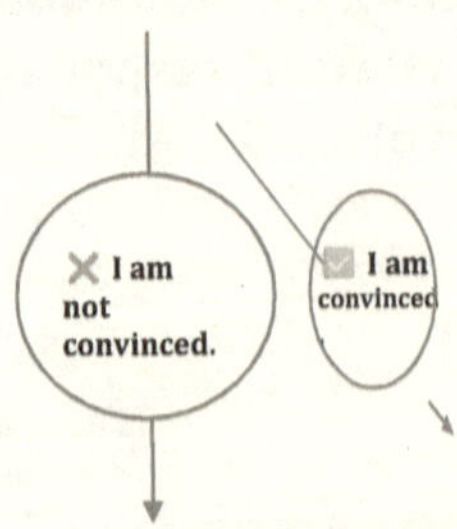

7. The Natural Differences Among People: People Are Not All the Same

Fundamentally, people's appearances, beliefs, speech, and behavior vary significantly across societies, cultures, and even genders. This diversity is both fascinating and inevitable. Even the way people think can differ, as they may employ different forms of logic to process ideas. If you find someone else's appearance, beliefs, speech, or behavior surprising, it's just as likely that they find yours equally unusual.

These differences are not mere peculiarities but invaluable treasures of human culture and civilization. Each one embodies a rich historical and spiritual heritage—often shaped by both wisdom and mistakes. For instance, in some cultures, marriage between cousins is seen as taboo, while in others, it is considered a sacred union "tied in the heavens."

Biological factors also play a role in shaping human diversity. Certain hormones may dominate in some individuals, influencing their temperament in ways beyond their control. Some people are naturally strong, while others are delicate. Some are highly extroverted, others deeply introverted, and many fall somewhere in between. Some individuals may display aggression without intending to, while others are naturally sensitive and reserved. Men may lean toward pragmatism, while women may be more attuned to emotional and intellectual subtleties, often requiring greater patience.

Physical differences extend across societies as well. Consider peanut allergies—unknown in some cultures but a serious concern in others, necessitating clear labeling on food products. In some regions, spicy foods are a staple, while in others, even a small amount of spice can trigger allergic reactions. These variations remind us that what seems "normal" is highly relative.

Ultimately, as long as people recognize these differences as natural and respect one another's interests, rights, and beliefs, they create an excellent opportunity for the growth of human consciousness. Even in cases where mutual understanding is absent among certain individuals or groups, these differences can still be interpreted through other perspectives outlined in this framework.

As Niels Bohr famously said, "Opposites are not contradictions but complements." A person with a strong temperament may thrive as a firefighter, professional athlete, soldier, or police officer, playing a crucial role in society. Even those whose extreme religious beliefs may seem irrational to us can function like the genetic codes of culture—preserving traditions that might one day serve as stabilizing forces in times of crisis.

Remember, you have undergone profound changes throughout your life and will continue to evolve. There may have been times when you thought or acted in ways that, if you encountered them today in someone else, would surprise or even upset you.

Therefore: Recognize that people are not all the same. Seek out the hidden opportunities within these differences and embrace the richness of diversity. As long as individuals respect each other's interests, rights, and beliefs, there is no reason to take offense. And in cases where mutual understanding is lacking, refer to the other insights within this framework for deeper comprehension.

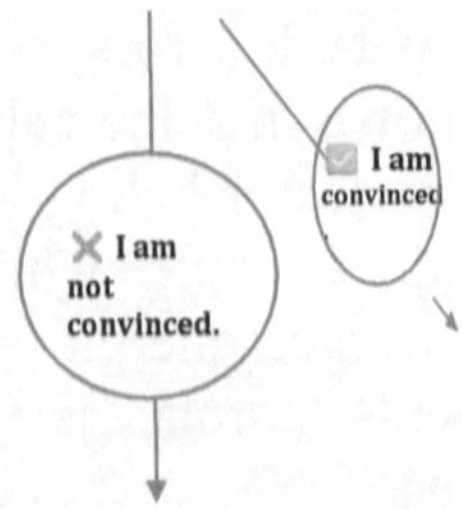

8. The Central Role of Love and Compassion: Love and Compassion

Love and compassion are pivotal in awakening our full-scale nature. They serve as transformative forces, shifting our happiness from an unconscious state to a conscious one. These qualities guide us in cultivating our abilities harmoniously, nurturing both ourselves and others with love, compassion, and a deep respect for their wholeness and irreplaceable essence.

If you find yourself feeling annoyance toward someone—who, in truth, is but a reflection of your truest and deepest self—activate the instinct of love and compassion within you. To do this, strip away the extraneous factors—the "accidental" or "derivative," as a philosopher might say—that surround the person. How? Picture them as a child, in old age, in sickness, in poverty, or experiencing the loss of a loved one, standing in a state of profound vulnerability, desperately in need of help. When you remove the layers unrelated to their true essence, compassion will arise naturally.

Now, consider a loved one in the same painful circumstances. Would you still feel hurt? Most likely, love and compassion—divine forces—will illuminate your path, dissolving your pain and bringing healing.

Therefore: Approach others with love and compassion, and imagine the suffering they endure, being on the edge of helplessness and homelessness. In doing so, you will uncover your genuine care and love for all, and any resentments will melt away.

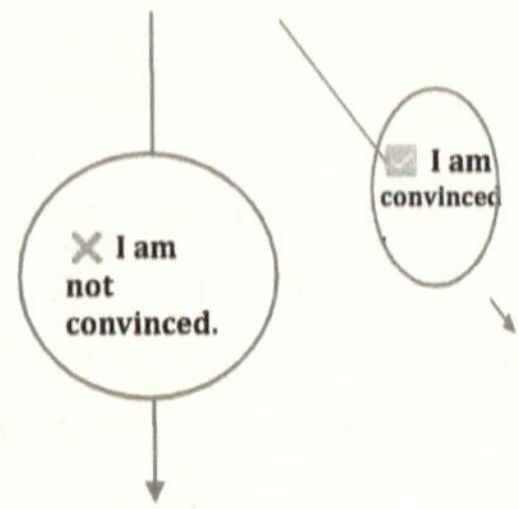

9. Action without Expectation: Volunteer Services

The tenth insight of NUSAWVO reveals that "voluntary service to our fellow beings and all other creatures, without expectation, without demand, and without discrimination" is an essential aspect of our full-scale nature.

This quality may have been overlooked, and it is crucial that we make a conscious effort to activate it now. If you are ready, the first step begins here: apologize to the person you feel has wronged you—even if you cannot find a clear reason for doing so. And offer them a gift. What harm could there be in this?

If you truly believe in voluntary service without exception, without demand, and without discrimination, that person, too, is a part of your fellow humanity. Perform this act to the best of your ability, knowing that it is unlikely you will not see some positive outcome from it.

Your goal is not to embarrass or pressure the person into changing their behavior. Rather, you are manifesting a higher aspect of your own nature, allowing you to release ownership of the suffering.

Therefore: By activating one of the most significant aspects of your transcendental nature— "unconditional, undiscriminating, universal voluntary service"—and by apologizing and offering a gift to the one you hold a grudge against, you facilitate your own growth. From that moment forward, the resentment will no longer belong to you. You will have transformed, shedding your old skin and leaving that resentment behind as you continue your evolution.

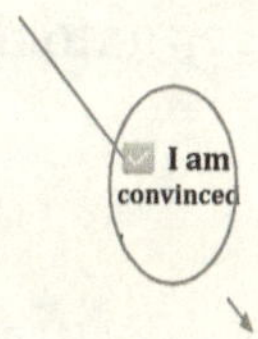

9.1. If Resolution is Found

If your resentment has dissolved, and you are now experiencing joy, love, and peace within yourself, it may be time for a very private—and we emphasize, very private—conversation with that person. Approach them with kindness, humility, and politeness. Acknowledge your own contribution to the situation and share your feelings about the resentment you once held. There may be an opportunity to uncover the root of your grievances, and through this, your relationship with them could deepen and improve on a more meaningful level.

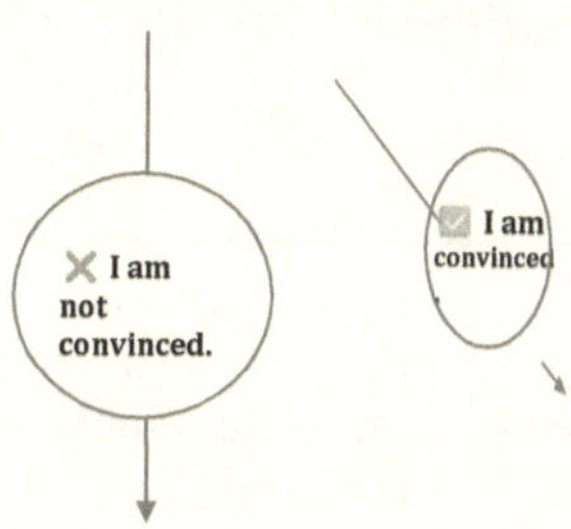

10. Letting Go and Surrendering to Meaning

If, after reviewing the nine items above, you find that the resentment still lingers in your heart and you haven't achieved the desired result, consider trying this alternative approach: Assuming you believe in "meaning" in your own way, take a moment—even just for the sake of trying something new—and follow the NUSAWVO method. Suspend your volition, and if you wish, engage in the special NUSAWVO ritual (along with the hymn: "O God"). Once you've done so, surrender the matter to meaning. Whatever happens next is not for you to control; the outcome is irrelevant, for whatever unfolds is the best. - (see Note 135)

Therefore: Suspend your volition, perform the NUSAWVO dance and prayer, and leave the matter to meaning. What happens after that is not important because it is good for everyone.

Summary of the Circular Diagram Discussion:

After reflecting on the circular diagram and wishing happiness to those who have irritated you, the problem—whether personal or social—will eventually resolve. If it's personal, the issue will be resolved sooner or later. If it's social, you will no longer harbor hatred or malice toward that society and will understand it better. You might even find ways to solve its problems. Ultimately, if nothing changes, you will let go and move forward, as there's no deadlock—either a solution will appear, or you'll create one.

Note 135*- Letting go means surrendering attachment to the results of your actions and trusting that things will unfold as they are meant to. It is a powerful practice that allows the divine or universal plan to manifest in its highest form, beyond your personal control. (Debbie Ford).*

When resentment is gone, the situation becomes as follows:

1. **There is no issue**: The problem dissolves.
2. **The whole world is consciousness, and others are my truest, deepest self**: A realization of interconnectedness.
3. **The unique essence of people is separate from their appearance, beliefs, speech, and behavior**: Understanding the deeper nature of individuals.
4. **The law of attraction works**: The best events are drawn to me and others.
5. **I am likely wrong about others**: Recognizing that others are often right in many ways—this reflects conditional persistence in interests, rights, and truth.
6. **Gratitude for my opportunities**: Understanding that not everyone has had the same chances I've had and recognizing my past mistakes.
7. **Love and compassion are my natural traits**: These qualities help me forget grievances by understanding the suffering of others.
8. **Voluntary service strengthens my nature**: Engaging in service, especially for those I've offended, activates my true self.
9. **Surrendering to meaning**: By suspending my volition and performing the NUSAWVO rituals, I entrust the outcome to God, with no attachment to the result.
10. **Happiness becomes agentic and self-aware**: I am at peace with myself and the world.

- I fulfill my full-scale nature.
- I am a lover, and nothing is a threat to me.
- I am grateful to the universe, others, and God.
- I hope for a brighter future.
- I am astonished and elated by existence.
- I see no threats or resentments—only opportunities.
- I find and create opportunities for myself and others.
- Ultimately, I float in the charm of existence, wishing happiness for everyone and all existence.

Summary of the Ninth Insight:

The Ninth Insight highlights that human beings are inherently inclined toward happiness—a happiness that becomes truly meaningful only when shared with others. However, this natural inclination can be obscured by six major anomalies (perversions) such as greed, ignorance, fear, unethical market practices, xenophobia, religious extremism, and petty resentments.

The insight offers solutions for the first five challenges through various perspectives and visions. The final challenge—minor resentments—is addressed by a circular diagram designed to alleviate suffering. This approach not only resolves small grievances but also clarifies the broader challenges at hand. Ultimately, the goal is to ensure that the inherent desire for happiness and benevolence for all remains unwavering—without expectation, exception, or discrimination.

On a larger scale, these insights do not negate the necessity of national defense or military deterrence. However, when peace is cultivated at the foundational level of human thought and behavior, excessive desires, unethical market practices, and hostile ideologies will naturally diminish. As discord decreases, international conflicts will likely subside, leading to a reorientation of military forces toward humanitarian efforts.

The core of this insight is captured in its full title: *"Wishing for the Happiness of Humanity and All Existence: Without Expectation, Without Exception, Without Differentiation, and Universally."* This means our wish for happiness must extend beyond humanity to all of existence, recognizing that humans are not separate from the universe. Ecological destruction serves as a stark reminder of this interconnectedness, and the growing awareness of this crisis calls for urgent action before irreversible damage is done.

The insight also raises critical ethical questions:

- Is it justifiable to hunt, breed, and raise animals for consumption in light of modern alternatives and scientific knowledge?
- Is it ethical to subject animals to inhumane experiments in the name of progress?
- Can we find alternatives to harmful pesticides in agriculture?

These concerns must be brought to conscious awareness and addressed through honest inquiry and active steps to remedy the injustices.

Finally, the symbol of this insight is part of NUSAWVO's smile, which will be fully revealed in the Twelfth Insight.

The Tenth Insight

“Voluntary Service for Everyone”

"It is not enough to want; you must act."
— Hyrum Smith

"The only ones who are truly happy are those who have sought to serve and found a way."
— Albert Schweitzer

"Greatness is not merely in words;
Two hundred speeches are not worth half a deed."
— Ferdowsi

"Always greet each other with a smile, for a smile is the beginning of love."
— Mother Teresa

"Return swiftly to righteousness and avoid mistakes.
He who is lazy in doing good finds comfort in wrongdoing."
— The Buddha

"If we lack peace in life, it is because we have forgotten
that we are all connected."
— Mother Teresa

"Spread love wherever you go.
Let everyone you meet feel your kindness."
— Mother Teresa

"You can be rich even in poverty."
— Mother Teresa

"O you whose hand can still extend-arise,
Do good before all power from you flies.
Act now, before the hour turns too late,
And you stand helpless at the gate of fate."

– Saadi

Tenth Insight: Voluntary Service to Everyone

(*Voluntary Service to Humanity and the Entire Being: Without Expectation, Without Exception, Without Discrimination, and Universal*)

The tenth insight is the sixth and final pillar of NUSAWVO's core. It highlights a fundamental principle: **voluntary service to all**—which can also be understood as charity in its purest form. This aspect of our nature is deeply spiritual, connecting us not only to our true selves but also to all of humanity and existence.

At the same time, voluntary service is the most profound form of prayer and a direct means of communicating with meaning (God). Our success and happiness depend on embracing and actively engaging in this natural characteristic.

We must step beyond the narrow confines of personal indulgences—whether for ourselves or our immediate families—and open our eyes to the world around us. Countless individuals need our immediate support. Recognizing this truth and taking action activates one of the most essential aspects of our being: **spirituality**.

If we believe in meaning—God—we must understand that experiencing and connecting with it is not found in abstraction but in the simple, meaningful acts of service to others.

Evolutionary and Spiritual Foundations of Voluntary Service

As explored in the third insight, our evolutionary roots trace back to the same ancestors as the great apes. Many of these creatures—especially our closest relatives, chimpanzees and bonobos—are inherently social and compassionate. Notably, no bonobo has ever been observed killing another bonobo in the wild. Chimpanzees, too, display profound emotions, forming deep bonds through love, affection, and care. Looking back at our evolutionary history, we see that the forces shaping our development over the past five million years have, in an astonishing way, worked toward deepening peace, friendship, love, and compassion. Human drivers such as organization, cooperation, solidarity, family bonds, and social unity have reinforced these values. Beyond this, another remarkable truth emerges: our cognitive and cultural evolution has reached a level where we can perceive the unity of the universe and experience meaning (God). This realization leads us toward an understanding of **pure spirituality**, revealing the love-based foundation of existence.

Thus, from both an evolutionary and spiritual perspective, human nature is deeply rooted in love, compassion, and voluntary service. Charity and selfless service are the most tangible expressions of these innate qualities.

Activating Voluntary Service in Consciousness

Like any other natural ability, voluntary service and charity must enter the realm of human consciousness and be actively cultivated. While this process may naturally unfold over time, cultural and societal influences often obstruct it. Some forces discourage voluntary service, redirect it toward corrupt or misguided paths, deem it useless, or suppress it entirely. Therefore, awakening and nurturing this aspect of our nature is essential. Voluntary service takes many forms, from the simplest gestures to grand efforts. It can be as small as offering a flower, a kind word, a warm smile, or a moment of empathy—or as profound as providing extensive aid to individuals, neighbors, colleagues, the sick, the elderly, the injured, other living beings, and the environment.

Building charitable institutions—such as schools, associations, clinics, research centers, or even ethical businesses - (see Note 136) that create employment opportunities—is another meaningful way to engage in voluntary service.

Note 136- *"If you give a man a fish, you feed him for a day. But if you teach him how to fish, you feed him for a lifetime."*

Ultimately, in the process of giving and serving, both the giver and the receiver benefit from the profound spiritual connection that is formed. This leads to the emergence of a dynamic society built on solidarity, empathy, and mutual support.

Parents who actively embody and encourage universal voluntary service in their own lives not only enrich their relationships but also plant the seeds of happiness in their children, ensuring their well-being for generations to come.

Among the ancient Indian wisdom stories, there is one about an old Hindu monk who sat silently by a river, repeating his mantra. Nearby, a scorpion crawled along a tree. Suddenly, a branch broke, and the scorpion fell into the river. Without hesitation, the monk bent down and rescued it, only to be stung in the process. Undeterred, he continued chanting his mantra.

A short while later, the scorpion fell into the river again. Once more, the monk reached out and saved it, receiving another sting in return.

A villager, who had been observing quietly, approached the monk with a puzzled expression. With mild frustration, he said, "Master, I've seen you save that scorpion multiple times, but it stings you every time. Why not just let it go?"

The monk smiled and replied, "Brother, the scorpion's nature is to sting; it does not do so on purpose."

The villager asked again, "But if you know its nature, why not leave it to its fate?"

The monk answered, "Brother, I do not act on purpose either. I am human—my nature is to rescue."

Summary of the Tenth Insight

Spirituality, love, compassion, and their manifestations—such as charity and voluntary service—are inseparable aspects of our true nature. It is essential to emphasize, remind, and awaken these qualities so they may emerge from potential into reality, lighting up the world. In the NUSAWVO school, we recognize the profound importance of this insight. We assert that this noble human trait should not be confined to family teachings and cultural practices alone but should be an integral part of formal education. It should be taught practically in schools as a mandatory part of the curriculum. This could include inviting philanthropists to speak, organizing visits to charitable and "economic-charitable" institutions, and similar initiatives.

A clear example of this insight is the practice of dedicating a portion of one's income to charity, such as the "Tithe" (one-tenth) or **Zakat**. This principle is widely upheld in both religious and non-religious traditions. In the NUSAWVO school, we emphasize that at least 10 percent of an individual's or family's income—or at least 10 percent of their time—should be dedicated to charity and voluntary service. However, it is crucial that this be done discreetly, away from public attention, as a private act between the giver and the receiver. During the drafting of the Universal Declaration of Human Rights, a letter was sent to Gandhi, asking world leaders to share their views on the declaration. Gandhi responded:

"Experience tells me that the proclamation of human duties is much more important than the proclamation of his rights."

As this insight expresses, its focus extends beyond humanity to encompass all of existence. In this context, no part of being is excluded. We believe that voluntary service should extend to all creatures and to our environment as a whole.

The symbol of this insight is a component of **NUSAWVO's Smile**, which will be revealed at the conclusion of the Twelfth Insight.

The Eleventh Insight

"The Power of Love and Compassion"

Whatever I say to capture love,
To shape it, frame it, lift it above,
The moment I touch its boundless flame,
I stand in silence—lost, ashamed.
— Rumi

O wise ones, O wise ones, seek love's embrace,
O lovers, O lovers, behold its grace.
Here we stand, and so do you,
Bound by love, forever true.
— Rumi

I'll say it openly, with joy untold,
Love's true servant, free and bold.
No ties to the earth, no chains above,
I dwell in the boundless realm of love.
— Hafiz

I was dead, and love gave breath,
Turned my sorrow into mirth.
Love took the throne within my soul,
And crowned me sovereign, whole.
— Rumi

"Love, and only love, carried me to the vast sorrow of lives.
Love, and only love, makes you familiar with the warmth of an apple."
— Sohrab Sepehri

In eternity, a spark of goodness was born.
Love arose, and fire engulfed the world.
— Hafiz

The world's rotation is driven by hunger and love.
— Schiller

Love is the fruit of all seasons, within reach of every hand.
— Mother Teresa

"It is a hidden secret from which love arises,
A name beyond ruby lips and the delicate hair beneath the ears."
— Hafiz

The Eleventh Insight: The Power of Love and Compassion - The Law of Attraction

In the Third Insight, we discovered that the NUSAWVO school views the transition from unconscious (or passive) happiness to conscious (or active) happiness as dependent on the realization of our full-scale nature. We learned that within this school's proposed model, our full-scale nature—a general subject—is fulfilled through six intrinsic features, known as the NUSAWVO Core. Additionally, we realized that the power of love and compassion, along with the law of attraction, serves as the activating force behind this set.

From the Fourth to the Tenth Insights, we explored the full-scale nature of human beings and examined the six fundamental characteristics that constitute NUSAWVO. Now, in this insight, we turn our attention to love and compassion, seeking to answer a fundamental question:

What is love, truly?

The Fundamental Nature of Love

The topic of love is simple, for no experience in life is ever separate from it. The very fact that we are here, reading or writing these words, is due to love. Why? Because across countless generations, our ancestors chose connection over enmity, closeness over division, and love over hatred. The result? After hundreds of millions of years, life continues to generously host us. If even a single pair of our ancestors had rejected love, at least one of us—perhaps none—would exist today as either writer or reader.

Love manifests in many forms: maternal love, paternal love, sibling love, romantic love, sexual love, love for life, love for the divine, love for humanity… Yet, at its core, all love is one. If we insist on defining it, perhaps the simplest way is through negation: What would the world be without love?

Without love and compassion, no relationship grounded in human affection or solidarity could form. No man would fall in love with a woman. No mother would nurse her child. No families would exist. No children would care for their aging parents. No nurse would tend to the sick. No worker would endure hardship. In short, nothing would function.

Even pleasure—if disconnected from love—would fade into emptiness.

Love in Nature and Spirituality

Does love exist among all beings? Yes. It manifests at different evolutionary levels and in various forms—parent-child bonds, herd loyalty, sibling affection, romantic attraction, and beyond.

Is love connected to spirituality? Love and spirituality are intertwined. Sometimes, love seems to be the source of spirituality; at other times, spirituality appears to give birth to love. Ultimately, they may be one and the same. When we experience meaning, a profound and pure love envelops us. Likewise, when we encounter selfless individuals—those who, with no expectation of recognition, dedicate their lives to humanity, charity, or the well-being of other creatures—we witness spiritual love in its purest form.

Is Love to Be Learned?

Yes. Although love is a natural trait of all living beings—especially humans, as natural as walking or speaking—we know that humans raised by animals struggle to walk on two legs or to speak a language. In the same way, love, compassion, and responsibility must be awakened and activated within a person.

This awakening occurs through observing deep human connections—among parents, friends, spiritual teachers, and communities—and through the transmission of loving and responsible beliefs. Love is intrinsic to human nature, yet like any natural ability, it requires learning, activation, and cultivation.

The Science of Love: Hormones and Emotion

Is love dependent on hormones such as oxytocin? Undoubtedly, yes. However, hormones do not define love itself; they merely regulate its intensity. They amplify or diminish the emotional aspects of love—but they are not its essence.

The Relationship Between Love and Compassion

Compassion refines and balances love, guiding it from the extremes of passion toward moderation, wisdom, and responsibility. In truth, love and compassion are complementary forces—one fuels warmth and connection, while the other brings clarity and depth. Neither can exist fully without the other.

Love, Compassion, and the Law of Attraction

What is true compassion? Compassion is the ability to see others as a reflection of our deepest, truest self—to recognize in them our brothers, sisters, children, parents, and loved ones. It is the capacity to perceive the sorrow and suffering of others, to empathize deeply with their plight, and to understand their pain as our own. True compassion means acknowledging the greatness within every human being, no matter their current circumstances, and honoring that greatness as a mirror of our own essence.

In alignment with this understanding, the NUSAWVO school views love as the primary force that activates every aspect of our full-scale nature. Love is the foundation of this realization—an unconditional love that emanates from our unity with existence and our connection to the divine.

There is a deeper dimension to love and compassion: the law of attraction, particularly in its positive form. While earlier insights have touched on this concept, what follows may feel repetitive—but it is a necessary repetition to truly understand the power it holds.

The law of attraction can be seen as a newly defined force, one that blends aspects of gravity and magnetism but with its own distinct polarity. Much like gravity, which pulls objects toward a central point, the law of attraction draws experiences, circumstances, and situations in alignment with the core of our thoughts, emotions, and actions. However, unlike magnetism, where opposite poles attract, the law of attraction works differently: positive energies attract positive experiences, and negative energies attract negative experiences. This creates a dynamic flow that mirrors our inner state.

Much like gravity distorts the fabric of space-time, the law of attraction reshapes the very fabric of our reality. It influences the events and conditions around us according to the dominant energies—positive or negative—within us. This force operates within the context of our core beliefs and thoughts: the framework of our ideas—whether positive or negative—determines what manifests in our lives.

The law of attraction, in this sense, has two aspects: positive and negative. It is a universal principle of action and reaction, one that responds to our actions and intentions.

- When we act with good intentions, aligned with our higher nature, the law of attraction manifests positively in our lives.
- When we stray from our true nature—allowing malice, selfishness, or negativity to dominate our thoughts and actions—the law of attraction reflects this misalignment, drawing undesirable experiences into our reality.

In simpler terms, when we prioritize our own interests or beliefs over those of others, or when we dwell on negative thoughts, we unknowingly activate the negative side of the law of attraction, bringing more hardship, conflict, and suffering into our lives.

Love and the Law of Attraction: Two Sides of the Same Coin

Love could be seen as the emotional dimension, while the law of attraction (in its positive aspect) represents the mental and logical counterpart of the same universal truth. In other words, the law of attraction (positively applied) is simply the practical manifestation of love and compassion in our reality.

In Zoroastrianism, this concept may be interpreted as "Asha." Asha aligns with a broad sense of cosmic order and righteousness. While Asha shares similarities with Karma and the Law of Attraction, it transcends the personal, focusing more on the universal moral principles that govern the natural and cosmic order.

In contrast, the Law of Attraction and Karma reflect a more personal understanding: they teach that what we encounter in life is a direct reflection of our own intentions, thoughts, speech, and actions. The law of attraction functions like a mirror: it does not create reality from nothing but merely reflects our inner world. If you smile at the mirror, it cannot help but smile back.

How the Law of Attraction Operates

The mechanism of the law of attraction is simple: **You attract whatever dominates your thoughts and influences your behavior.**

For example:

- If someone tells you, **"Don't think about pickles,"** your mind does not process the negation; instead, the core idea—

"pickles"—takes center stage, triggering an automatic reaction (such as salivation).

- If someone warns you, **"Be careful not to hurt someone,"** your mind fixates on the idea of **hurting**, increasing the likelihood of an accident.
- If someone says, **"Don't be angry!"**, the mind still latches onto **anger**, making it more likely that irritation will take hold.
- If you are told, **"Don't be afraid of aggressive people,"** your mind fixates on **fear**, making you more anxious and likely to attract confrontations.

However, contrast this with a **positive** framing:

- If instead, you hear, **"All people are your own, treat them with love and compassion,"** your mind aligns with **kindness and connection**. In this state, **even an aggressive person who crosses your path may be transformed into an opportunity for healing and service**. With your usual cheerful smile and inner calm, you can **handle the situation with wisdom and grace**, offering goodness even to those who have lost their way—because in the end, **they too are your truest, deepest self**.

The Ultimate Measure: Seeking the Greater Good

An essential aspect of the law of attraction (or karma, or even Asha, to some extent) is the true intention behind our thoughts, words, and actions. The central question is:

How much happiness and goodness does your intention bring—to yourself, to others, and to the world at large?

When your actions are guided by the well-being of all, the law of attraction will return joy and fulfillment to both you and the world around you. However, if your motivation is solely driven by your own beliefs, interests, and rights, disregarding others, then while the law of attraction may still provide you with what you seek, the balance will ultimately shift toward the negative. In the end, you will find yourself unfulfilled, at a loss. The law of attraction does not reward or punish—it simply reflects. It mirrors what we cultivate within ourselves, whether love and harmony or division and discord.

Final Thoughts

Love, compassion, and the law of attraction are **not separate forces**; they are different expressions of the **same universal reality**. When we truly understand this, we see that our **thoughts, intentions, and actions** shape the world—not through mystical intervention, but through the **direct, natural, and inevitable unfolding of cause and effect**.

Thus, in every moment, the choice is ours:

- To **live in love and attract goodness**
- Or to **live in fear and suffer the consequences**

The Law of Attraction and Its Delicate Precision

Even if you doubt the absolute effectiveness of this law, you cannot deny its influence—especially in the realm of **human relationships**. The dynamics of human interaction are extraordinarily **precise, sensitive, and accountable**. Every **small act of malice** sets in motion **equivalent and parallel reactions**, none of which remain hidden from the grand balance of existence.

To ensure that the **law of attraction** works in your favor, you must first **carefully and wholeheartedly** choose the **central core of your thought**. This means examining your intention with complete **awareness and benevolence**, reviewing it as many times as necessary. Imagine that in the **world of existence, no transaction occurs without an invoice**—every desire, every action, comes with a cost.

So, ask yourself:

- **Are you prepared to pay the price?**
- **Does this intention serve the greater good—for both you and others?**
- **Is it truly worth the portion of your life that it will consume?**

If the answer is **yes**, then allow that **noble thought**, one that **aligns harmoniously with every aspect of your full-scale nature—the NUSAWVO core—to take root in your being**. Let it **energize you, uplift you, and fill you with joy**. In this state, the **law of attraction aligns all your inner forces**, guiding them **effortlessly toward realization**.

Summary of the Eleventh Insight

- **Love is the life force itself.** It is the **objective interpretation of spirituality**—beyond mere words and logic. Without love, neither **human history** nor the **concept of humanity** holds any meaning.
- **Compassion is an essential aspect of love**—but it is an aspect infused with **fortitude, responsibility, moderation, and wisdom**. Love and compassion are **two wings of the same bird**.
- **Love has another face—the law of attraction in its positive aspect.**
- In the **NUSAWVO school**, love and its **practical counterpart, the law of attraction**, serve as the **activating force and balancing mechanism** of **full-scale nature** and its six intrinsic characteristics, known collectively as **NUSAWVO**.

Looking Ahead: The Twelfth Insight

In the next discussion, we will explore the **twelfth insight** of this school—**"self-aware happiness."**

The **symbol of love** in this school is as follows:

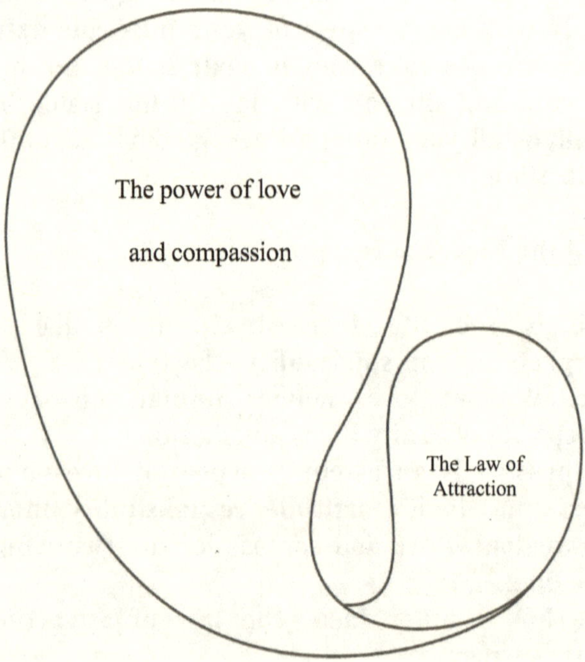

The Twelfth Insight

"Self-Aware Happiness"

Maybe life is not the celebration you wished for,
But now that you are invited to it, dance as beautifully as you can.
— Unknown

When life is sweet, be thankful and celebrate.
When life is bitter, be grateful and grow.
— Shauna Niequist

To see a world in a grain of sand,
And heaven in a wildflower,
Hold infinity in the palm of your hand,
And eternity in an hour.
— William Blake

The Twelfth Insight: Self-Aware Happiness

As mentioned in previous insights, this school believes that by embracing the power of love and compassion, and through the activation of the six "active gems of happiness," known as "NUSAWVO core," we can fulfill our full-scale nature. In doing so, we take command of our life- (see Note 137), our destiny, and our happiness. However, even while walking the path of happiness, it is possible that we remain unaware of it. That is, we may be so immersed in the flow of daily life and its demands that we lose touch with the joy, wonder, and intoxication that should arise from living in this mysterious world. As a result, we may not feel as happy as we truly should. This condition reflects a clear interpretation: simply being on the path of happiness does not automatically mean we are aware of or experiencing it. This principle is not exclusive to happiness alone; it applies to any aspect of ourselves that lies outside our conscious awareness. For example, many people possess a characteristic, talent, or quality without realizing it. To truly benefit from it, that characteristic must first enter the field of our awareness, allowing us to recognize and embrace it.

Note 137- *It is evident that many other factors contribute to human happiness. Among these, we believe that the awareness of our unity with the entire existence—what we refer to as "consciousness" (the subject of the second insight)—plays a significant role. This awareness reduces the anxiety associated with loneliness and the fear of death. Additionally, our direct and passionate relationship with "meaning" (God)—a concept explored in both the first and last insights—elevates our thinking, fills us with love, compassion, and humility, and naturally eliminates unrighteousness and violence. It is assumed that the seeker of truth, before engaging with the present insight, is already familiar with these concepts through a review of the previous insights. Furthermore, these concepts are presented in the form of five noble traits, which will be further elaborated in this insight, offering new perspectives on them. Of course, there are many other factors that influence human happiness, and we humbly leave the discovery of these elements to the truth seekers themselves.*

Happiness, like any other aspect of ourselves, requires conscious awareness. This is the foundation of "self-aware happiness," a concept born from the need to bridge this gap and bring happiness into our field of self-awareness.

This discussion introduces five noble traits that help us become aware of our own happiness, transforming it from mere "happiness" into "self-aware happiness." It is important to understand that this gift—"happiness that comes to human awareness," or "self-aware happiness"—is not exclusive to a select group of so-called intellectuals, sages, or chosen ones. It is a precious gem that belongs to all people, and we all deserve it, whether we are aware of it or not.

The five noble traits are as follows:

- **Thanksgiving (Appreciation, Gratitude, Thankfulness)**: The Thanksgiver (T)
- **Hope**: Hopeful (H)
- **Astonishment and Elation**: Astonished and Elated (A)
- **Opportunity Finding and Opportunity Making**: Opportunity Finder and Opportunity Maker (O)
- **The Seeking of Meaning (Spirituality)**: The Seeker of Meaning (S)

As you can see, the first letters of these five traits form the acronym "THAOS." If this does not resonate with you, feel free to choose a different letter combination that fits your preference. In the following sections, we will explore these noble traits in more detail.

The First Trait: Thanksgiving (Appreciation, Gratitude, Thankfulness) — The Thanksgiver (T)

Without gratitude and appreciation, a person cannot truly recognize their happiness or understand its value. As a wise person once said, *"I asked life, 'Why are you so hard?' Life smiled and replied, 'You people never know the value of comfort.'"*

In this school, by adding the attributes of "thankfulness" or "gratitude" to the definition of a "happy human," we are reminded that it is essential to express gratitude for the blessings we receive from the universe. Even if we don't benefit much from the 20 foundations outlined in the pillars of human happiness, or if other factors are missing entirely, we can still be thankful for everything we have and everything we are. We can be grateful to nature, to the people who have shaped history, to the spirit of humanity that, through perseverance and hardship, has brought us to this level of comfort, welfare, security, and opportunity. We can express gratitude to God, to the soul of the universe (consciousness), to our ancestors, to our mothers, fathers, siblings, spouses, children, friends, colleagues, and neighbors. We can also be grateful for the social order, even if it doesn't align perfectly with our ideals in every way, as well as for our employees and employers.

Without gratitude and thankfulness, we cannot truly appreciate our happiness (i.e., seeing the glass as half-full). Without this awareness, we won't have the hope needed to improve our situation, because in order to improve something, we must first accept it. If we don't, we risk destroying rather than improving it. Consider this: people were unaware of the beauty of nature around them until artists captured it in their paintings. Only then did people begin to appreciate that beauty. Similarly, when someone sees a beautiful view today, they may exclaim, "What an amazing view! It's like a painting."

Without gratitude, a person cannot recognize or value their happiness. Instead, they focus on their perceived shortcomings, wrongly believing they are unhappy. This mindset stifles creativity and the drive to improve, leading to the destruction of one's own happiness and, potentially, harming the happiness of others who are working toward the same goal.

An artist once said, *"I am sick of those who see the misfortune of others and are thankful for their own happiness!"* This statement might be wrongly interpreted as suggesting that we should curse the earth and time when witnessing others' misery. However, it can rightly be interpreted as a reminder that we should be grateful for our own happiness and, more importantly, strive to share it with those who are less fortunate.

In the heart of this trait, it is good to prepare a list of the strengths in our life. We will see that, in many ways, our situation is better than that of even one generation before us.

The Second Trait: Hope – Hopeful (H)

Hope is the light that guides us through life's uncertainty—hope for the future, hope for improvement, hope for the continuation of today's blessings, and hope that the bitter days will eventually pass. It is the realization that every seemingly bitter experience carries the potential for growth. Hope is not just about wishing for a better outcome, but about maintaining optimism, even when the odds are stacked against us. It is the essence of life itself. Life is not about dwelling on death—it is about embracing what is, and looking forward to what can be.

As André Zidé wisely observed, "Dying is not a difficult task at all, because in the end, everyone succeeds at this work!" Yet, life is something that needs to be cherished, especially by those who remain hopeful. Hopeful individuals, regardless of their health or circumstances, continue to appreciate the joy of living, remain grateful for their present happiness, and strive for a better and more meaningful future. Life belongs to the hopeful, and it is everyone's responsibility to carry this torch of hope, passing it on to the next generation.

From the perspective of this school's philosophy, everything is essentially an expression of consciousness, or perhaps an illusion—albeit one that persists. So, why not be happy and hopeful? *Que sera, sera.* If there's something we can do to improve our situation or avert an adverse event, we should act without hesitation. But after that, we must never despair. For someone who has recognized the value of existence and its infinite mysteries, no challenge feels like the end of the road. Life—indeed, consciousness—has no beginning or end. It is eternal.

Victor Hugo famously said, "The only word that God has written on every man's forehead is hope." Similarly, Shana Nyquist, a renowned American author, reminds us: "When life is sweet, be thankful and celebrate. When life is bitter, be grateful and grow."

All of this is shared without even referencing "meaning" (God). The connection with meaning—entrusting the flow of things to that mysterious force—adds a dimension that transcends the realm of these concepts.

A wise person once shared a diagram that offers a playful way to address worry. When your mind is heavy with concerns, write them down and compare them to the diagram below. It's unlikely that you won't find yourself smiling.

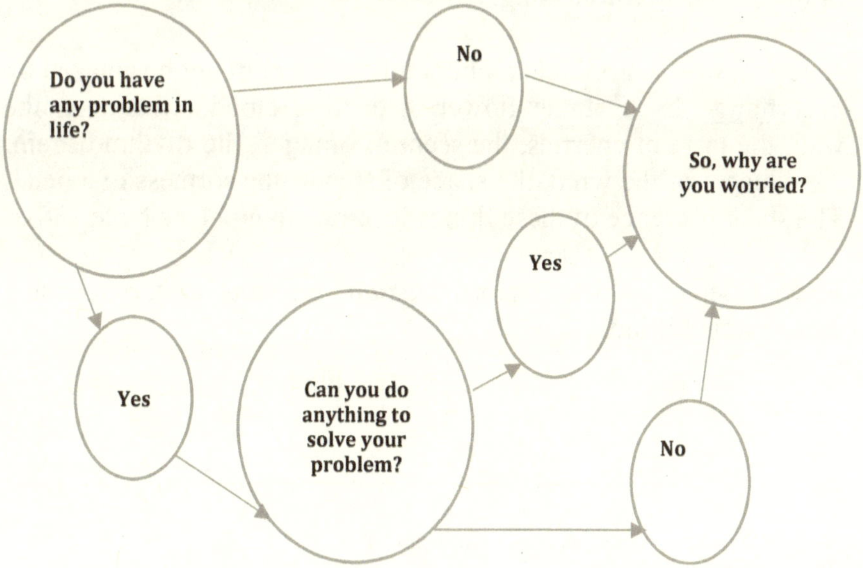

The Third Trait: Astonishment and Elation – Astonished and Elated (A)

It is nearly impossible to grasp the grandeur of existence and not be astonished. Existence is awe-inspiring, a boundless wonder that fills the heart with amazement. Simply the fact that there *is* existence—that we *are* here, that we *perceive* it—is astonishing in itself.

To hear the sound of rain, birds, and flowing streams. To inhale the fragrance of lemon leaves. To witness the sun setting over the highest branches while shadows gently embrace the earth. To hold one another in laughter, to embrace in shared sorrow. To be moved by great music, to feel an overwhelming sense of oneness that dissolves all distances. To know that there *is* a world and we are *in* it, and that one day, there will be a world in which we will *not* be, yet we will still remain a part of it. To reflect on a time when there *was* a world and we were *not*, and to consider that once, there was no world at all—yet, in some inexplicable way, everything *was*. All of this is utterly astonishing.

A person who is grateful and hopeful cannot help but be amazed by everything—by a single flower, a drifting cloud, a stain on the wall, the taste of cherries, the scent of oranges, the rhythm of rain, the whisper of the wind, the shape of stones, the softness of a petal. The sheer presence of these things is miraculous. - (see Note 138)

Paulo Coelho, in *The Second Maktub*, recounts a story about a monk named Stindel:

Note 138-*To see a world in a grain of sand,*
And heaven in a wildflower,
Hold infinity in the palm of your hand,
And eternity in an hour.
— William Blake

One day, the daughter of a friend asked me, *"Father, isn't it truly strange that I exist?"*

Coelho continues:

Children instinctively know that life is a miracle. We adults know it too—because deep down, we are still children. This part of us never truly dies. We might suppress it, forcing ourselves into the rigid seriousness of adulthood, but as long as we live, that child remains within us. It is better, then, to trust in its wonder. Once we learn life's lessons, we must merge the childlike excitement of discovery with the wisdom we have gained. As Jesus Christ said, "You must be born again."

To exist in this world and to be aware of one's presence in it is not only astonishing but profoundly blissful—almost intoxicating. Contrary to what Shariati once wrote (though his intention was different, as we have seen), we hold that one who does not exult in this life, beneath this sky, has not truly *been*. For simply *being here*, participating in existence, is an incomparable and rapturous experience. This feeling is neither indulgent nor self-serving; it is natural, spiritual, and pure. It arises from the depths of our very being, dissolving the illusions of separation—and even the illusion of death itself.

A happy, hopeful, and astonished human being is one who is overwhelmed by the sheer experience of being—who takes pride in the joy of existence. This happiness is noble and unburdened; it does not come at the expense of others, nor is it built on deprivation or misfortune. It is the fruit of goodwill, a natural state of elation that is meant to be shared with the world.

The Fourth Trait: Opportunity Finding and Opportunity Making – The Opportunity Finder and Opportunity Maker (O)

Human beings are both opportunity finders and opportunity makers. Our ancestors—and the ancestors before them—embodied this trait, even at the dawn of life, when they existed as resilient, determined single-celled organisms. Braving vast oceans and enduring the scorching, tumultuous conditions of early Earth, they sought and created opportunities to survive and evolve.

For humanity, nothing is ever truly a disaster. Every crisis carries within it the seed of growth. - (see Note 139) Without the ability to recognize opportunities, life itself would not have emerged from its origins. And Without the instinct to seek and create opportunities, our ancestors would have perished, left to the indifference of nature. Life, by its very essence, seeks opportunity, and human beings—its most intelligent and enigmatic manifestation—are its most gifted seekers and creators of opportunity.

A wise person once said:
"If you come across a great door locked before you, do not fear and do not despair. If it were meant to be permanently closed, there would be a wall instead." (Or simply: *Every lock has a key.*)

Francis Bacon, the English philosopher and statesman, remarked:
"To live happily, we must create favorable circumstances, not wait for them."

Note 139- *The Chinese word for "crisis" (危机, wēijī) consists of two parts: 危 (wēi), meaning "danger," and 机 (jī), which signifies a pivotal moment or turning point. In every crisis, there exists not only risk but also the potential for change and transformation.*

Abraham Lincoln affirmed:
"The best way to predict the future is to create it."

William James, the philosopher and psychologist, declared:
"The greatest discovery of our generation is that a person can change their life by changing their mindset."

Viktor Frankl, the renowned psychiatrist, wrote in *Man's Search for Ultimate Meaning*:
"To be human is to continually face situations that are both opportunities and struggles at the same time. The path to overcoming struggle is to recognize the meaning within the situation, for it is this meaning that grants us the opportunity to transcend it. Every situation calls out to us—we must first listen, then respond, and ultimately rise to the challenge."

It is clear, then, that work and entrepreneurship are direct expressions of opportunity-seeking and opportunity-making. It is through action—through working and creating—that our ancient human faculties are awakened and sharpened. It is in this process that our raw, untested imaginations are tempered by the realities of life, giving rise to true creativity. Without this process, our fantasies remain nothing more than lavish illusions, lacking substance or foundation. This truth applies to raising children as well. Exposing them to life's realities is essential for cultivating well-adjusted, resilient individuals. Shielding them from hardship can be as damaging as the tale of the well-meaning observer who, upon seeing a butterfly struggle to emerge from its cocoon, decided to assist by gently widening the opening. The butterfly emerged effortlessly—but its transformation was incomplete. Deprived of the struggle that would strengthen its wings, it was left too weak to fly.

We must go further—we must remove the words *problem* and *bad luck* from our vocabulary entirely. For you, as a *self-aware and joyful human being*, there is no such thing as misfortune. Every challenge is an opportunity for growth. In the end, you will either find a way—or you will make one.

The Fifth Trait: The Seeker of Meaning (Spirituality)

Spirituality is not separate from the wholeness of human existence. As explored in the second insight, the true essence of a human being is fundamentally grounded in consciousness. Thus, the perceived gap between ourselves and others—or even between ourselves and the world—does not truly exist. We are all one.

By realizing this truth, our sense of loneliness, separation, and anxiety dissolves. We come to see that others, as well as all things in existence, are not external to us but rather extensions of our very being. This realization grants us a profound form of immortality—not in a physical sense, but in understanding that death is merely the dispersion of our concentrated consciousness into the vast ocean of pure consciousness. This understanding is the essence of spirituality.

Spirituality, in this light, is not an abstract or mystical notion but a natural and logical conscience that emerges from the awareness of our unity with existence. It is the recognition of our intrinsic connection to the universe—an eternal bond with neither beginning nor end.

Moreover, as discussed in the first insight, we uncovered the fundamental reality underlying existence: meaning. We realized that we can establish a deep, reciprocal relationship with meaning itself. By entrusting our sorrows, pains, and defeats to meaning, we open ourselves to receive hope, joy, mercy, and love in return. This relationship is yet another expression of spirituality.

Additionally, in the fourth insight, we identified spirituality as one of the four fundamental aspects of our full-scale nature. Activating this dimension plays a crucial role in the transition from *unaware happiness* to *self-aware happiness*.

Summary of the Twelfth Insight

The twelfth insight naturally extends from the previous ones. In earlier discussions, we explored the concept of *active happiness* and examined how it manifests through:

- The *full-scale nature* of human beings
- The six insights forming the core of *NUSAWVO*
- The transformative power of love and compassion

These elements were introduced as pathways to achieving *active-agentic happiness*.

In this insight, five noble traits—collectively known as *THAOS*—are introduced as key aspects of self-aware happiness. Through these traits, the NUSAWVO school completes its exploration of human happiness by presenting the concept of the *self-aware, agentic, happy human*.

In the next discussion, we will turn our attention to the final insight of this school—*meaning*.

The symbol of the twelfth insight is a flower with five petals, each representing one of the five noble traits. This flower, simple yet beautiful, serves as a symbol of a self-aware, happy human being.

The choice of this symbol reflects the journey outlined in the NUSAWVO school:

- From the fourth to the eleventh insights, an individual embarks on an inner journey, integrating the full-scale nature of human existence, the six insights that form the NUSAWVO core, and the power of love and compassion—NUSAWVO's *smile and its companions*.
- Through this journey, one gains true agency over their happiness.
- By understanding the five noble traits (*THAOS*), one attains self-awareness of happiness.

- In the end, this journey leads back to the original simplicity and unaltered state of being—blooming like a flower, radiating joy, and intoxicating the world with the fragrance of meaning.

The symbol is represented as follows:

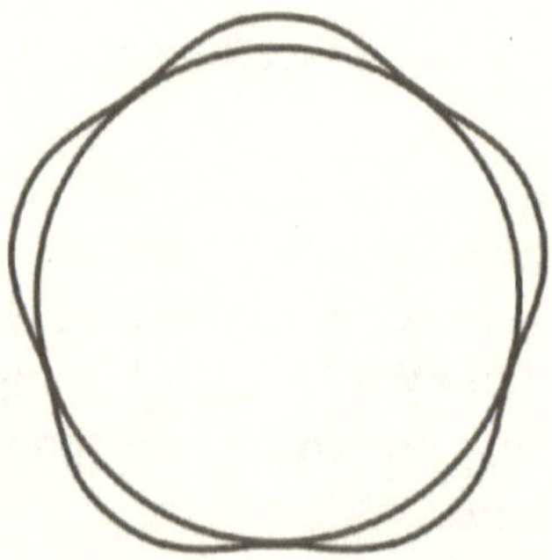

Now, we turn our attention to the symbolic connection between the third insight (*the happy human*) and the insights that relate to it—namely:

- The fourth insight (*the full-scale nature of human*),
- The fifth through tenth insights (*the six insights forming the NUSAWVO core*),
- The eleventh insight (*the power of love and compassion, along with the law of attraction*), and
- The twelfth insight (*self-aware happiness*).

The symbolic connection among these insights is represented as follows:

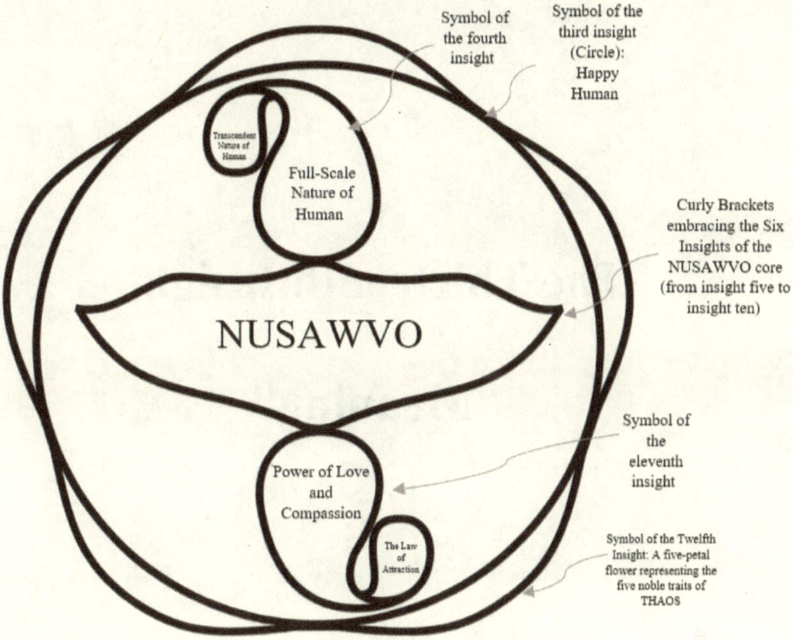

The above symbol can be simplified as shown below, capturing the true essence of this school—a "smile" and a "flower".

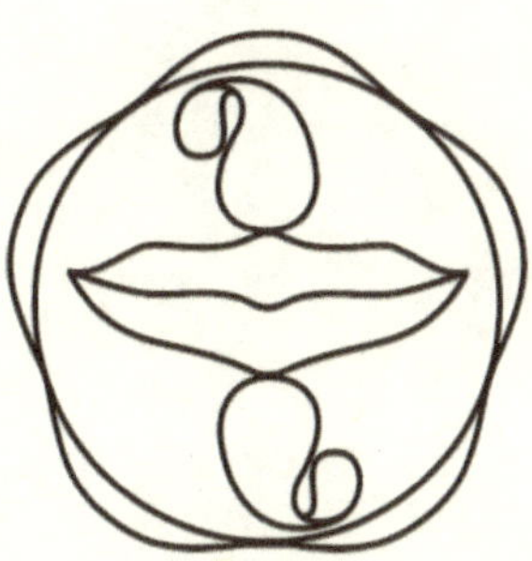

The Thirteenth Insight

"Meaning"

"It is not our task to unravel the secret of the rose;
Perhaps our task is simply to float within its charm."
— Sohrab Sepehri

"The grandeur of your beauty cannot be contained by earth or sky;
And yet, I marvel at how you reside within my heart."
— Jami

"God is with us."
— Unknown

"Sometimes, the only thing you can do is refrain from wondering, refrain from thinking.
No imagination, no surprise—just take a deep breath and trust that everything will unfold as it should."
— Unknown

"Look back with gratitude and thank God.
Look ahead with trust and have faith in God."
— Unknown

"God's path is serene and peaceful."
— Unknown

The Thirteenth Insight: "Meaning"

The thirteenth insight of this school, much like the first, centers on "meaning" (God). Since the foundational approach of this school toward "meaning" is based on silence, humility, and heartfelt faith, no new points are added here beyond the implicit explanations found in the first insight and the brief references scattered throughout the other insights. In essence, the appearance of this final insight serves as a reaffirmation that the core of this school—and its most fundamental pillar—rests in the belief in "meaning."

Just as nothing possesses true consistency without its essence, it must be stated that this school too lacks consistency without its essence. More precisely, we can say that outside of "meaning," the other subjects discussed in this school are relatively and conditionally valid.

One might ask: Why is this the case? What sets "meaning" apart from the other topics explored within this school, making them, by comparison, conditional and relative? The answer lies in the fact that everything discussed here regarding "meaning" is fundamentally different and incomparable to other subjects. What we speak of in relation to "meaning" is not merely theoretical or conceptual. These ideas represent reports from a singular, simple, and unique experience—**experienced as such by the heart, whether real or not—not as a privilege, but as something quietly accessible to all.** Though these reports are imperfect and incomplete, the experience itself remains distinct. This experience transcends mental interpretations and imaginations, remaining separate from such matters. (In fact, it was this very experience that inspired the completion of this book.) In contrast, the other subjects within this school are undeniably products of our minds and emotions. While these concepts may prove helpful and resolve certain issues, they are limited to the specific conditions under which they arose and developed. In this Context, two key aspects must be noted: first, we do not know how these ideas will function or be helpful under different conditions or on a global scale; second, with inevitable changes in circumstances, human beings, and culture over time, it remains uncertain how these ideas will adapt to new realities or what their outcomes will be.

Therefore, at the conclusion of this book, we return to the foundational pillar of this school— "meaning"—and hope that we remain grounded in this relationship, free from the distractions of our mental constructs and untested notions. May the grace of our connection with "Meaning" guide us through every stage of life.

The symbol of this insight mirrors the circle from the first insight, representing the connection between us and God.

And finally, the symbol of this school will be as follows:

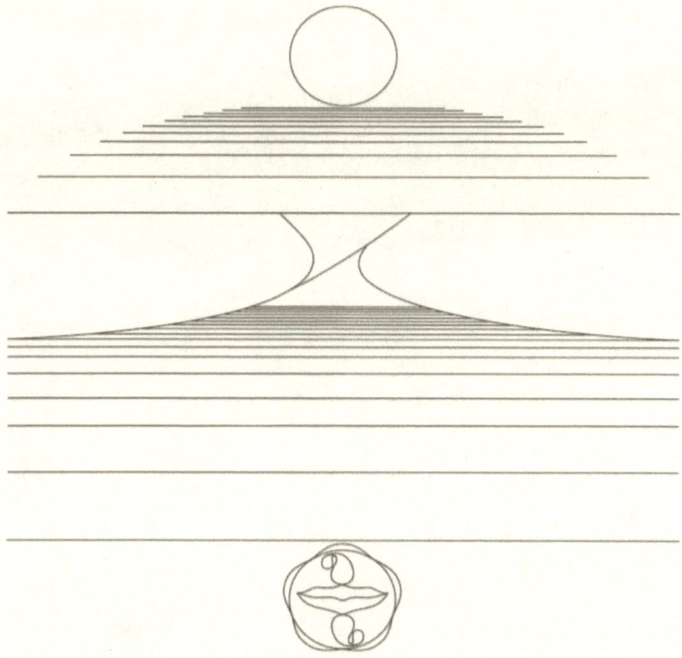

Farewell

"Nothing in the world is more powerful than thoughts whose time has come to be implemented."
— ***Victor Hugo***

"The best way to predict the future is to create it."
— ***Abraham Lincoln***

"If you want to plan for one year, plant rice; for ten years, plant a tree. But for the humanization of society, a century is needed."
— ***Confucius***

Goodbye

Truck drivers in Iran are known for their tastefulness and kind-hearted nature. From time to time, you come across a simple phrase on the back of a truck—one so profound it seems to hold the wisdom of an entire library. On the back of one such truck, it was written:

"Be with God, and reign as king,
Without Him, do as you wish—anything."

At least once in life, everyone asks themselves: "When I say goodbye to this world, what will be the last thing I say to my child?" These final words are perhaps the most precious gift a parent can leave behind. There are countless possible answers to this question, yet some may choose to say exactly what was written on that truck.

It is only natural, then, that the final words of this book—this school of thought—are also captured in that very phrase. And yet, the opportunity for farewell has not truly come.

As we reach the end of this book, it is time to say goodbye. But you, dear friend, who have walked this path with us, must know: If you have accepted the assumptions and claims within these pages, you will not receive a reward for it. On the contrary, you will find yourself in a predicament—one that could be called human self-awareness or natural conscience!

(And now, for the sake of argument—though this is merely an assumption—let us suppose you have taken some interest in these ideas!)

By now, you have come to recognize the true nature of the universe in both its dimensions. You understand that the essence of all things—mountains, oceans, galaxies, cockroaches, elephants, and of course, you and me—is consciousness. In another realm of existence, we are not separate but one, a singular whole. It is only in this realm of reality that we appear under different names and forms. Knowing this, you can no longer perceive others as truly "other"—for their pain is your pain, and resentment toward them loses its meaning. In this way, the essence of spirituality has been revealed to you.

You now know that your existence in this realm of reality is as a concentrated form of consciousness—an "I"—yet, at the same time, you are that vast, oceanic consciousness itself. With this understanding, the mystery of death is unraveled: death is simply the expansion and diffusion of your concentrated "I" back into your true self, the pure consciousness—the soul of the world. With this awareness, the loss of loved ones becomes more bearable, and the fear of death fades, allowing you to embrace life with greater intimacy and reverence.

You see that you are not merely a guest in this world; you are its fruit, its outcome—its very child. In a sense, you are this world.

You have recognized the extraordinary nature of your being and the astonishing effects of harmony within yourself. Love and compassion have become your allies, and the so-called "law of attraction" is no longer a mystical concept but a practical reality. You have encountered truth in its broadest and deepest sense, and it is unlikely that you will let go of this irreplaceable treasure easily.

You have understood the nature of perversions and the hidden dangers within them, and how they can lead one astray from natural happiness—even from humanity itself. Likewise, you have come to recognize the abysses that exist within these perversions—deep chasms of disillusionment and suffering that distort the soul's path and inhibit its true potential.

You have found the overarching meaning of life, and with this clarity, your search has become effortless and natural. As a result, your energy is freed. Work is no longer an obligation but a fulfillment of your overflowing vitality. It is not merely a task but a joy—perhaps you are even an entrepreneur, for you now understand that work shapes your abilities, refines your being, and introduces you to the real world and true human connections.

You have discovered the joy of serving others and how this deeply spiritual experience is linked to your own wholeness. Service heals wounds, nourishes the soul, and allows the essence of your humanity to bloom.

You are aware of the boundless possibilities that life has placed before you, and you cannot remain indifferent to those who have worked selflessly to pave the way.

You have recognized your innate happiness, knowing how to deepen and share it, making it richer with each moment.

You no longer see problems as obstacles or failures, but as opportunities for growth, learning, and contribution. The word "problem" has long vanished from your vocabulary, replaced by challenge, opportunity, and lesson.

The sacredness of family has revealed itself to you in all its dimensions, and you will not trade this invaluable bond for any illusion or mirage.

You love your children and understand their vulnerability, innocence, and need for protection. Their physical and mental well-being is of utmost importance to you. With open-mindedness, foresight, balance, and trust, you nurture their growth, allowing them to benefit from the healing embrace of nature. You are not oblivious to their upbringing, nor do you take it lightly.

You are now familiar with the purest, most innocent, and untainted concept of God that can be conceived. You understand that there is nothing fearsome, judgmental, or transactional in this relationship—only a profound unfolding that enriches you moment by moment, inspiring you to serve others and the world itself. This relationship allows you to surrender your sorrows and burdens in the darkest moments, placing your trust in the natural flow of existence.

Hope has become your way of life. With the awareness of your spiritual nature—within the framework of what we call consciousness—and through your intimate connection with meaning (God), your only occupation is wonder, joy, and unwavering hope. Everything else is but a passing entertainment.

And so, truly, you now possess all that you need to know. This is your human consciousness—your natural conscience.

It is no surprise, then, that in your personal philosophy, phrases like "I didn't know," "They didn't tell me, "Or "It wasn't meant for me" have long disappeared. With self-awareness, you no longer blame anyone—not even yourself—for the past or the present, for you know that your present choices and future direction rest solely in your hands. You see the past not as a fixed reality, but as an interpretation—one that you can reshape, even redefine, if you dare.

Thus, my dear friend, you are in quite a predicament. Because now, you have everything you need to create more success, more joy, and more meaning in your life.

The final choice is entirely yours.

In such a state, there are usually only two options:

1. To embrace this newfound happiness and share it with others.
2. ____________________________ We don't know! Believe us, we truly don't know what the second option could be!

Once, a wise mother, as she bid farewell to life, left her children with these words: *"Be happy, my dears!"*

A simple phrase, yet if you look deeply, it is the most beautiful blessing ever spoken.

We, too, have no greater wish for you than this: Dear one, be happy. Live long, with dignity and health.

If you find errors in this book, be kind—for you are already a spiritual being. Treat this book as a corridor in your journey of seeking. Walk through it but light your own way.

And truly—one last thing: if you remember only one thing from this book, let it be its smile. This smile is yours. Keep it as a memory—it will do no harm! Take the flower, too; it may serve you well.

Finally, do not withhold your knowledge from us, for as Shakti Gawain once said: *"At this time, our awareness is undergoing a great leap, and in this process, evolution takes a long step."*

And without a doubt, that long step somehow depends on you—yes, you!

The following email is at your disposal for this purpose: nusawvo@gmail.com

Be happy and prosperous.

Appendix One

Appendix One: Understanding the Universe (From a Scientific Perspective- (see Note 140)

In this appendix, we will first explore the ordinary realm of the universe, which we refer to as the first realm. We will then introduce another realm—metaphorically called the second realm—and, based on this distinction, recognize that the universe can symbolically be understood as a dual reality.

Since a deeper understanding of any phenomenon often comes through examining its history, we will adopt this approach here and throughout the book, striving to study phenomena in relation to their historical development whenever possible.

Note 140*- In Appendices, 2, and 3, we will explore the universe through philosophical and mystical lenses.*

The First Realm of the Universe-Existence

To explore the universe—existence—in the first realm, we begin by examining our closest relative in this world: our home, Earth. From there, we trace the history of life and the evolution of humans, reflecting on our role as questioners of the universe. Finally, we turn our attention to the past and the boundless nature of the universe itself. - (see Note 141)

The First Realm of Existence – The History of Earth

The study of Earth's history initially relied on analyzing geological layers and comparing the fossils embedded within them. Over time, additional methods emerged, including measuring the salinity of ocean water, examining sedimentation processes, studying the cooling of Earth's crust, and analyzing the decay of radioactive materials trapped in rocks.

Among these approaches, radiometric dating—the last method—proved to be the most precise and definitive. First introduced in the early 20th century, it reached a high level of accuracy by the mid-20th century with the development of advanced spectrometers.

Note 141- *"Science will never fully unveil nature's ultimate secret, for in the end, we ourselves are part of the mystery we seek to understand."*

-Max Planck

Dating Earth's History and the Formation of the Solar System

To accurately trace Earth's history, it is essential to focus on precise scientific methods rather than broad generalizations. One such method is radiometric dating, which relies on the predictable decay of radioactive elements.

Radioactive materials naturally decay over time, transforming into other elements at a measurable rate. The time required for half of the atoms in a given sample of a radioactive element to decay into its daughter product is called its **half-life**. These half-lives vary significantly among elements, ranging from fractions of a second to several billion years. For example, the half-life of **Carbon-14** is approximately **5,730 years**, while **Uranium-238** has a half-life of about **4.47 billion years**.

Rocks from different geological layers often contain radioactive elements such as **uranium, rubidium, thorium, and potassium-40**.- (see Note 142) By measuring the ratio of the original radioactive isotope to its decay product and knowing the element's half-life, scientists can determine the age of a rock sample. These dating techniques have been applied to Earth's deepest rock layers, using various isotopes to ensure accuracy and consistency.

Since all current astrophysical theories and empirical evidence suggest that Earth and the rest of the Solar System formed simultaneously, additional data from meteorites and lunar samples provide further confirmation. Radiometric dating of meteorites and moon rocks has yielded consistent results, leading to the widely accepted conclusion that Earth and the Solar System are approximately **4.6 billion years old**.

Note 142*- Through radioactive decay, uranium transforms into lead, rubidium into strontium, thorium into lead, and radioactive potassium into argon.*

This finding is now firmly established and universally accepted within the scientific community.

The Formation of the Solar System

Scientific hypotheses suggest that the Sun and the planetary system, including Earth, formed from the remnants of earlier massive stars. Long before the birth of the Solar System, one or more giant stars occupied this region of the galaxy. After exhausting their nuclear fuel, these stars collapsed and exploded in a dramatic **supernova** event, dispersing their constituent elements into interstellar space. - (see Note 143)

The resulting dense cloud of gas and dust, enriched with heavy elements, began contracting under the influence of **gravity**, eventually forming a rotating disk. At the center of this disk, a dense core emerged—the proto-Sun. As the core grew in mass, its gravitational pull increased, drawing in surrounding gas and dust until it ignited nuclear fusion, marking the birth of our Sun.

Meanwhile, the remaining material in the surrounding disk gradually coalesced into small **planetesimals**—early building blocks of planets. These objects repeatedly collided and merged, growing into larger planetary bodies over millions of years. Through this process of accretion, the planets of our Solar System, including Earth, took shape.

Note 143- *As scientific knowledge continues to grow, it is possible that new theories regarding the origins of our solar system may emerge.*

The First Realm of the Universe: The History of Life

In the beginning, the Earth was a molten mass that gradually cooled. Over time, oceans and seas formed. Then, after hundreds of millions of years, driven by a high affinity for the elements to combine—particularly carbon—and assisted by various sources of energy, - (see Note 144) the first glimpses of life appeared in the form of single-celled organisms. This was not a sudden event but rather a gradual process that took place in a vast laboratory, the Earth, over a span of hundreds of millions of years. The oldest microscopic fossils and traces of life discovered so far are approximately 4.1 billion years old. These traces seem to be remnants of bacteria that lived near cracks in the ocean floor, where molten material emerged and provided iron for them to feed on. Until 550 million years ago, life was largely confined to the seas and oceans and existed primarily as unicellular and protocellular organisms.

Note 144- *Today, under strictly controlled laboratory conditions and by applying various forms of energy—such as electrical discharges that simulate natural lightning—within sealed containers with water and simple gases thought to have been present in the early oceans (carbon dioxide, methane, ammonia, and hydrogen), researchers can produce amino acids—the building blocks of proteins—as well as sugars, alcohols, and fats. These compounds are closely related to life and can form in just a few days.*

In 1953, American chemists Stanley Miller and Harold Urey first conducted this experiment, showing that complex organic molecules could arise from simple inorganic compounds under conditions similar to those believed to exist on early Earth. Since then, it has been successfully replicated in laboratories around the world.

In subsequent experiments using different materials, even nucleotides—the building blocks of DNA—have been synthesized. Additionally, some of the most complex organic molecules associated with life have been found in meteorites, suggesting that these celestial bodies may have played a role in the prebiotic chemistry that contributed to the origins of life on Earth.

From that time onward, life expanded rapidly across the oceans and onto land, flourishing in a dazzling variety of forms. This surge in life was due to the creation of favorable environmental conditions, as well as the operation of two key factors: "life forces" (self-preservation and reproduction) and "evolutionary factors" (genetic mutations and natural selection).

Over the past 550 million years, the Earth has experienced five major mass extinctions, triggered by events such as extreme climate changes, volcanic eruptions, and asteroid collisions. The most recent extinction occurred 66 million years ago, when a massive meteorite struck the Earth, wiping out the dinosaurs and many other species. Despite these catastrophic extinctions, which have wiped out up to 99% of all life, Earth is now home to five major kingdoms of life: protozoa, bacteria, fungi, plants, and animals. Among the animals, the five classes of vertebrates—fishes, amphibians, reptiles, birds, and mammals—are most closely related to us in terms of evolution.

Among these groups, early mammals, considered our common ancestors, first appeared around 200 million years ago. However, it wasn't until the extinction of the dinosaurs 66 million years ago that mammals began to expand. The mammals that survived this cataclysm, by taking full advantage of the post-extinction environmental conditions and through the process of evolutionary factors (genetic mutations and natural selection), spread across the planet, diversifying into a wide range of forms.

Around 14 million years ago, hominids—our common ancestors with orangutans, gorillas, chimpanzees, and bonobos—emerged, marking a significant step in human evolution. Our direct human ancestors, however, did not appear until about 5 million years ago, when they branched off from the ancestors of chimpanzees and bonobos and began an independent evolutionary path. These early human ancestors were bipedal in a limited way and used very simple tools for daily tasks. Since then, human evolution has been marked by numerous changes, resulting in a variety of human subspecies, including the relatively recent ones such as *Homo floresiensis*, *Homo erectus*, Denisovans, *Homo heidelbergensis*, and Neanderthals. These subspecies evolved through a combination of interbreeding and extinction, giving way to the emergence of *Homo sapiens* around 300,000 years ago. This species was distinguished from other human subspecies by evolutionary and biological features resembling those of modern humans. These early humans, who lived in the plains of Africa, used stone tools and had mastered the use of fire. Over time, they migrated to other continents and, through a combination of interbreeding with subspecies such as the Denisovans and Neanderthals, became the dominant human species across the planet. Today, the various groups of humans are the descendants of this early *Homo sapiens*. - (see Note 145)

While human biological evolution will never cease, for 300,000 years ago, what has primarily shaped human destiny has been cultural rather than biological. However, the development of human culture and civilization remained slow and primitive until about 10,000 years ago, when the advent of livestock domestication and agriculture accelerated the pace of civilization. The invention of writing a few thousand years ago marked a significant turning point, sparking a rapid growth in sciences, arts, and technologies, while also strengthening philosophical and spiritual thought. This, in turn, created a fertile ground for the growth of ideas and the emergence of diverse material and spiritual phenomena.

In recent centuries, particularly in the 20th and 21st centuries, the development of advanced theoretical and technical tools has enabled humans to look back at their origins and trace their evolutionary story. The study of the Earth's age and the history of life has been among the remarkable achievements of this period.

Note 145- *As we have repeatedly emphasized, scientific research has shown that humans are not divided into distinct races. All humans belong to a single race.*

In the third insight, we have focused on the evolution and transformation of humans over the past few million years. In the following appendices, we will explore the history of the universe and following that, we will examine key moments in recorded human history.

The First Realm of the Universe: The History of the Universe

Three key discoveries in cosmology have guided physicists in revisiting the history of our universe and formulating the "Big Bang" theory. The first of these was the result of Edwin Hubble's groundbreaking work in the 1920s. Hubble's observations, along with those of subsequent scientists, conclusively showed that the universe is expanding. This discovery laid the foundation for understanding the dynamic nature of the cosmos.

The second significant discovery came in the 1960s with the detection of the Cosmic Microwave Background (CMB) radiation. This faint radiation, a remnant from the early universe, provided critical insights into the conditions that must have prevailed shortly after the Big Bang, helping physicists to model the universe's early moments more accurately.

The third discovery stemmed from research into the "abundance ratio of light elements" in the universe. Cosmological models predicted the relative amounts of elements like hydrogen, helium, and lithium in the early universe. Interestingly, only those models whose assumptions about the universe's origin were consistent with the first two findings—the expansion of the universe and the presence of CMB radiation—accurately predicted the observed abundance ratios. This confirmed the validity of the Big Bang theory. Together, these three discoveries have enabled scientists to estimate the age of the universe and to form a coherent understanding of its origin and evolution.

It is fascinating to consider how these interconnected findings provide a glimpse into the earliest moments of our universe's existence. - (see Note 146) It is beneficial to examine some details in this context.

1 - The Reality of Universe Expansion

In the 1920s, American astronomer Edwin Hubble conducted observations on a large number of galaxies using the 2.5-meter Mount Wilson telescope. By analyzing the Doppler effect—the same principle used by police radar to measure vehicle speeds—he discovered that the light emitted from these galaxies was shifted toward the red part of the spectrum- (see Note 147) (a phenomenon known as redshift). This indicated that the galaxies were moving away from us—and, in fact, from each other.

Note 146- *These findings are not only supported by solid, well-established theoretical models but are also backed by extensive experimental data from a wide range of sources. These include satellites, telescopes, lasers, gravitational wave detectors, radio telescopes, and particle accelerators. The precision of this knowledge is so advanced that it has enabled us, for example, to launch the* Rosetta *spacecraft, which traveled through the solar system for 10 years, eventually landing on the comet 67P/Churyumov–Gerasimenko. This comet is 6.4 billion kilometers away from Earth and is traveling at a speed of 135,000 kilometers per hour. The* Rosetta *mission allowed scientists to study the comet's composition and gain valuable insights into the materials that make up these ancient objects.*

Note 147- *If a star is moving toward Earth, the light waves it emits become compressed, similar to an accordion being squeezed. This compression shortens the wavelength of the light, shifting its color slightly toward the blue end of the spectrum—a phenomenon known as blueshift. Conversely, if a star is moving away from Earth, its light waves are stretched, increasing their wavelength and shifting the color slightly toward the red end of the spectrum, known as redshift. The greater the shift in wavelength, the faster the star is moving relative to us. By measuring these frequency changes, scientists can determine the velocity at which celestial objects are approaching or receding from Earth.*

The implication of this discovery was profound: space itself (more precisely, space-time) is expanding. Hubble further determined that the farther away a galaxy is, the faster it is receding. - (see Note 148)

Note 148- *The Earth and the Milky Way are certainly not the center of the universe. To understand why, imagine the universe as a rising loaf of bread with raisins scattered throughout it. As the bread expands, all the raisins move away from one another. In this analogy, the raisins represent galaxies, and just as no single raisin is the center of the loaf, no galaxy—including the Milky Way—holds a central position in the universe. The apparent motion of other galaxies moving away from us is not due to our galaxy being at the center, but rather due to the expansion of space itself.*

Furthermore, the expansion effect is more pronounced over greater distances. The farther apart two galaxies (raisins) are, the faster they appear to move away from each other. To illustrate this, imagine a hypothetical cake that starts with a diameter of 10 meters and is placed in a giant oven. Suppose three raisins are positioned in a straight line on the cake:

- ***Raisin A*** *at the far left*
- ***Raisin B*** *in the middle, 1 meter away from A*
- ***Raisin C*** *at the far right*

Now, as the cake expands, its size increases tenfold over an hour. As a result:

- *The distance between **A and C** expands from 10 meters to 100 meters. The speed at which they move apart is calculated as:*

  (100−10) ÷1=90 meters per hour

- *The distance between **A and B** expands from 1 meter to 10 meters, meaning B moves away from A at a speed of:*

  (10−1) ÷1=9 meters per hour

This illustrates that the farther apart two points are, the faster they appear to move away from each other—precisely what we observe with galaxies in the expanding universe.

He quantified this relationship, finding that the velocity of a galaxy's recession is proportional to its distance from us. The constant of proportionality in this relationship became known as *Hubble's constant* (now referred to as the *Hubble parameter*). By conceptually reversing this expansion—like playing a movie in reverse—one could trace the universe back to a state where all galaxies were once much closer together, eventually converging at a single point. This idea laid the foundation for estimating the age of the universe. Hubble initially attempted to determine the universe's age using this method, but an error in one of his key measurements led him to calculate an age of only 1.8 billion years—less than the age of the Earth itself. However, with more precise measurements and improved techniques, modern cosmology has refined this estimate. Today, the age of the universe is calculated to be approximately **13.8 billion years.**

2. Cosmic Microwave Background Radiation-

Continuing the study of the universe's age and origin, in the mid-20th century, American physicist George Gamow, building upon Edwin Hubble's discovery of cosmic expansion, proposed a compelling idea. He reasoned that if we were to conceptually reverse the expansion of the universe—tracing it backward in time—all matter and energy would converge, ultimately reaching a single, extremely dense and hot state- (see Note 149).

Note 149- *A primary example supporting this assumption is the behavior of gases under compression. As we know, when a gas is compressed, its temperature increases due to the conservation of energy and the reduction in available volume. This phenomenon is particularly evident in the cores of collapsing supernovae. During the final stages of a massive star's life, when gravitational forces cause the core to collapse, the temperature can soar to* ***several billion degrees Celsius****. Under these extreme conditions, nuclear reactions occur at an unprecedented rate, leading to the formation of heavy elements. In some cases, the density at the core of a collapsing star can reach* ***400 billion times the density of water****. This immense compression and heating play a crucial role in triggering the explosive supernova event, which disperses newly formed elements into space, contributing to the cosmic cycle of matter.*

This moment, which marked the birth of the universe, would have occurred billions of years ago in an environment of unimaginable heat and density. Gamow predicted that remnants of this primordial event—specifically, residual radiation from the early universe—should still be detectable.

Despite the passage of billions of years, he theorized that a faint glow of this ancient heat would persist throughout the cosmos in the form of background radiation. He estimated that by now, due to the universe's expansion and cooling, this radiation would have a temperature of approximately **5 Kelvin**. In 1965, American physicists Arno Penzias and Robert Wilson accidentally discovered this radiation while working with a radio antenna at Bell Labs. They observed an unexplained, uniform microwave signal coming from every direction in the sky, which turned out to be the **Cosmic Microwave Background (CMB)**—the very radiation Gamow had predicted.

The measured temperature of this radiation was **2.7 Kelvin**, remarkably close to Gamow's estimate. Since then, advancements in observational technology have significantly enhanced our ability to study the CMB. Satellite missions such as **COBE (Cosmic Background Explorer), WMAP (Wilkinson Microwave Anisotropy Probe), and Planck** have provided increasingly detailed maps of this radiation. These observations have allowed scientists to refine our understanding of the universe's early conditions, confirming the predictions of the **Big Bang theory** with astonishing precision.

3-The Ratio of Light Elements in the Universe

In addition to the **expansion of the universe** and the **cosmic microwave background radiation**, another key piece of evidence supporting the **Big Bang theory** is the observed **abundance of light elements** in the universe.

Observations show that the universe consists of approximately **75% hydrogen, 25% helium**, and trace amounts of **lithium** and

beryllium. These ratios are not random; they align precisely with predictions made by Big Bang nucleosynthesis (BBN), the process that took place within the first few minutes after the universe's birth.

According to the Big Bang model, during the first **three minutes**, the universe was an extremely hot, dense plasma with a temperature of about **one billion degrees Kelvin**. In this environment, **nuclear fusion** occurred, leading to the formation of light elements. As the universe expanded and cooled, further fusion reactions ceased, locking in the observed elemental abundances.

Remarkably, no alternative cosmological model has successfully reproduced these observed ratios. The fact that this specific composition is seen **not just in stars or galaxies, but across the entire cosmos**, provides strong evidence for the **Big Bang theory** as the origin of the universe.

The Birth of the Universe and the Limits of Our Observations

Based on extensive astrophysical data, physicists widely agree that if we **reverse the expansion of the universe**—like playing a movie backward—all matter and energy would converge into an increasingly dense and hot state.

Tracing this process back **13.8 billion years**, we reach a point where only **hydrogen and helium** existed. Before that, going even further back, we encounter a state where even these elements had not yet formed, and the universe was an opaque, hot plasma. This marks the earliest phase we can directly study—what we call the **Big Bang era**.

However, it is important to be cautious in how we interpret this. The **universe is not a finite structure** but is considered **infinite or unbounded**, with our **"observable universe"** being just a small fraction of its full extent—comparable to a single grain- (see Note 150) of sand relative to the entire Earth. Therefore, rather than picturing the universe as emerging from a single point—like the center of a black hole- (see Note 151)—it is more accurate to consider that **before 13.8 billion years ago, space-time itself did not exist** in its current form. Instead, the universe may have been in a state of **quantum non-locality**, sometimes described as a **quantum vacuum** in modern physics. This interpretation aligns with certain cosmological models suggesting that before the Big Bang, the universe existed as a **probabilistic quantum state**, outside of classical space-time. When the **Big Bang occurred**, it did not happen at a single location but across **all of space simultaneously**, giving rise to the structure of space-time as we perceive it today.

Note 150- *A more precise way to interpret the vastness of the universe is to consider that if we* ***scale down the entire observable universe*** *to the size of a* ***subatomic particle****, the universe beyond our observable horizon would still be* ***infinitely larger****. This is because* ***the universe itself is infinite****, and all observational data suggest that it has* ***no detectable curvature****—meaning it extends endlessly in all directions.*

Note 151- *The Canadian physicist* ***Hubert Reeves*** *has pointed out that the term* ***"Big Bang"*** *simply refers to the fact that* ***13.8 billion years ago, the universe was in an extremely hot and dense state****—far beyond human imagination. However, contrary to common misconceptions, the universe did* ***not*** *begin as a single point. This idea appears in some early writings, but modern cosmology provides strong evidence that* ***the universe has always been infinite****. Reeves emphasizes that the* ***Big Bang occurred across an already infinite thing****, not as an explosion from a single point, making descriptions like a* ***"fireball"*** *misleading.*

This is why the universe appears **flat, isotropic, and homogeneous** on large scales—no matter where we look, it looks roughly the same in all directions. - (see Note 152) There are several hypotheses regarding the early universe, one of the most well-known being the **inflationary hypothesis**, proposed by **Alan Guth**, an American cosmologist. According to this hypothesis, immediately after the birth of the universe, **space underwent an exponential expansion** within a **tiny fraction of a second**—inflating from an almost infinitesimally small state to a vast, essentially infinite scale.

This hypothesis leads to conclusions that align with the statement above: whether one says that **space-time was born infinitely large from the very beginning** or that it **expanded from an incredibly small size to an immense scale in an instant**, the practical implications remain the same. According to **Einstein's theory of relativity**, space and time are fundamentally interconnected, meaning that **before the birth of our universe, there was no pre-existing space in which to define a point or location**. Therefore, in all respects, the **Big Bang did not happen at a single point in space but rather occurred everywhere at once within an already infinite framework**. The state of the universe **before its birth remains unknown**. It may have been **"nothing"** in the classical sense, or it could have been a **quantum vacuum**, as some models suggest. The American physicist **Edward Tryon** once proposed that **"the universe is simply a quantum fluctuation that emerged from a vacuum"**, meaning that quantum mechanics allows for spontaneous energy fluctuations that could give rise to a universe.

Note 152- *The universe has* ***no boundaries*** *in any direction. If it did, there would have to be an edge somewhere—but as the* ***Roman philosopher Lucretius*** *pointed out, something can only have a limit if* ***something else exists beyond it to impose that boundary****. Since no such external "edge" has been found, the universe remains boundless and ever-expanding.*

Many physicists argue that the universe **emerged from "nothing" yet** still maintains a fundamental balance. This idea is supported by the fact that when we sum the **total energy and matter content of the universe**—taking into account **Einstein's mass-energy equivalence equation** ($E = mc^2$)—and compare it with the **negative gravitational energy** of the universe, the total sum appears to be **zero**. Additionally, the universe has **no net electric charge**, and the total angular momentum (the sum of rotational motion of all cosmic structures) also appears to be **zero**.

Thus, the universe may be a **self-contained system with a net energy of zero—an elegant, self-sustaining balance that emerged from a quantum state of nothingness**.

Regardless of what existed before the birth of the universe, it has undergone **tremendous transformations** since its formation. Before the Big Bang, the four fundamental forces—**gravity, electromagnetism, the strong nuclear force, and the weak nuclear force**—were unified in a state of **symmetry**. There were no distinct particles or structures, just a high-energy quantum field in an **extremely hot, dense state**. For unknown reasons, this **symmetry broke** approximately **13.8 billion years ago**, triggering the birth of the universe. The universe did not expand from a single point but instead emerged **everywhere at once** in an infinite expanse, or, as the **inflationary hypothesis** suggests, it underwent an **exponential expansion** in an incredibly short fraction of a second. In the earliest moments following the Big Bang, the four fundamental forces **separated**, various **quantum fields** stabilized, and the first **wave-particle excitations** appeared. Almost immediately, the universe entered a period of rapid **expansion (which is distinct from inflation)**.

Within just a few minutes, the first **atomic nuclei** of **hydrogen, helium, and trace amounts of lithium** formed through a process called **Big Bang nucleosynthesis**. However, for the next **380,000 years**, the universe remained in a **hot, dense plasma state**—a turbulent sea of ionized particles where **free electrons scattered photons, making the universe opaque**.

As the universe expanded and **cooled to about 3000 K**, electrons slowed down enough to **combine with nuclei**, forming the first neutral atoms. This event, known as **recombination**, allowed photons to travel freely for the first time, making the universe **transparent** to radiation.

The photons released at this moment continue to travel through space and are detected today as the **Cosmic Microwave Background (CMB) radiation**, an afterglow of the Big Bang that we observe in all directions.

Following this, gravitational forces began **pulling together vast clouds of hydrogen**, leading to the formation of the **first generation of stars and galaxies**. The Hubble Space Telescope, and more recently the **James Webb Space Telescope**, have captured images of these **early galaxies, which are over 13.5 billion years old - (see Note 153)**—essentially snapshots of the universe in its infancy.

Note 153- *It took approximately* ***300 million years*** *after the Big Bang for the first galaxies to form. While galaxies of this age can be found throughout the universe, if we want to observe galaxies in their infancy, we must look at the most distant regions of space, approximately* ***13.5 billion light-years*** *away. Observing these distant galaxies allows us to glimpse the early universe, essentially witnessing a* ***300-million-year-old cosmos*** *at the time when the first galaxies began to take shape.*

Over time, as the universe continued to evolve, stars of different sizes and masses formed. **Smaller stars** like our Sun- (see Note 154) primarily **fused hydrogen into helium**, generating energy and light. However, **more massive stars**, which contained significantly larger amounts of hydrogen, underwent **faster and more intense nuclear fusion**, producing elements heavier than helium **all the way up to iron**. In the most **massive stars**, nuclear fusion continued beyond iron in extreme conditions, but as their fuel supply depleted, they **collapsed under their own gravity - (see Note 155)**. This rapid collapse triggered a **supernova explosion**, - (see Note 156) where temperatures soared into the **billions of degrees**. In these violent explosions, elements **heavier than iron, such as gold, uranium, and platinum**, were formed and dispersed into space. What remained of the core of these massive stars often became a **neutron star**—a city-sized remnant with a mass several times greater than the Sun—or, in the case of even larger stars, a **black hole**.

Note 154- *It is estimated that the Sun has a lifespan of about* ***10 billion years****, and it is currently about halfway through its life. However, stars with significantly more mass, such as those 10 times heavier than our Sun, reach the end of their lifecycle in only tens of millions of years—far shorter than the Sun's expected lifespan.*

Note 155- *For stars massive enough—about* ***40 times the mass of the Sun****—their ultimate fate is to collapse into a black hole. The intense gravitational pull of the black hole becomes so strong that even light cannot escape. The concept of black holes was first theorized in the* ***18th century****, and their observation in the real universe became possible in the* ***1990s****. Since then, the* ***Hubble Space Telescope****,* ***Chandra X-Ray Observatory****, and various radio telescopes have detected numerous black holes. In fact, it is believed that most galaxies, including our own, harbor a* ***supermassive black hole*** *at their centers.*

Note 156- *Supernova* ***explosions*** *occur throughout galaxies, and it is estimated that a supernova is observed in each galaxy—including ours—roughly once every* ***100 years****. The most recent supernova explosion in our galaxy was observed in* ***1604*** *(Kepler's supernova), and these explosive events continue to provide critical insights into the lifecycle of stars and the chemical enrichment of the universe.*

The **elements ejected by these ancient supernovae** eventually formed **new cosmic clouds**, which became the raw material for the **next generations of stars**. In fact, our **Sun and its planetary system—including Earth—were formed from the remnants of these earlier stellar explosions**. In other words, if Earth is our **mother**, then our **great mother** is not the Sun, but the **massive stars that lived, died, and exploded billions of years ago**, enriching the cosmos with the heavy elements necessary for planets—and eventually life itself.

This brings us to three fundamental topics that extend beyond the story of cosmic evolution:

1. **The Observable Universe** – The part of the universe we can see, limited by the speed of light.
2. **Parallel worlds** – The Theoretical possibility of multiple universes coexisting beyond our perception.
3. **String Theory** – A framework in modern physics that attempts to unify all forces and particles at a fundamental level.

These topics push the boundaries of our understanding and continue to inspire scientific exploration.

The First Realm of the Universe – The Observable Universe

To enter this discussion, we must ask an important question: If the universe is infinitely vast—meaning that from every angle we observe, the number of stars is infinite—then why isn't the entire night sky as bright as day? In an infinite and uniform universe, the light from an infinite number of stars should reach our eyes from every direction, making the night sky completely illuminated.

The answer is not that starlight weakens with distance, because in an infinite universe, the number of stars also increases with distance, balancing out the reduction in brightness. Nor is it because cosmic dust clouds absorb the light, as over an infinite time, these clouds would also heat up and radiate light like stars themselves. The correct explanation is that, while the universe is boundless, its age is finite. The universe had a beginning—approximately 13.8 billion years ago—meaning that light from regions farther than 13.8 billion light-years has not yet reached us. This leads us to the concept of the "observable universe."

While the universe itself may be infinite, the portion we can observe and interact with is limited. This is because no causal effect—such as light or gravity—can travel faster than the speed of light, which is 300,000 kilometers per second. Consequently, the farthest cosmic signals that reach us today originate from a sphere with a radius of 13.8 billion light-years. If the universe were static (i.e., not expanding), this would mean our observable universe would be only 13.8 billion light-years in size, and over time, light from more distant regions would gradually reach us.

However, because the universe is expanding—where the repulsive force (dark energy) exceeds gravitational attraction—the situation is different. Although it appears that we are at the center of a sphere with a 13.8 billion light-year radius, the reality is that, over the past 13.8 billion years, the universe itself has expanded. A region that was 13.8 billion light-years away at the moment of the Big Bang has now receded much farther. Using Hubble's constant and other calculations, scientists have determined that the current observable universe extends to a radius of 46.5 billion light-years, meaning we see a universe with a total diameter of 93 billion light-years. However, the light we receive from these distant regions corresponds to a time when the universe was much younger—just after the beginning of cosmic expansion.

Importantly, there is no real edge to the universe, and we are not at its center. If we could instantaneously travel to a location 40 billion light-years from the Milky Way, we would see distant galaxies that were previously unobservable from Earth, while looking back toward our original position, we would see our own galaxy as it appeared billions of years ago. This means that the observable universe shifts depending on the observer's location, much like passengers on a ship who, seeing no edge to the ocean, mistakenly assume they are at its center.

Furthermore, the edges of the observable universe are receding from us at speeds approaching the speed of light. As a result, in the future, light and other causal signals from objects currently at the edge of our observable universe (about 13.8–16 billion light-years away) will eventually become undetectable. Even galaxies visible today will gradually disappear from our view, and their gravitational influence on us will also vanish.

Composition and Future of the Observable Universe

The observable universe contains approximately two trillion (two thousand billion) galaxies, and remarkably, the number of stars in these galaxies is vastly greater than the total number of grains of sand on Earth. The composition of this universe includes:

- **5% ordinary matter** (atoms and molecules that make up stars, planets, and life)
- **27% dark matter** (an unknown, invisible form of matter detected through its gravitational effects)
- **68% dark energy** (a mysterious force accelerating cosmic expansion)

Dark matter, though unseen, can be traced by its gravitational influence. It surrounds galaxies and provides additional mass that prevents them from dispersing due to rotation. Dark energy, on the other hand, is an unknown form of energy embedded within space itself, causing galaxies to accelerate away from one another. Einstein initially introduced this concept in 1917 as the "cosmological constant," later dismissed it as his "greatest blunder," but modern observations confirm its existence.

According to Einstein's theory of relativity, the speed of light in static space is limited to 300,000 km/s. However, in an expanding universe, light from regions beyond a certain distance (estimated at 14–16 billion light-years) will never reach us, as space itself expands faster than light can travel. This can be understood through the analogy of a rising raisin bread: as the dough expands, raisins move apart—if two raisins are far enough, their separation increases faster than any signal could travel between them.

The Event Horizon – The Ultimate Limit of Observability

A crucial calculation helps determine the boundary beyond which causal connection with the universe is permanently lost. This boundary is called the **Event Horizon** and is derived using the **Hubble's law** formula:

$$v = H_0 \cdot d$$

Where:

- v is the recession velocity of a galaxy (in km/s)
- H_0 is the Hubble constant (currently estimated at 70 km/s per megaparsec)
- d is the distance between two galaxies (in megaparsecs, where 1 megaparsec = 3,262,000 light-years)

To determine the distance at which a galaxy moves away from us at the speed of light (v =300,000 km/s), we solve for d:

d=300,000/70=4285 megaparsecs
Converting this distance to light-years:
4285×3,262,000≈14 billion light
Thus, any galaxy located beyond approximately **14 billion light-years** is forever beyond our reach. In more refined calculations, considering the time-dependent nature of Hubble's parameter, this distance increases to around **16 billion light-years**.
Over the next two hundred million years, as the universe ages to 14 billion years, we may briefly observe additional galaxies, but afterward, due to ongoing expansion, even visible galaxies will start disappearing. In **hundreds of billions of years**, only the 36 galaxies gravitationally bound to the Milky Way (our "Local

Group") will remain visible, while the rest of the universe will fade into darkness.

The Mystery of Dark Matter

There are many hypotheses regarding dark matter. Some scientists propose it consists of common yet dim celestial objects like brown dwarfs and black holes. Others suggest that dark matter is composed of an entirely new type of exotic particle, such as **Weakly Interacting Massive Particles (WIMPs)**, which only interact through gravity and the weak nuclear force.

Regardless of its nature, dark matter remains one of the biggest unsolved mysteries in modern cosmology.

The First Realm of the Universe – Parallel Worlds and the Multiverse

The idea of parallel worlds has existed for a long time, but it was not widely accepted within the scientific community. Today, with the emergence of theories such as **string theory, eternal inflation, and the Many-Worlds interpretation of quantum mechanics**, the concept of a **Multiverse** has gained significant traction in modern physics. Although this hypothesis is not currently testable—and may remain beyond direct experimentation for a long time—it nonetheless has many prominent supporters.

In this framework, our universe can be compared to a **single bubble within a vast cosmic ocean**—an ocean where new bubbles (universes) continuously form, each potentially governed by different physical laws. If this hypothesis is correct, regions of space beyond our own may still be undergoing inflation, giving rise to entirely new universes. Other cosmological models suggest that quantum fluctuations or gravitational collapses within black holes could also serve as cosmic seeds for fresh universes.

Some of the world's leading physicists propose that an **uncountable number of Big Bangs** may have already taken place across an infinite cosmic landscape, with new ones continuously emerging. The reason we observe a universe with such finely tuned physical laws is because only such a universe permits conscious observers like us. This perspective, known as the **Anthropic Principle**, does not prove the multiverse's existence but helps explain why we find ourselves in a universe capable of supporting life.

Theoretical physicist **Edward P. Tryon** famously stated: *"In response to the question about the Big Bang, I humbly say that our universe is just one of those things that happen once in a while."*
He suggested that while the spontaneous creation of a universe like ours may seem improbable, the sheer number of possible attempts could make it inevitable.

Astrophysicist **Martin Rees** supports the idea that an enormous number of universes may exist, each with its own distinct physical properties. He compares this to shopping in a vast clothing store: *"If the number of clothes in the store is large enough, you will not be surprised to find clothes in your size."*

Similarly, if the number of universes is immense, some will inevitably have the precise conditions necessary for life. We simply exist in one of these universes.

Rees also explores the idea—originally proposed by physicist **Lee Smolin**—that every collapsing black hole may contain the seed of a **new universe**, potentially expanding into its own distinct cosmos.

Ultimately, while the **Multiverse hypothesis** remains theoretical, it has gained substantial credibility among physicists. The idea suggests that universes continuously give birth to other universes, forming an **eternal and self-perpetuating system**. This means that our universe may have emerged from a pre-existing one, or perhaps our universe itself is giving rise to countless new universes.

This idea does not contradict the hypothesis that our universe came from "nothing" because, in such models, each new universe is causally disconnected from the previous one. In essence, each new universe emerges as a **quantum event**, without direct causal ties to its predecessor.

NUSAWVO School's Perspective

While standard physics suggests that these universes are **causally disconnected**, the **NUSAWVO school** offers a **deeper perspective** in the discussion of the **"second realm of existence."** Beyond material causality, all universes—no matter how physically separate—are fundamentally linked through **consciousness**.

The whole of existence, regardless of how many worlds it contains, is an **inseparable unity**—a **pure, undivided consciousness** that transcends individual realities. This means that while traditional physics may describe separate universes, in the ultimate sense, **they are not truly separate**. Consciousness serves as the **underlying fabric** that connects all possible realms, beyond the limits of causality and space-time.

The First Realm of the Universe – String Theory

For much of history, it was believed that matter was composed of **indivisible** units called **atoms**, thought to be the smallest building blocks of reality. However, the advent of **particle accelerators** shattered this notion, revealing that atoms themselves consist of smaller components: **protons, neutrons, and electrons**. Further advancements in physics have uncovered an even deeper structure—protons and neutrons are composed of even smaller entities such as **quarks, neutrinos, mesons, leptons, and other subatomic particles**.

To explain the **diversity and behavior** of these fundamental components, several theoretical models have emerged. Among them, **string theory** has gained significant support in recent years.

According to this theory, what we perceive as **subatomic particles** are not **point-like objects** but rather **infinitesimally small, vibrating strings**. These strings are so minute that current observational tools can only detect them as points. The **vibrational patterns** of these strings determine the nature of the particles they represent. By altering the **tension or frequency** of a string, its physical properties change—what appears as an **electron** in one vibrational mode could manifest as a **quark or neutrino** in another.

American physicist **Michio Kaku** eloquently describes this concept:
"In this new paradigm, the laws of physics, distilled from thousands of years of experience, are nothing but the harmonics that can be written for strings. The laws of chemistry are the melodies played on these strings. Hence, the universe is a cosmic symphony of vibrating strings, and consciousness—or the mind in its entirety—is a form of cosmic music resonating through the fabric of space-time."

He further expands on this analogy by quoting world-renowned violinist **Yehudi Menuhin**:
"Music creates order out of chaos: rhythm imposes accompaniment over disparity, melody imposes continuity over separation, and harmony imposes compatibility over disharmony."
(*As narrated by Michio Kaku in* Parallel Worlds)

As we will explore further, what we call **particles—or more precisely, wave-particles—are ultimately various forms of energy** that manifest as **mass** to our perception. Therefore, beyond any hypothesis attempting to unify the **fundamental forces and particles** of the universe, the world at its deepest level appears to be **energy—or more precisely, energy-consciousness—and perhaps, in the ultimate sense, only consciousness itself.**

A Review of the First Realm and an Introduction to the Second Realm of the Universe

In our exploration of the **first realm of the universe**, we examined the **history of Earth**, the trajectory of our **evolutionary past**, and various theories concerning the **origin and expansion** of the cosmos. We established that our universe **emerged 13.8 billion years ago** and has been expanding ever since. As a result of this relentless expansion, the farthest edges of the universe will eventually fade from view, as the stretching of space will prevent their light from ever reaching us.

By retracing **Earth's history**, we discovered that our planet and the **solar system** formed approximately **4.6 billion years ago** from interstellar dust—remnants of one or more ancient **supernovae**. Within a few hundred million years, Earth became the **cradle of the first flashes of life**. We then briefly reviewed the **evolutionary journey** of living organisms, the emergence of **humanity**, and nature's continuous process of **experimentation and refinement**.

In the **cosmology section**, we considered the possibility that **infinite worlds** may have existed before our own. It is conceivable that our universe **sprouted from the remnants of a prior cosmos**—or perhaps even from the depths of **"nothingness"** itself. Likewise, our universe may have given rise to **countless parallel universes**, and this process could continue **indefinitely into the future**.

However, for the purpose of our inquiry, what matters most is **our own universe**—one with a **distinctive history** and a **set of behaviors** that, while **partially understood**, still hold **profound mysteries**. Yet, what we have examined so far is only **part of the story**.

Up until now, we have focused on a **realm of the universe largely governed by causal relationships**—the principles of **Newtonian mechanics** and **Einstein's relativity**. But why **"largely"**? Because even within this realm, the processes underlying **the birth of our**

universe, as well as the behavior of **subatomic and atomic structures**—from **galaxies and stars** to our own **bodies**—do not always conform to **classical cause-and-effect laws**. Instead, they obey the principles of **quantum physics**, which operate under **unconventional, probabilistic, and non-causal laws**.

Since these **subatomic and atomic components** constitute the **macroscopic universe**, it follows that even the apparent **causal order** of the cosmos is underpinned by a **foundation that is loose, fluid, uncertain, and shaped by probabilities**. This realization, as we will explore further in our interpretation of the **nature of reality**, leads us to an **unconventional domain**—one that we refer to as **the second realm of the universe, or the "basis realm"**.

The Second Realm of the Universe – Scientific Knowledge

To explore the second realm from a scientific perspective, we must first establish a foundational understanding of core physical principles. The following sections will provide a brief overview of three major frameworks in physics:

1. **Newtonian Physics** – The foundation of classical mechanics and the deterministic view of the universe.
2. **Relativistic Physics** – The extension of classical mechanics that accounts for the nature of space-time and gravity.
3. **Quantum Physics** – The framework that governs the behavior of matter and energy at the smallest scales, revealing the probabilistic and indeterminate nature of reality.

Newtonian Physics

Newtonian physics was developed by **Isaac Newton**, the brilliant 17th-century English physicist, and it formed the cornerstone of classical mechanics. In this framework, the universe is composed of **small, solid, and indestructible material particles** that combine to form all physical matter. These particles interact according to deterministic laws, and their motion is governed by the **force of gravity**.

The **Newtonian worldview** envisions the universe as a **giant clockwork mechanism**, meticulously set in motion at the dawn of time—perhaps by a divine hand—and thereafter operating **with absolute predictability** according to the three laws of motion and the law of gravity.

Despite later refinements and the discovery of its limitations, Newtonian mechanics remains one of the most influential scientific theories in history. It continues to be widely used in engineering, astronomy, and many areas of applied physics.

Newton's Three Laws of Motion

1. **Law of Inertia:**
   An object at rest remains at rest, and an object in motion continues to move in a straight line at constant speed unless acted upon by an external force. This principle, derived from **Galileo's research**, challenged the ancient belief that motion required a continuous force to sustain it.
2. **Law of Acceleration:**
   The acceleration of an object is directly proportional to the **net force** applied to it and inversely proportional to its **mass**:

   $a=F/m$

   Where:

   - a is the acceleration of the object,
   - F is the applied force, and
   - m is the object's mass.
3. **Law of Action and Reaction:**
   For every action, there is an **equal and opposite reaction**.

For example, when you push against the ground with your feet, the ground exerts an equal force back on you.

Newton's Law of Universal Gravitation

Newton's law of gravity states that every mass in the universe attracts every other mass with a force directly proportional to the product of their masses and inversely proportional to the **square of the distance** between them:

$F=G(m1m2)/r^2$

Where:

- F is the gravitational force between two objects,
- m1 and m2 are their respective masses,
- r is the distance between their centers, and
- G is the universal gravitational constant.

This formula mathematically explains why an apple falls to the ground, just as it governs the motion of planets, stars, and galaxies.

However, Newtonian physics, though remarkably accurate within its domain, **fails to describe the behavior of objects at very high speeds, in extremely strong gravitational fields, or at the quantum level**. In these realms, **relativity and quantum physics** take over, leading us to the second realm of the universe.

Relativity in Physics

At the beginning of the 20th century, Newtonian mechanics was the dominant framework for describing motion, applying even to celestial bodies. However, certain phenomena—such as the anomalous precession of Mercury's orbit—revealed inconsistencies that Newtonian physics could not explain. Additionally, electromagnetic theory, particularly as formulated by Maxwell's equations, did not fully align with Newtonian mechanics, creating a conceptual division within physics.

A major breakthrough came in the 1880s with the Michelson-Morley experiment in the United States. This experiment attempted to detect differences in the speed of light due to Earth's motion through an assumed "luminiferous ether" but instead provided unexpected results: the speed of light remained constant in all directions. This discovery challenged classical physics and paved the way for a new understanding of motion and space-time.

In 1905, Albert Einstein resolved these contradictions by introducing the theory of **special relativity**. This theory reconciled Maxwell's electromagnetism with mechanics and incorporated the fundamental postulate that the speed of light in a vacuum is a universal constant: **approximately 300,000 kilometers per second**. It also provided a deeper explanation for Mercury's orbital irregularities, later fully accounted for in his **general theory of relativity**.

In the following sections, we will explore special relativity and later introduce general relativity.

Special Relativity

The theory of special relativity applies to **inertial reference frames**, which are either at rest or moving at a constant velocity. As its name suggests, the concept of relativity expresses that motion, space, and time are all **relative**—there is no absolute, universal spatial or temporal reference frame. In this framework, all physical measurements depend on the observer's state of motion.

This theory primarily concerns the effects of motion on speed, distance, and time. At low speeds, Newton's classical laws accurately describe motion. However, at very high velocities—close to the speed of light—special relativity reveals fundamental differences. The key insight is that time and space are not fixed; they change depending on the observer's relative motion.

This leads to **time dilation** and **length contraction**:

- A clock moving at high speed runs slower compared to one at rest.
- An object in motion appears shorter along its direction of travel when viewed from a stationary frame.

These effects are direct consequences of the fact that the speed of light in a vacuum is the same for all observers, regardless of their motion or the motion of the light source. - (see Note 157)

Understanding Time Dilation:

Time dilation, a key concept in Einstein's theory of relativity, describes how time slows down for an object moving at a significant fraction of the speed of light relative to an observer. This idea may seem counterintuitive at first, but it's a direct consequence of the principles governing the universe, specifically the behavior of fundamental forces.

These forces—gravity, electromagnetism, and the strong and weak nuclear forces—all propagate at the speed of light, **c** (300,000 km/s), which is the maximum speed at which information can travel. When an object moves at a high velocity, the distance these forces need to travel in the object's frame of reference increases. Since the forces still propagate at the constant speed of light, this results in an increase in the time required for interactions to complete, causing time to appear to slow down from the perspective of an external observer.

Note 157- *The constancy of the speed of light had already been discovered by Michelson and Morley, but Einstein revealed a deeper insight: this speed remains independent of the motion of its source. Whether the light-emitting source moves forward in the direction of the beam, backward against it, or perpendicular to it, every photon, once emitted, travels entirely independent of the source's speed and direction. Instead, it follows the space-time framework of the surrounding world. This is because a photon, having no mass, also has no inertia and therefore does not retain any sense of the motion of its source. Regardless of these factors, the speed of light in a vacuum always remains 300,000 km/s. However, its wavelength shifts depending on whether the photon moves toward or away from the observer—a phenomenon known as the Doppler effect.*

To help illustrate this, let's consider a tangible example: - (see Note 158)

Example: Time Dilation for an Object (i.e., an astronaut) Moving at 99% of Light Speed

Let's say we have an object moving at **99% of the speed of light** (0.99c), and we want to calculate the time dilation effect from the perspective of an observer at rest.

We can use the time dilation formula from special relativity:

The formula for time dilation is:

$\Delta t' = \gamma \Delta t$

This states that the **time interval $\Delta t'$ measured by an observer in the moving reference frame (moving clock)** will be **longer** than the time interval Δt measured by an observer in the rest frame (resting clock).

Note 158*- Perhaps the most compelling explanation of relativity comes from a student-made video that earned its creator a $400,000 prize. You can find it online under the title:*

"Student Wins $400,000 Award for Relativity Explanation."

The Key Factor: γ(gamma)

The factor γ(gamma) (the **Lorentz factor**) is defined as:

$$\gamma = \frac{1}{\sqrt{1 - \frac{v^2}{c^2}}}$$

where:

- v is the relative velocity of the moving frame with respect to the stationary frame,
- c is the speed of light.

Using the time dilation formula:

$\Delta t' = \gamma \Delta t$

Where for an object moving at 99% of light speed (v=0.99c):

$$\gamma = \frac{1}{\sqrt{1 - \frac{(0.99c)^2}{c^2}}} = \frac{1}{\sqrt{1-0.9801}} = \frac{1}{\sqrt{0.0199}} = \frac{1}{0.1411} \approx 7.09$$

So, the **time experienced by the astronaut** is **dilated** (slowed down) by a factor of **7.09** compared to the stationary observer on Earth.

Thus, while **1 year** passes for the **stationary observer (Earth)**, the astronaut will experience only:

$\Delta t' = \frac{1}{7.09} \approx 0.1411$ years≈51.5 days

This means that for every 1 second that passes for the observer at rest, only **0.1411 seconds** will pass for the object moving at 99% of the speed of light. In other words, **time slows down significantly** for the moving object from the perspective of the stationary observer.

So, putting everything together:

- **1 year on Earth** corresponds to **51.5 days for the astronaut** moving at 99% the speed of light.

Why is this important?

This example illustrates that as objects approach the speed of light, their internal processes (clocks, biological reactions, etc.) slow down relative to a stationary observer. While this effect is small at low speeds, it becomes extreme as you get closer to the speed of light. For example, at 99.9% of light speed, the time dilation would be even more pronounced.

Einstein's Mass-Energy Equivalence: $E = mc^2$

Einstein's most famous equation,

$E = mc^2$

demonstrates the profound equivalence of mass and energy. Here:

- E is energy,
- m is mass,
- c is the speed of light.

This equation shows that even a small amount of mass can be converted into a **tremendous** amount of energy, as seen in nuclear reactions.

For example:

- **Nuclear bombs** release energy by converting a tiny fraction of mass into energy.
- **Stars** generate energy through nuclear fusion.
- **Particle collisions** in accelerators create new particles by converting energy into mass.

Practical Example: Increased Mass at High Speed

A fascinating consequence of $E = mc^2$ is that an object's **mass increases as it gains energy**. - (see Note 159) If a spaceship were somehow able to keep accelerating toward the speed of light (assuming an unlimited fuel source), its mass would continually increase. The closer it got to c, the more energy would be required to continue accelerating it. At $v \approx c$ the required energy would approach infinity—meaning **no material object can ever reach the speed of light-** (see Note 160) because it would require an infinite amount of energy.

Note 159- *To better understand this relationship, it is helpful to examine two fundamental relativistic equations related to moving objects. In these equations,* ***v*** *represents the object's velocity,* ***m₀*** *is its rest mass, and* ***m*** *is its relativistic mass when in motion. Although these two formulas may appear distinct, they essentially describe the same concept from different perspectives within the framework of special relativity. One equation expresses the energy-mass relationship, while the other highlights how an object's mass increases with velocity as it approaches the speed of light.*

$$E = \frac{m_0 c^2}{\sqrt{\left(1 - \frac{v^2}{c^2}\right)}}$$

$$m = \frac{m_0}{\sqrt{\left(1 - \frac{v^2}{c^2}\right)}}$$

Note 160- *"Light, along with its constituent particles—photons—and the three other fundamental forces and their associated particles, are considered material according to physical definitions. However, they do not have intrinsic mass on their own. Nevertheless, according to the equation $E = mc^2$, their mass and energy equivalents can be calculated."*

This principle is also observable in everyday devices, such as **rechargeable batteries**. When a battery is charged, its energy content increases, and **its mass also increases slightly**, though this change is extremely small and usually undetectable without precise instruments.

Conclusion

Einstein's special relativity transformed our understanding of space, time, and motion. It demonstrated that:

- **Time and length are not absolute but depend on the observer's motion.**
- **The speed of light is a universal constant, independent of the observer or source.**
- **Mass and energy are interchangeable, leading to powerful applications in physics.**
- **No object with mass can reach the speed of light due to infinite energy requirements.**

Special relativity laid the foundation for **modern physics, cosmology, and technology**, influencing everything from GPS satellites to nuclear power and quantum field theories.

General Relativity

The theory of general relativity, developed by Einstein in 1915, came a decade after his theory of special relativity.

In special relativity, Einstein focused on inertial systems—those to which no external force is applied, meaning they are either at rest or in motion (as described by Newton's first law). In contrast, general relativity addresses gravity and the effects of force on objects that are subjected to acceleration. Einstein's key insight was that space and time are not separate entities but instead form a unified structure known as spacetime, which encompasses the entire universe. - (see Note 161) This spacetime behaves like an elastic rubber sphere (or an elastic space), bending around objects with mass. The greater the mass of an object, the more significant the curvature. For example, near massive stars, including our Sun, this curvature is so pronounced that even light is bent. This effect is dominant across all four fundamental forces.

Einstein also realized that accelerating bodies produce effects equivalent to gravity. Therefore, there is no distinction between the gravitational force felt by passengers aboard an accelerating spaceship (or the additional force felt when an elevator starts) and the gravitational force experienced on the Earth's surface. In terms of their effects, both are the same. The greater the mass of an object, the stronger its gravitational pull, or in other words, the more intense the curvature of spacetime around it. From the perspective of an observer outside the object's gravitational field, clocks on or near the object will run slower than those outside. For instance, a clock located on the Sun would run 4,200 parts per billion slower compared to one in space, whereas a clock on Earth would be 1.4 parts per billion slower than one in space.

Note 161- *"For the first time in 1738, the English philosopher David Hume addressed the issue of the unity of time and motion. Later, Einstein acknowledged that without engaging with these philosophical reflections, he may never have succeeded in developing the theory of relativity."*

An interesting and practical example of how gravity (and speed) affect time can be seen in GPS satellites. Twenty-four satellites orbit Earth at a distance of 20,000 kilometers and a speed of 14,000 kilometers per hour. Each satellite carries an atomic clock with an accuracy of one nanosecond (one-billionth of a second). For precise positioning, the time on the satellites must be accurate within 20 to 30 nanoseconds. Since the satellites are farther from Earth, the effect of Earth's gravity on them is weaker, meaning time passes **45,000 nanoseconds per day faster** for them than on Earth, as predicted by general relativity. However, due to their high speed, special relativity causes their clocks to run **7,000 nanoseconds per day slower**. The combined effect results in the satellite clocks running **38,000 nanoseconds (38 microseconds) ahead per day** compared to clocks on Earth. Without accounting for both general and special relativity, GPS accuracy would degrade significantly, leading to errors of up to 10 kilometers in just two minutes. To counter this, the satellites are equipped with clocks that are pre-adjusted to be 38,000 nanoseconds slower than those on Earth.

Einstein's theories of special and general relativity have been confirmed by numerous experiments over the past century. One of the most notable confirmations came in 2015, when the LIGO detectors in the United States observed gravitational waves—ripples in spacetime—caused by the merger of two black holes, each roughly 30 times the mass of the Sun, 1.3 billion light years away. During the merger, about three times the mass of the Sun was converted into gravitational waves, which traveled through spacetime and were detected by LIGO. This discovery provided further confirmation of Einstein's predictions.

General relativity has become the most successful theory for describing large-scale phenomena such as the expanding universe, supernovae, black holes, galaxies, and stars, as well as the behavior of matter and energy at high velocities. These theories naturally include Newton's laws under low-speed conditions, giving consistent results. A fundamental insight of these theories is the introduction of the spacetime structure, which replaces the separate concepts of "space" and "time." Additionally, Einstein's famous equation $E = mc^2$, which expresses the equivalence of matter and energy, remains one of the most profound discoveries in understanding the physical nature of the universe.

Although general relativity is the most successful theory for describing large-scale phenomena, another theory, quantum mechanics, has emerged to explain the behavior of very small particles, such as atoms. Interestingly, Einstein also contributed to the development of quantum theory. The theories of relativity and quantum mechanics are based on different mathematical frameworks and physical principles, addressing different realms of reality. Together, however, they form the basis of our complete understanding of the physical world.

In the following sections, we will explore topics related to quantum physics.

Quantum Theory

To describe the nature and behavior of extremely small entities such as atoms and subatomic particles, a revolutionary framework called **quantum theory** was developed. The German physicist **Max Planck** introduced the concept of **the quantum** while studying radiant energy. Through his experiments, he discovered that radiant energy, which propagates in waves, is **absorbed or emitted in discrete packets** rather than in a continuous flow. These fundamental units of energy, later known as **quanta**, laid the foundation for quantum mechanics.

Planck formulated the following equation to describe the relationship between the energy of a quantum and the frequency of its associated wave:

$E=h\nu$

where:

- **E** is the quantum energy (the energy contained in a single packet),
- **h** is Planck's constant, - (see Note 162)
- **ν** (nu) is the frequency of the wave (measured in vibrations per second).

This equation indicates that the energy of each quantum is **directly proportional** to the frequency of the emitted radiation—higher frequencies correspond to higher energy quanta.

A few years later, this equation was expanded to incorporate the wavelength (**λ**) of light:

$E=hc/\lambda$

where:

- **c** is the speed of light in a vacuum,
- **λ** (lambda) is the wavelength of light.

At the time, a puzzling phenomenon known as the **photoelectric effect**—in which light striking certain materials caused the ejection of electrons—could not be explained using classical physics. In **1905**, Albert **Einstein**, building upon Planck's quantum hypothesis, proposed that **light itself consists of discrete packets** - (see Note **163) of energy**, later termed **photons**.

Note 162- *Planck's constant plays a fundamental role in modern physics, and its presence in any formula signifies the influence of quantum mechanics.*

Note 163- *The concept of quantization arises from this principle: the absorption and emission of energy occur in discrete, indivisible units. What does this mean? Imagine buying eggs at a store—you can purchase them only in whole numbers (one, two, a dozen, etc.), never in fractions. This is a quantized process. In contrast, when buying milk, you can purchase any amount, such as 1.5 liters or 2.3 liters, making it a continuous quantity. Similarly, energy is not emitted or absorbed as a continuous stream but instead exists in discrete packets, known as quanta.*

He suggested that a light source emits a **finite number** of these energy packets, which move intact without splitting apart. If a photon has sufficient energy (i.e., a high enough frequency), it can transfer its energy to an electron, causing the electron to be ejected from the material and generating a **photoelectric current**. This groundbreaking insight earned Einstein the **Nobel Prize in Physics in 1921**. Einstein's work led to the realization that **photons act as the fundamental carriers of electromagnetic energy** across the entire spectrum—from **X-rays and visible light to infrared radiation and radio waves**. - (see Note 164)

Bohr's Quantum Model of the Atom

In **1913**, the Danish physicist **Niels Bohr** introduced the first quantum theory of atomic structure, proposing the following principles:

1. **Electrons orbit -** (see Note **165) the atomic nucleus in stable, quantized orbits -** (see Note **166)** without radiating energy.

Note 164*- Visible light, X-rays, infrared waves, and heat waves are all forms of electromagnetic radiation. Their only difference lies in their frequency—the number of vibrations per second—which directly determines the energy of their quanta (photons). Radio waves have the lowest frequency, followed by heat and infrared rays. As the frequency increases, we move through the visible spectrum (red, orange, yellow, and violet light), then to ultraviolet rays, X-rays, and finally, gamma rays. Gamma rays, which originate from radioactive materials like radium and certain cosmic elements, have the highest frequency and energy.*

Note 165*- These orbitals are not limited to simple circular or oval shapes; they can also take on more complex forms, such as spindle-like structures and other intricate geometries.*

Note 166*- Although the classical concept of planetary-like electron orbits has been entirely replaced in modern physics, electrons do possess an intrinsic form of rotation known as* ***spin****. However, this spin is not a literal spinning motion in space but rather a fundamental quantum property that describes an intrinsic angular momentum. In fact, all fundamental particles in the universe are associated with a specific spin value, which plays a crucial role in defining their behavior and interactions.*

2. **Only specific, allowed orbits exist**, corresponding to discrete energy levels.
3. **When an electron transitions - (see Note 167) between these orbits, it absorbs or emits a photon** with energy equal to the difference between the two levels.

Wave-Particle Duality and the Birth of Quantum Mechanics

In **1924**, French physicist **Louis de Broglie** extended quantum ideas by proposing that **all matter exhibits wave-like properties**, not just photons. - (see Note 168)

Note 167- *This transition is known as a* ***"quantum leap."*** *Unlike classical motion, where objects move continuously through space, an electron makes a sudden, non-classical jump from one orbital to another. If we were to assume—incorrectly—that the electron behaves purely as a particle in orbitals, it would seem to vanish from one orbital and instantly reappear in another without passing through the space in between. However, if we consider the electron as a* ***wave*** *or a* ***wave-particle duality****, its transition from one orbital to another simply reflects a change in its* ***wave function****.*

What remains certain is that an electron's transition between orbitals is ***discrete and quantized****, not continuous or gradual. This separation between orbitals and the electron's inability to occupy the space between them leads to the formation of distinct spectral lines—a unique "barcode" for each element. This principle is the foundation of* ***spectroscopy****, which allows scientists to precisely identify elements, whether on Earth, in the Sun, or anywhere in the universe.*

Note 168- *The interactions between elements of the universe occur through the four fundamental forces:* ***electromagnetism, gravity, the strong nuclear force, and the weak nuclear force****. Each of these forces is mediated by specific quantum particles:*

- ***Electromagnetic force****: Its quantum carrier is the* ***photon****, which manifests as light across different frequencies.*
- ***Weak nuclear force****: This force is mediated by the* ***W and Z bosons****.*
- ***Strong nuclear force****: Its quantum carrier is the* ***gluon****, which binds quarks together within protons and neutrons.*
- ***Gravity****: The theoretical quantum carrier of gravity is the* ***graviton****, but it has not yet been experimentally detected.*

Since Einstein's relativity equation relates energy and mass:

$E = mc^2$

and Planck's equation relates energy and frequency:

$E=h\nu$

de Broglie combined these concepts, introducing a relationship between **momentum (p) and wavelength (λ)**:

$p=E/c=h/\lambda$

which simplifies to the famous **de Broglie wavelength equation**:

$\lambda=h/p$

This equation suggests that **every particle, including electrons and atoms, exhibits wavelike behavior**—a concept later confirmed in experiments.

Schrödinger's Wave Equation and the Probabilistic Nature of Particles

In **1926**, Austrian physicist **Erwin Schrödinger** formulated a mathematical framework known as the **Schrödinger wave equation**, which accurately described the wave behavior of electrons in atoms. In this model:

- **Electrons do not follow fixed planetary orbits** but instead exist as probability waves within certain regions called **orbitals**.
- The **wave function**, denoted by the Greek letter ψ (psi), describes the likelihood of finding an electron at a given location.

Shortly afterward, **Max Born** introduced the interpretation that the **square of the wave function represents the probability density**

of an electron's presence. This meant that **electrons could never be pinpointed precisely**, only described probabilistically—a radical departure from classical physics.

Heisenberg's Uncertainty Principle

In **1927**, **Werner Heisenberg** formulated his **uncertainty principle**, stating that:

"It is impossible to simultaneously determine both the precise position and momentum of a subatomic particle with absolute accuracy."

This principle, inherent to quantum mechanics, contrasts with classical physics, where such properties could be measured precisely. The **act of observation (Measurement) -** (see Note **169)** **itself disturbs the system**, leading to fundamental indeterminacy in nature. - (see Note 170)

Note 169- Observation *does not mean that simply looking at an electron stimulates it or forces it to exhibit its particle-like nature. If mere visual observation had such an effect, we would never be able to observe the phenomenon of light interference—something we will discuss in detail later in the context of the double-slit experiment. In reality, what causes the collapse of the wave function (i.e., the loss of wave-like behavior and the disappearance of interference patterns) is not human observation itself but rather the interaction with a measuring device. Unfortunately, some educational videos on the internet mistakenly suggest that human observation alone alters quantum behavior, which is a significant misinterpretation.*

Why is this a misinterpretation? Because ***our eyes are receivers, not transmitters****. The light that reaches us is either emitted directly by an object (such as an excited electron releasing a photon, similar to an electric lamp) or is reflected after striking an object from another source. Our vision depends on receiving this light, not on actively influencing the observed phenomenon. Believing that sight works by emitting rays from the eyes—like laser beams or radar—is a misconception that has long been debunked.*

It is more precise to use terms like ***"detection"*** *or* ***"measurement"*** *instead of* ***"observation"*** *in quantum mechanics. This is because when an experimenter measures a quantum system, they use instruments that interact with the wave-*

(continued from Note 169)- *particle, transferring or extracting energy in the process. This exchange of energy—whether in the form of an electromagnetic field, photons, or ion flow—causes the wave function to collapse, forcing the wave-particle (such as an electron) to exhibit its particle-like nature.*

In nature, wave-particles are constantly interacting with forces, fields, and other particles. They do not need an observer to exist or behave in a certain way. However, from a philosophical perspective, the way the human mind interprets information—whether waves, particles, or the world around us—introduces another fascinating discussion. If we consider the entire structure of reality, including both the observer and the observed, as a manifestation of ***pure consciousness****, we enter a deeper philosophical exploration, which we will address in due course.*

Note 170- *If we attempt to use classical physics concepts to understand this issue, we can observe that even with macroscopic parameters, our measurements influence the object being measured. For example, measuring the temperature of an object causes the thermometer to absorb or release energy, thus affecting the object's temperature. Similarly, measuring tire pressure will inevitably result in a slight reduction of pressure. In the subatomic world, however, the situation is even more profound. When we measure or track a photon, we effectively "destroy" it, because the measuring device absorbs its energy. Therefore, it is not surprising that, in experiments such as the two-slit experiment (which will be discussed later), any form of measurement leads to the loss of certain features of the system.*

Let's consider this more closely: how do we find the location of an object? We do so by shining light on it and recording the reflected light, which allows us to track its position. Similarly, how do we determine the speed of an object? By observing it over time, measuring the distance it travels, and then calculating the speed by dividing the distance by the time elapsed.

This method, however, doesn't work for atomic and subatomic components, because these components exist as waves—and even more strangely, as probability waves. As soon as we send a photon with sufficient energy to identify the position of a wave-particle (such as an electron), the collision with the photon causes the wave function of the electron to collapse. As a result, the electron manifests as a particle, but its momentum characteristics are lost in the process. Furthermore, if the electron is inside an atom, it may even jump to a higher orbit or escape into free space, and in this case, no meaningful measurement can be made.

The Copenhagen Interpretation

The **Copenhagen interpretation**, developed by **Bohr and Heisenberg**, remains one of the most widely accepted explanations of quantum mechanics. It asserts that:

- **Wave-particle duality**: Particles behave as both **waves and particles**, but only **one aspect is observable at a time**.
- **Superposition**: Until measured (observed), **quantum systems exist in multiple possible states simultaneously**.
- **Collapse of the wave function**: The act of **measurement forces a quantum system into a definite state**.

Quantum Entanglement and the EPR Paradox

Another strange quantum phenomenon is **quantum entanglement**, first discussed by **Einstein, Podolsky, and Rosen (EPR)** in 1935.

Entangled particles remain **instantaneously connected**, no matter how far apart they are, meaning that a change in one particle's state **immediately affects the other**—a phenomenon Einstein famously called "**spooky action at a distance**."

Despite Einstein's skepticism, entanglement has been **experimentally confirmed**, raising profound questions about the **nonlocal nature of reality**.

Final Thoughts

Quantum mechanics has fundamentally altered our understanding of reality, revealing a world governed by **probabilities, uncertainties, and dualities**. The strange and counterintuitive nature of quantum physics has given rise to **modern technologies**, including **lasers, semiconductors, quantum computing, and quantum cryptography**. Two key experiments that illustrate quantum strangeness—**the double-slit experiment** (demonstrating wave-particle duality) and **quantum entanglement**—will be discussed in the following sections.

Wave-Particle Duality and the Double-Slit Experiment

Until the advent of quantum theory in the early 20th century, the nature of light was a subject of ongoing debate. Isaac Newton and many scientists believed that light consisted of tiny particles (corpuscles). In contrast, other researchers, relying on Thomas Young's **double-slit experiment** in the early 19th century, demonstrated that light behaves as a wave.

With the emergence of quantum theory, it was revealed that light exhibits **wave-particle duality**, meaning it behaves as a wave under normal circumstances but manifests as a particle when subjected to energy exchange or measurement. This dual nature is not exclusive to light; it applies to all quantum entities, including **electrons, protons, neutrons, and even entire atoms and molecules**. A more precise way to describe such entities is to consider them **wave-particles** or **probability waves**—quantum objects that exhibit both localized (particle-like) and delocalized (wave-like) characteristics.

Each quantum entity has an associated **wave function**, a mathematical representation that describes its probabilistic spread in space. The wave function dictates the likelihood of finding the particle at different locations, with some regions having a higher probability than others. When a quantum entity interacts with an energy field—such as a measuring device—its **wave function collapses** instantaneously, meaning the particle's probabilistic spread contracts into a definite location. This collapse occurs regardless of the speed of light limitation and reflects the fundamental nonlocality of quantum mechanics.

The Double-Slit Experiment: A Gateway to Quantum Reality
The double-slit experiment was designed to investigate the wave behavior of light, electrons, and other quantum particles. It also demonstrates how measurement influences quantum states.

In this experiment:

• When light or electrons pass through two slits without being measured, they produce an interference pattern, characteristic of wave behavior.

• However, if a detector is placed at the slits to determine which slit a particle passes through, the interference pattern disappears, and the particles behave as individual entities, suggesting their wave function has collapsed upon measurement.

Richard Feynman, one of the pioneers of quantum physics, famously stated that "all of quantum mechanics can be understood through the double-slit experiment." The experiment has since been performed on various quantum objects—including electrons, protons, entire atoms, and even large molecules—demonstrating that wave-particle duality is a universal property of matter.

While it might not seem surprising that massless photons can exhibit wave-like behavior, the fact that electrons, protons, and even complex molecules (which have mass) behave in the same way is a profound discovery. This reinforces the idea that all physical interactions, whether in nature or within our own bodies, occur within the framework of wave-particle duality.

The Conscious Universe Hypothesis

A striking implication of wave-particle duality is the apparent awareness of quantum particles regarding their environment. The fact that a quantum entity's wave function collapses instantaneously upon measurement suggests that it "knows" when it is being detected. This has led some researchers to explore the idea that the universe exhibits an inherent form of consciousness or information processing at all levels, even at the smallest scales of existence.

By studying the double-slit experiment across various quantum systems, we continue to unveil the deep and mysterious nature of reality—one where the measurer, the act of detection, and the physical world are intricately interconnected. To further explore this phenomenon, we focus on the electron as the subject of the experiment while emphasizing that the double-slit experiment can be conducted on all subatomic particles, atoms, and even molecules.

Experiment 1: Single-Slit with Macroscopic Particles

We create a very narrow **single slit** on a sheet and place a **sensitive screen** at an appropriate distance behind it. Using a gun, we fire small bullets or marbles toward the slit. Some particles collide with the edges of the slit and scatter, while others pass through and create a **single line pattern** on the screen.

This experiment demonstrates the expected behavior of **macroscopic particles** when encountering a **single slit**—they either pass through or scatter but do not exhibit wave-like properties.

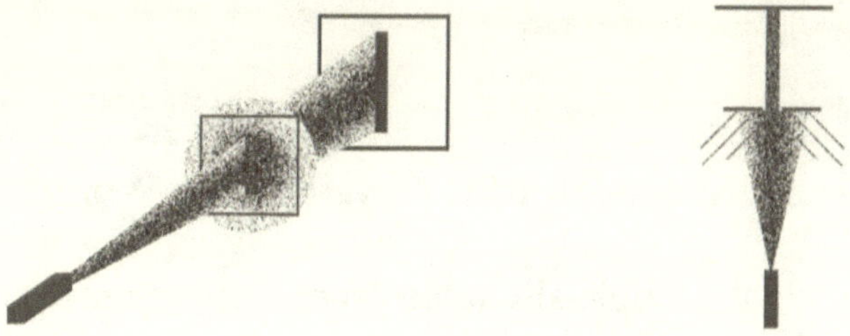

Experiment 2: Double-Slit with Macroscopic Particles

We now modify the setup by introducing **two parallel slits** instead of one. When bullets or marbles are fired toward the slits, the pattern that forms on the sensitive screen consists of **two distinct lines**, corresponding to the two slits.

This experiment confirms that macroscopic objects behave **classically**—they do not interfere with themselves but instead travel through one slit or the other, producing a straightforward **two-line pattern**.

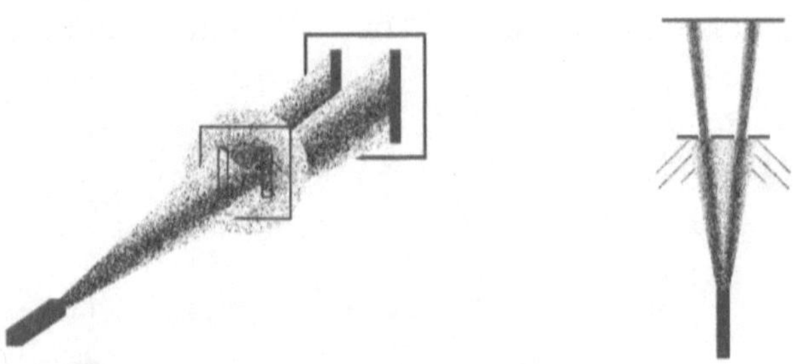

Experiment 3: Single-Slit with Waves (Water Waves)

Next, we take the **single-slit setup** from Experiment 1 but place it in a **waveable medium**, such as water. We then generate waves that pass through the slit. On the screen, a **single line pattern** appears, similar to Experiment 1.

This demonstrates that, like particles, waves passing through a **single slit** form a **single diffraction pattern**, confirming that a single opening does not create interference.

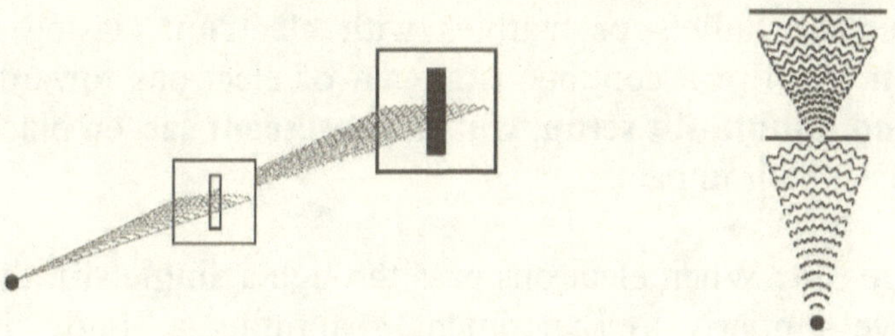

Experiment 4: Double-Slit with Waves (Water Waves)

Now, we modify the setup by introducing **two slits**, just as in Experiment 2, but keep the system in a **waveable medium** (e.g., water). When waves are generated, they pass through both slits and interfere with each other.

- At some points, the waves **reinforce each other** (constructive interference), creating bright bands.
- At other points, they **cancel each other out** (destructive interference), creating dark bands.

Instead of just two lines, a **series of alternating bright and dark bands** (interference pattern) appear on the screen. This experiment highlights the **fundamental difference between the behavior of particles (Experiment 2) and waves (Experiment 4)**.

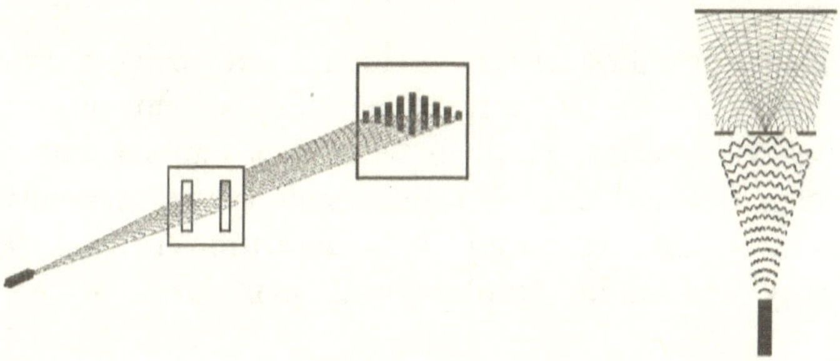

Experiment 5: Double-Slit with Electrons

Now, we replace bullets or marbles with **electrons**. Using an **electron gun**, we fire a continuous stream of electrons toward a **single-slit and double-slit setup**, with a **fluorescent screen** placed behind to detect their impact.

- **Single Slit:** When electrons pass through a **single slit**, they behave similarly to Experiment 3, forming a **single line pattern**, just like individual particles.
- **Double Slit:** When electrons pass through **two slits**, the result is striking: instead of two lines (as in Experiment 2), **an interference pattern appears**, similar to Experiment 4.

This means that:

1. Electrons behave like a **wave**, passing through both slits simultaneously and interfering with themselves.
2. When electrons hit the screen, they **collapse into particles**, forming bright spots where they land.
3. The **overall pattern** remains an interference pattern, proving that **electrons exhibit wave-like behavior**.

The most intense (brightest) band appears at the center of the interference pattern, where **constructive interference is strongest**, while the bands on either side gradually fade due to weaker interference effects.

Conclusion: The Dual Nature of Electrons

This experiment provides **direct evidence of wave-particle duality**. Electrons behave like waves when they are not observed (measured) but collapse into particles when they interact with the detection screen. This is identical to the behavior of **water waves in Experiment 4**, demonstrating that **quantum entities, like electrons, obey the same fundamental principles of wave interference**.

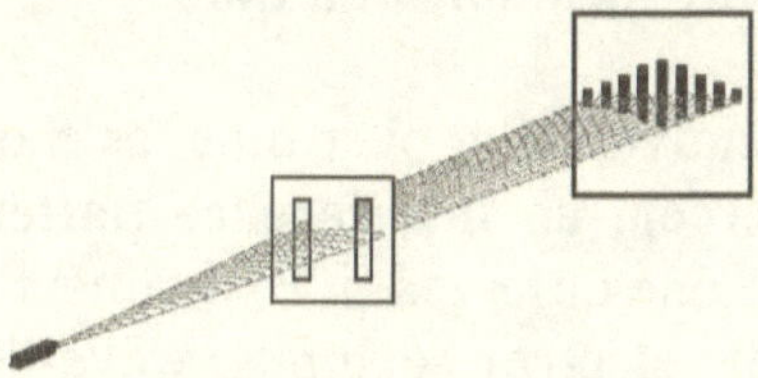

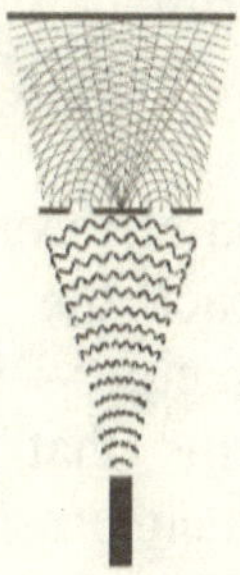

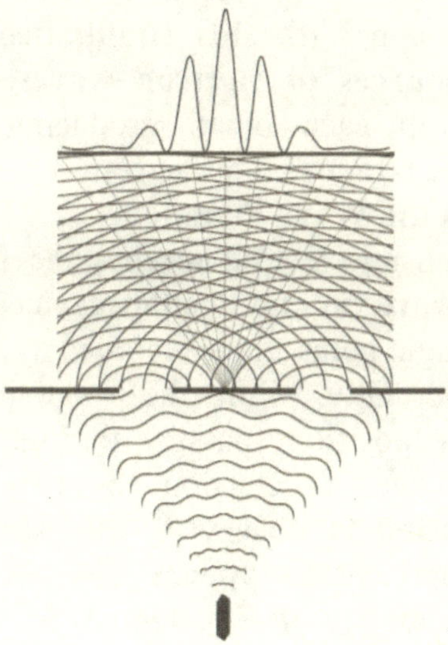

Experiment 6: Single-Electron Interference

In this experiment, we repeat **Experiment 5**, but instead of firing a continuous stream of electrons, we fire them **one at a time**.

Despite each electron being fired individually, over time, as more electrons reach the **fluorescent screen**, an **interference pattern** gradually emerges—identical to the one observed in Experiment 5. This confirms that even **a single electron** exhibits wave-like behavior when passing through the **double slit**, interfering with itself as if it traveled through **both slits simultaneously**.

Key Observations and Analysis

1. **Wave Behavior**: Each **individual** electron does not simply travel through one slit or the other, as a classical particle would. Instead, it behaves as a **wave**, spreading out and passing through **both slits simultaneously**, creating **two coherent sources of electron waves**. These two waves interfere with each other, producing the characteristic **interference fringes** on the screen.
2. **Collapse of the Wave Function**:
    - When an electron reaches the **fluorescent screen**, its **wave function collapses**, meaning it **localizes** to a single point and is recorded as a **particle**.
    - Importantly, while the **wave** initially exists in a **superposition** (passing through both slits), at the moment of detection (by the screen), **only one location is observed**—the other associated wave component **disappears**.
3. **The Consistency with Experiment 5**:
    - The result is identical to Experiment 5, proving that **interference is not caused by electrons interacting with each other**, but rather is an **intrinsic property** of **each individual electron's wave-like nature**.

- Even though electrons are emitted one at a time, their **individual wave functions still interfere** in a manner consistent with quantum mechanics, reinforcing the idea that the **wave function is real** and **not just a statistical abstraction**.

Conclusion: The Fundamental Nature of Wave-Particle Duality

Experiments **5 and 6** demonstrate the same fundamental principle: **electrons exhibit both wave-like and particle-like behavior simultaneously**. Each electron travels as a **wave**, interferes with itself, and then **collapses into a particle** upon touching the screen. The only difference between the two experiments is that in **Experiment 5**, a continuous stream of electrons produces the interference pattern more quickly, while in **Experiment 6**, the pattern builds up **electron by electron** over time.

This experiment serves as one of the most striking demonstrations of **quantum mechanics**, challenging classical intuitions about the nature of reality and reinforcing the principle that **wave-particle duality** is a fundamental feature of the universe.

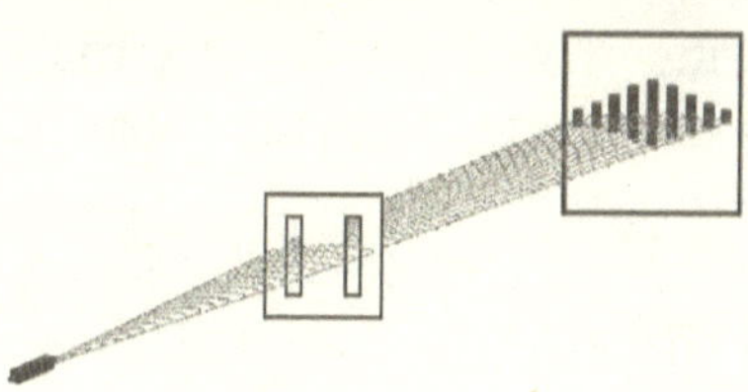

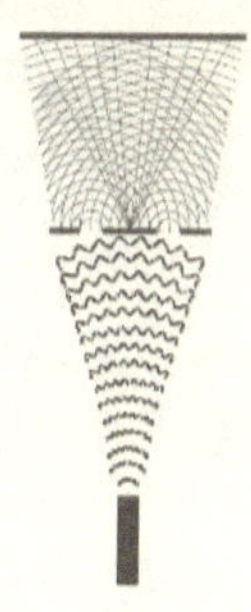

Experiment 7: The Effect of Measurement on Wave-Particle Duality

This experiment follows the same setup as Experiment 6, with one key modification: detectors are placed at the slits to determine which slit each electron passes through. These detectors interact with the electrons by exchanging energy through mechanisms such as electromagnetic fields, ion flow, or photon interaction.

Procedure & Observations

- As in Experiment 6, electrons are fired one by one toward the double slit.
- Unlike before, the detectors now allow us to determine the path of each electron.
- The interference pattern disappears. Instead of multiple bright and dark interference fringes, we observe only two distinct lines on the screen—one corresponding to each slit.

Analysis: Why Does the Interference Pattern Disappear?

1. **Wave Function Collapse Due to Measurement:**
    - When an electron passes through the slits undisturbed (as in Experiment 6), its wave function remains intact, allowing for self-interference and the formation of an interference pattern.
    - However, measuring which slit the electron passes through involves an energy exchange that disturbs its wave function.
    - This disturbance causes the wave function to collapse, forcing the electron into a definite state. Instead of behaving as a wave, it behaves like a classical particle, passing through only one slit.
2. **Loss of Wave Interference:**
    - Since each electron now travels through a single slit, there is no second associated wave to interfere with.

  - The electron follows a classical trajectory, producing only two bands—one for each slit—like the pattern observed in Experiment 2 with bullets.

Key Conclusion: The Role of Measurement in Quantum Mechanics

This experiment highlights a fundamental principle of quantum mechanics:

- **Measurement alters reality.**
- **Without measurement** (Experiment 6), the electron behaves as a wave and interferes with itself.
- **With measurement** (Experiment 7), the wave function collapses, and the electron behaves as a classical particle.

This striking result demonstrates the **quantum observer effect**, showing that the mere act of observation changes the outcome of an experiment. It reinforces the concept that quantum objects do not have definite properties until they are measured—a fundamental aspect of wave-particle duality and quantum mechanics.

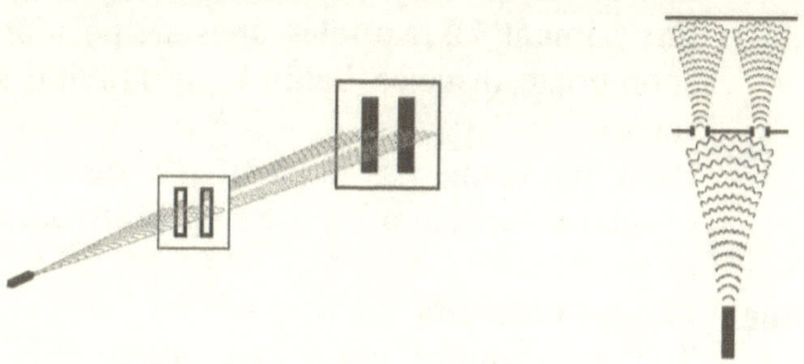

Experiment 8: Wheeler's Delayed-Choice Experiment

Between **1978 and 1984**, the American physicist **John Wheeler** proposed a series of thought experiments known as **Wheeler's Delayed-Choice Experiments**. These experiments, though conceptually more sophisticated, fundamentally align with the **previous quantum double-slit tests** and confirm the predictions of **standard quantum mechanics**.

One version of these experiments is examined here. This test is a **hybrid of Experiments 6 and 7**, incorporating a more **intelligent design** while **yielding the same fundamental quantum results**.

Although Wheeler originally designed it for **light (photons)**, the experiment is also applicable to **electrons** and other **quantum particles**.

Experimental Setup

1. **Moving Sensitive Screen**
   - A **sensitive detection screen** is positioned **after the double slit** but can move at **very high speeds**, including both **before and after the photon-wave passes through the slits**.
2. **Lens Placement**
   - Immediately after the **double slit**, a **lens** is placed to induce a slight deviation in the photon's path.
3. **Telescopic Detection System**
   - Instead of placing **detectors at the slits** (as in **Experiment 7**), two **telescopes** are positioned at an appropriate distance **behind the moving sensitive screen**.
   - Each **telescope** has a **sensitive detector at its eyepiece**, capable of registering **individual photon impacts**.
4. **Single-Photon Emission**
   - The light source emits **one photon at a time**, ensuring that each photon behaves **as a quantum wave-particle**.

Experimental Scenarios & Observations

First Scenario: The Sensitive Screen is Present

- A **single photon** is emitted from **Region I** and travels toward **Region II**, encountering the **double slit**.
- As in **Experiment 6**, the photon **passes through both slits simultaneously** (as a wave), then moves through the **lens** and enters **Region III**.
- It then **strikes the sensitive screen** in **Region IV**.
- When this process is repeated for a large number of photons, an **interference pattern** emerges—just as in **Experiment 6**—indicating that **each photon has traveled as a wave and interfered with itself** before being measured.

Second Scenario: The Sensitive Screen is Removed

- The **sensitive screen is removed**, allowing the photon wave-particle to continue propagating into **Region V**.
- This can be done **either before or after** the photon has passed through the **double slit**, but **the timing does not affect the result**.
- Without the screen, the **photon wave function does not collapse immediately**; instead, it **enters both telescopes simultaneously -Region VI**.
- Eventually, the photon **collapses onto one of the two sensitive screens** inside the **telescopes** with a **50/50 probability**.
- Instead of forming an **interference pattern**, the photon is detected **as a particle**, localized on one of the two screens.

Analysis: The Delayed-Choice Phenomenon

This experiment demonstrates a remarkable quantum principle:

1. The Role of Measurement in Determining Observable Behavior

- If the sensitive screen is present, the photon behaves in a wave-like manner and produces an interference pattern.
- If the screen is removed, the photon behaves like a particle and no interference is observed.

2. The "Delayed-Choice" Insight

- Intriguingly, the choice to insert or remove the screen can be made after the photon has already passed through the slits.
- Classically, one might assume that the photon must have already "decided" its behavior—wave-like or particle-like—at the moment it passed the slits.
- However, the experiment shows that the outcome depends on the measurement made afterward suggesting that the quantum system **remains in a superposed, indeterminate state** until it is finally measured.
- This does not mean the past is changed. Rather, it indicates that prior to measurement, **no definite state existed in classical terms**, and it is only through the act of measurement that one possibility becomes actualized.

Comparison to Experiment 7

In Experiment 7, measurement occurs directly at the slits (using detectors), collapsing the wave function early and causing the photon to behave like a particle, producing two distinct bands.
In Experiment 8, no measurement is made at the slits, allowing the photon's wave function to remain in superposition. The final

behavior—wave-like or particle-like—depends on whether the screen is present or removed, even if that decision occurs after the photon has passed the slits.

Key Conclusion: Quantum Reality and the Role of Measurement

- A quantum particle does not possess definite properties (such as "wave-like" or "particle-like" behavior) until it is measured.
- The act of measurement determines the outcome—not by altering the past, but by **resolving a prior state of indeterminacy.**
- Wheeler's delayed-choice experiment suggests that quantum systems are fundamentally probabilistic and contextual. Their properties are **not fixed in advance** but are shaped by the interaction with the measuring apparatus.
- This supports the Copenhagen interpretation of quantum mechanics, while cautioning against the misleading idea that the future "changes" the past.

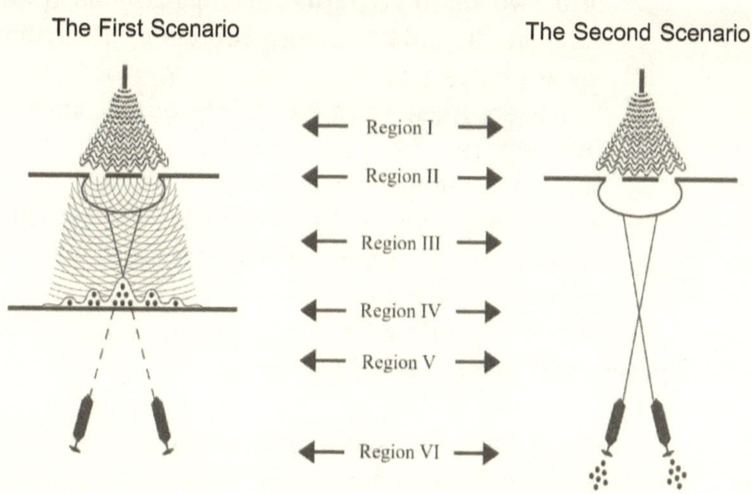

Experiment 9: Cosmic-Scale Double-Slit Experiment & Instantaneous Wave Function Collapse

This experiment explores the phenomenon of **wave function collapse on a cosmic scale**, demonstrating that **quantum measurement effects can occur instantaneously, independent of the speed of light**. This thought experiment was also conceived by **John Wheeler** as an extension of his **delayed-choice quantum experiments**.

Concept & Setup

1. **Cosmic Light Source (Quasar)**
    - Consider a **quasar**—a highly luminous celestial object located **several billion light-years away** from Earth.
    - The light emitted by the quasar travels toward Earth over this vast cosmic distance.
2. **Gravitational Lensing Effect**
    - On its way to Earth, the **quasar's light encounters a massive galaxy**, which acts as a **gravitational lens** due to its intense gravitational field.
    - This effect causes the **incoming light waves to split into two distinct paths**, bending around **opposite sides of the galaxy** before continuing their journey toward Earth.
    - When observed from Earth, the quasar appears **as two separate sources of light** (even though it is actually a single distant object).
    - This setup **naturally replicates a double-slit experiment on a cosmic scale** because each photon emitted from the quasar must have passed through **both gravitationally lensed paths simultaneously**.
3. **Telescopic Detection & Observation**
    - Two telescopes are positioned on Earth, each pointed at one of the **two apparent quasar images**.
    - Since both telescopes are actually observing the **same quasar**, every **single photon-wave arriving**

at Earth still shares a single wave function, just as in a conventional **double-slit experiment**.

Experimental Scenarios & Observations

First Scenario: Direct Photon Detection

- Each photon arriving at Earth **enters one of the two telescopes** and **directly impacts its respective detector**.
- Upon interaction with the sensitive detector, the **photon's wave function collapses**, registering it as a **localized particle** at a single position on the detector.
- Over time, as more photons arrive and collapse independently, the two telescopes together **produce two distinct images of the quasar**, corresponding to **two separate points of light** in the sky.
- This is analogous to the **particle-like behavior** seen in **Experiment 7**, where placing detectors at the slits **prevented the formation of an interference pattern**.

Second Scenario: Interferometric Measurement

- Instead of detecting the photons separately, we **combine the light from both telescopes** using an **interferometer** and direct the merged light onto a **single common sensitive screen**.
- Because the quasar's light originated from a **single coherent source**, its split photon-waves maintain **quantum coherence** even after traveling billions of light-years.
- As a result, each **individual photon interferes with itself**, just like in a standard **double-slit experiment**.
- On the **sensitive screen**, instead of observing two separate points of light (as in the first scenario), we see an **interference pattern**.
- Over time, as more and more photons are detected, the **full interference pattern emerges**, exactly as seen in **Experiment 6**.

Key Implications & Quantum Insights

1. Wave Function Collapse Is Instantaneous and Non-Local

- The photon's behavior is only determined at the moment of measurement. This means that its prior state—whether wave-like or particle-like—**was not fixed** during its journey, even if that journey lasted billions of years.
- This suggests that quantum information is not constrained by classical limitations such as the speed of light. Wave function collapse appears to occur instantaneously, regardless of distance.

2. Quantum Mechanics Applies Even at Cosmic Scales

- Even across vast astronomical distances, quantum principles—such as superposition, interference, and measurement-induced collapse—remain valid.
- These experiments reaffirm that quantum mechanics is not restricted to the microscopic domain but governs behavior on the grandest scales of the cosmos.

3. Delayed-Choice Experiments Reveal the Role of Measurement in Finalizing Quantum States

- In delayed-choice scenarios, the decision to measure a photon as a wave or a particle can be made long after it has left its source—for example, billions of years after a photon has departed a distant quasar.
- If we choose to perform an interference measurement, it reveals that the photon's wave-like behavior remained **unresolved** during its entire journey, only being defined upon final measurement.
- This does **not** mean the photon "actually traveled as a wave" the whole time in a classical sense. Rather, it remained in a superposed state until a specific measurement finalized one outcome.

- These results challenge classical intuitions about time and causality, showing that in quantum mechanics, **a system's behavior may remain fundamentally undefined until measured—even if the relevant events occurred in the distant past.**

Comparison to Previous Experiments

Experiment 6

- **Setup:** Electrons pass through a double slit without detectors.
- **Measurement:** An interference pattern forms.
- **Conclusion:** Electrons display wave-like behavior when not measured for path information.

Experiment 7

- **Setup:** Detectors are placed at the slits to observe electron paths.
- **Measurement:** Two distinct bands appear instead of interference.
- **Conclusion:** Measurement collapses the wave function into a particle-like state.

Experiment 8

- **Setup:** A delayed-choice experiment using telescopes after the particle passes the slits.
- **Measurement:** The electron's behavior depends on the later measurement step.
- **Conclusion:** The system remains in superposition until measured; **the outcome is not defined by the earlier event, but by the measurement context.**

Experiment 9

- **Setup:** A cosmic-scale double-slit experiment using gravitational lensing over billions of years.
- **Measurement:** An interference pattern is observed only when light paths are recombined.
- **Conclusion:** Measurement collapses the wave function even across vast cosmic timescales—highlighting that quantum indeterminacy persists until resolved by measurement.

Final Thought: Quantum Measurement & the Fabric of Reality

This experiment demonstrates that **the quantum wave function remains undetermined until a measurement is made**—even if that measurement is made **long after the photon has traveled across cosmic distances**.

- If we **observe the photons separately**, they **act like particles** and produce two distinct quasar images.
- If we **interfere them**, they **act like waves**, forming an interference pattern, as if they had been in superposition all along.

This suggests that **the fundamental nature of reality may not be predefined**; rather, it is shaped by **the act of observation itself**. **Quantum mechanics defies classical causality**, revealing that **the past can remain undetermined until it is measured in the present**—even on a cosmic scale.

Wave-Particle Duality: Extending the Two-Slit Experiment

The **double-slit experiment** has not only been performed on **photons, electrons, and other atomic-scale particles**, confirming their dual wave-particle nature, but has also been successfully extended to **larger quantum systems** such as **atoms and complex molecules**.

Notably, even molecules as large as **fullerenes (C_{60} and C_{70}, also known as "buckyballs," composed of 60 or 70 carbon atoms, respectively) and clusters containing up to 800 atoms** have exhibited the same **wave-like interference patterns**—a striking confirmation that **wave-particle duality** extends beyond fundamental particles to composite systems.

However, as molecular size increases, their **de Broglie wavelength** becomes extremely small, making quantum interference experiments **practically impossible** for macroscopic objects. This is why **we do not observe everyday objects behaving as waves**—

their **quantum coherence rapidly decoheres due to environmental interactions**, reinforcing their classical particle-like nature.

Instantaneous Wave Function Collapse & Nonlocality

A crucial aspect of wave function collapse—the transition from **quantum superposition to a definite state upon measurement**—is that it occurs **instantaneously** across the entire wave function. This means that when a **quantum wave interacts with a field or another particle**, the **entire wavefunction collapses at once**, rather than propagating information through the wave medium or requiring any energy transfer.

In other words:

- **Wavefunction collapse does not involve any signal transmission or causal propagation.**
- **It occurs instantaneously, independent of spatial separation and the speed of light.**
- **No information physically travels within the wave; instead, the entire wave system "knows" of its measurement state at once.**

This **remarkable nonlocal behavior**—the fact that the wave function "self-recognizes" its collapse state everywhere simultaneously—is precisely what **Einstein famously referred to as "spooky action at a distance"**. It is also **fundamentally the same phenomenon observed in quantum entanglement**, where two entangled particles exhibit correlated behaviors **instantaneously, regardless of their separation.**

In 2014, an **experiment conducted at Griffith University (Australia)** provided further empirical support for this instantaneous **collapse of the wave function**, reinforcing the idea that **wavefunction reduction is not constrained by relativistic causality or light-speed limits.**

Beyond Physics: Ontological Implications of Wave-Particle Duality

With this, we conclude our investigation into **wave-particle duality**. However, since the **philosophical implications** of this phenomenon are often misinterpreted in **ontological discussions**, before transitioning into the discussion of **quantum entanglement** we will analyze these ideas here slightly from an ontological perspective although we will further analyze these ideas in **Appendix II of this book**.

The **mutual relationship between human cognition and the external world** is a key issue that must be examined carefully. By doing so, we can better understand **the implications of quantum mechanics on our perception of reality**, providing a foundation for deeper philosophical inquiries.

The Nature of Reality: Quantum Objects vs. Macroscopic Perception

We have established that **all fundamental components of the universe—particles, fields, and forces—exhibit both wave-like and particle-like behavior**. At **atomic and molecular scales**, whether an entity manifests as a wave or particle **depends entirely on the measurement apparatus used**. However, these dual characteristics **never appear simultaneously**—a system always exhibits **either** wave or particle behavior in any given experiment, but never at once. This fundamental feature of quantum mechanics is strikingly demonstrated by the **double-slit experiment**.

In **everyday macroscopic reality**, however, objects consistently appear as **localized particles**. This is due to **quantum decoherence**, a process in which interactions between particles and their surroundings (including thermal and electromagnetic environments) cause quantum superpositions to collapse into distinct classical states. As a result, macroscopic objects behave deterministically, and **our nervous system interprets them accordingly**.

Einstein's statement that **"tangible objects exist in well-defined states before we observe them"** aligns with this classical interpretation of reality. This perspective leads to the assumption that the **physical world exists independently of human consciousness**—an idea reinforced by **the concept of quantum decoherence**, which suggests that:

- The **universe's fundamental components** (particles, fields, and forces) have been interacting **independently of human perception** since the **Big Bang (~13.8 billion years ago)**.
- These interactions **do not require human observation to exist**; rather, humans are **a product of these quantum and physical forces** rather than their defining factor.
- The **relationship between the universe and human beings is analogous to that of a parent and child**—human existence **emerges from** the universe, rather than the universe emerging from human perception.

Regardless of whether we interpret **physical existence as objective reality or a construct of perception**, the fundamental **physical framework of the cosmos operates independently of human thought**.

The Assumption That the World Owes Its Existence to Human Perception

The idea that the world's existence depends on human perception and awareness can be examined from multiple perspectives. It is worth briefly addressing a few key points here.

1. Misinterpretation of the "Observation Process" in Quantum Experiments

A common misunderstanding in discussions of quantum mechanics—especially regarding the double-slit experiment—is the role of the observer in altering a system's behavior. Some interpretations incorrectly suggest that merely "looking at" an experiment affects the results. However, in reality, passive observation alone has no impact on quantum outcomes. If merely observing the double-slit experiment influenced its results, the interference pattern would vanish the moment someone glanced at it—yet this is not the case.

A more precise term than "observation" would be **"measurement" or "detection,"** as these processes involve physical interactions. Any measurement in quantum mechanics requires the use of devices that employ electromagnetic fields, filters, or particles such as ions. These components interact with the system being observed—such as photons or electrons—exchanging energy and altering the state of the quantum system. Importantly, these measurements often occur automatically, meaning the results can be recorded without any human presence or awareness at the time. This reinforces the idea that **the mere presence of an observer does not determine quantum outcomes—rather, the physical act of measurement does.**

2. The Conflict Between Materialism and the Boundless Nature of Thought

The dominance of **materialism** in physics—rooted in **Cartesian dualism** and its strict separation of matter and mind—has built rigid conceptual barriers that often hinder the exploration of reality's deeper nature. Some physicists who favor the primacy of consciousness over materialism eagerly embrace any sign of cracks in this materialist framework. However, in some cases, this enthusiasm leads to misinterpretations.

A key example of such a misunderstanding is **Schrödinger's Cat Paradox.** Erwin Schrödinger proposed this thought experiment in 1935 to illustrate the absurdity of applying quantum probability to macroscopic objects. His intent was **not** to suggest that the cat exists in a superposition of being both alive and dead until observed, but rather to highlight the **limits** of quantum mechanics when extrapolated to everyday objects.

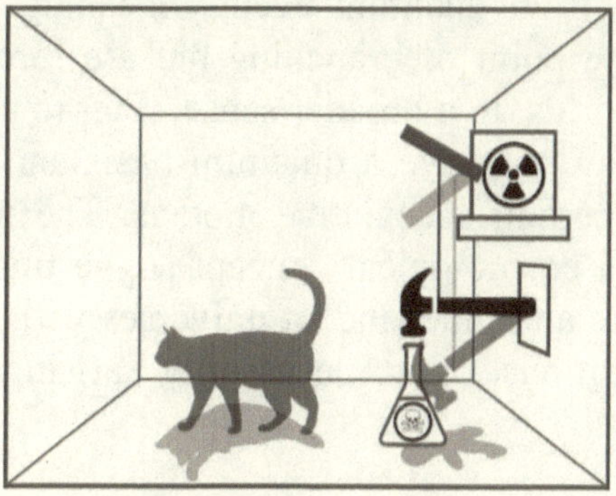

The experiment involves a cat in a sealed box with a radioactive source, a Geiger counter, a hammer, and a vial of cyanide gas. If the Geiger counter detects radiation (due to atomic decay), it triggers the hammer, breaking the vial and releasing the gas, killing the cat. The key question is: Before opening the box, is the cat alive or dead?

A common misinterpretation suggests that the cat exists in a quantum superposition of life and death until an observer collapses its wave function by looking inside the box. However, the reality is that the cat's fate is determined at the moment the radioactive atom decays and interacts with the detector. The collapse of the wave function happens due to this physical interaction, not because a human observes the system. The act of opening the box merely reveals an already determined outcome.

This misinterpretation extends to some versions of the **Many-Worlds Interpretation** (MWI), proposed in 1957 by Hugh Everett. According to MWI, instead of wave function collapse, every possible quantum outcome results in the universe "splitting" into separate, parallel realities. In the case of Schrödinger's cat, one universe contains a dead cat, while another contains a living one. These separate realities exist independently and never interact.

It is crucial to distinguish "Many-Worlds" from other ideas like "parallel worlds" and the "multiverse". In MWI, the different worlds emerge from quantum events, meaning they are causally connected at the point of branching but are forever isolated after that. Parallel worlds, in a broader sense, refer to coexisting realities that may or may not have a quantum basis and could interact or overlap under certain speculative models. The multiverse, on the other hand, is a cosmological concept suggesting the existence of entirely separate and independent universes with different physical laws or conditions, not necessarily arising from quantum branching.

It is important to note that even within the Copenhagen interpretation, the macroscopic world does not rely on measurement to manifest as particles. Instead, quantum decoherence—the process by which quantum superpositions interact with their environment—naturally explains why large-scale objects behave classically. Niels Bohr himself acknowledged a conceptual "wall" separating the atomic world from the macroscopic realm, where quantum effects are not directly observable. The wave functions of large objects have already collapsed due to continuous interactions with their surroundings.

3. **The Philosophical Question of Existence Without Observers**

One final reason some might argue that the universe owes its existence to human consciousness lies in philosophical considerations rather than physics. At the beginning of the universe, 13.8 billion years ago—and for hundreds of millions of years thereafter—there was likely no life, awareness, or consciousness. Earth itself only formed 4.6 billion years ago, and human history spans mere millions of years. Yet despite the lack of conscious observers for most of cosmic history, the universe evidently existed and evolved. This suggests that the world does not depend on human consciousness for its existence.

However, from a philosophical perspective, does an unobserved universe "exist" in any meaningful sense? If no conscious beings were present to perceive or interpret reality, would the universe still "exist" in any meaningful way, or would it be a realm untouched by recognition? This raises an elegant paradox: a universe without observers is, in some sense, meaningless—until an observer emerges.

Even today, if all conscious beings were removed from the cosmos, the physical universe would persist. However, without an observer with sensory faculties, neural processing, and a developed field of consciousness, the universe would remain an abstract concept, awaiting recognition. In this sense, existence without consciousness is incomplete.

The NUSAWVO Perspective: A World of Pure Consciousness

Returning to the fundamental insights of **NUSAWVO**, a different interpretation arises. It may be that the true nature of reality is **"pure consciousness,"** expressed through energy and manifested within the framework of physical laws. From this perspective, the evolution of the universe—beginning 13.8 billion years ago—has been the progressive emergence of intelligence **throughout all its components, from wave-particles to living beings including those with self-awareness.**

If this view is correct, then the material world is merely a **persistent illusion—one deeply embedded in the structure of consciousness itself.** However, even in this interpretation, the universe is **not** a product of human consciousness alone. Rather, consciousness itself is **the fundamental reality** that predates and transcends individual human awareness.

Thus, the relationship between humans and the universe is akin to **a dream dreaming of itself.** Our perception of reality may be **a projection of pure consciousness,** yet the universe does not originate from our minds alone. Instead, we arise **within** the broader consciousness that underlies existence itself.

In the **second appendix**, we will explore this discussion further from a philosophical standpoint, examining how the **appearance of**

reality is shaped by the sensory faculties and cognitive evolution of different beings. In doing so, we will consider how the absence of human observers might alter the world's perceptual reality, even as its fundamental essence persists.

Quantum Entanglement (EPR Paradox)

One of the most astonishing discoveries in quantum theory is the phenomenon known as "quantum entanglement," often referred to as the EPR paradox. Ironically, this concept emerged from Albert Einstein's attempt to disprove the probabilistic nature of quantum mechanics and to highlight its perceived inadequacies. The inherent randomness in the properties of subatomic particles was something Einstein found deeply unsettling. His famous remark, "God does not play dice with the world," reflects his skepticism regarding quantum uncertainty.

Einstein argued that if quantum mechanics were correct, then particles that were once in close interaction and later separated would maintain an instantaneous connection. That is, if a quantum property (such as spin, charge, or polarization) of one particle were measured, the corresponding property of the other particle—no matter how distant—would immediately adopt the opposite value. This apparent instantaneous correlation seemed to violate the fundamental principle that information cannot travel faster than the speed of light.

To challenge the randomness of quantum mechanics, Einstein, along with Boris Podolsky and Nathan Rosen, proposed a thought experiment in 1935—now known as the EPR paradox. They envisioned a scenario where a single particle decays into two photons that move in opposite directions. Because these photons originate from the same parent particle, they share the same wave function and remain entangled. One of their defining quantum properties is spin. Since the original particle had no net spin, the two photons must exhibit opposite spins to conserve angular momentum. However, according to quantum theory, the spin state of each photon remains indeterminate until it is measured—at which point, it collapses into a definite state. If one photon is measured and found to have an upward spin, the other photon must instantaneously adopt a downward spin, even if the two are light-years apart.

For Einstein, such a phenomenon seemed paradoxical. He believed that two possibilities existed:

1. The assumption of randomness in quantum mechanics was incorrect, and particles carried hidden variables that predetermined their properties at the moment of their creation.
2. Information about the measurement of one particle was transmitted to the other at an infinite speed, violating the principle that nothing can travel faster than light.

Since faster-than-light communication contradicts relativity, Einstein concluded that the first case must be correct—implying that quantum mechanics was incomplete and that hidden variables governed particle behavior. This led to a long-standing debate between proponents of quantum mechanics and those who, like Einstein, believed in the necessity of deterministic hidden variables.

The debate was settled in 1982 when French physicist Alain Aspect conducted an experiment using detectors 13 meters apart. His results provided strong empirical evidence for quantum entanglement and confirmed the probabilistic nature of quantum properties. The experiment was repeated in 1997 with detectors 11 kilometers apart and again in 2008 with detectors 18 kilometers apart at the University of Geneva, Switzerland. Each time, the results upheld quantum mechanics, demonstrating that entangled particles exhibit instantaneous correlations independent of the speed of light.

To summarize, quantum entanglement implies that at the quantum level, entangled particles remain fundamentally connected regardless of distance. No physical information or energy transfer is required for this connection to manifest. From a physical standpoint, since the entire universe was entangled at the moment of its birth 13.8 billion years ago, it suggests that there may be two fundamental levels of interaction: the traditional causal level (which we refer to as the "first realm" in the NUSAWVO school) and a deeper level beyond force, information transfer, or energy exchange (which NUSAWVO designates as the "second realm" or "basis realm").

However, it is crucial to revisit Einstein's perspective on the impossibility of faster-than-light information transfer. Although the EPR experiments proved Einstein wrong regarding the impossibility of instantaneous correlations, he was correct in asserting that faster-than-light transmission of meaningful information remains impossible. While quantum entanglement allows one to receive random information from across the universe, it does not enable the transmission of structured, non-random messages.

In other words, an EPR transmitter sends only a random signal. Every time an observer measures the spin of a photon; they are observing a purely probabilistic outcome. This randomness prevents the use of quantum entanglement for meaningful communication. To illustrate this, consider a scenario where the two photons were not entangled. In that case, measuring the spin of the second photon would yield a completely random result, independent of the first photon's spin—akin to flipping two separate coins, each with a 50% chance of landing heads or tails. Thus, entanglement establishes an intrinsic connection between particles, but it does not allow for faster-than-light communication of useful information.

Summary of Quantum Principles

1. **Quantization of the World's Elements:**

All fundamental constituents of existence at atomic and subatomic scales exhibit quantization, meaning they exist in discrete states rather than continuous forms.

2. **Wave-Particle Duality:**

At the atomic and subatomic levels, particles exhibit both wave-like and particle-like behavior. Their presence is described probabilistically until they interact with a system through energy exchange, at which point they manifest as particles. This transition, known as wavefunction collapse, occurs instantaneously across space, regardless of the speed of light. While information transfer is still constrained by relativity, the nonlocal nature of quantum mechanics suggests an underlying interconnectedness that defies classical intuition.

3. **Quantum Leap:**

Electrons occupy specific energy levels (orbitals) within an atom. When absorbing or emitting energy, they transition between orbitals in a discrete, non-continuous manner—vanishing from one state and appearing in another instantaneously.

4. **Uncertainty Principle:**

The Heisenberg Uncertainty Principle states that it is fundamentally impossible to simultaneously measure certain pairs of properties, such as position and momentum, with absolute precision at the quantum level.

5. **Quantum Entanglement:**

When two particles become entangled, they remain interconnected as a unified system, regardless of spatial separation. Any change in one particle instantaneously correlates with the other, without requiring classical information transfer.

Review of the Second Realm (Basis Realm)

The principles of the "second realm" of existence can be outlined as follows:

1. **The Wave-Energy Structure of Reality:**

From the fundamental equations $E=mc^2$ and $\lambda=h/p$, it follows that the fabric of reality consists of energy and wave-like structures.

2. **The Primacy of Consciousness in Physical Phenomena:**

At the quantum level, intelligence and consciousness influence the existence and behavior of fundamental wave-particles. More profoundly, each individual wave-particle is not only aware of its surrounding fields but also exhibits a form of instantaneous awareness that does not require the transmission of information. When a wave-particle interacts with a force or another wave-particle (i.e., undergoes an energy exchange), it instantaneously collapses into a definite state, a process known as **wave function collapse**.

3. **Nonlocal Awareness in Quantum Entanglement:**

The phenomenon of entanglement demonstrates a form of interconnected awareness independent of space and time constraints. Regardless of the vast distances (even spanning billions of light-years), entangled particles react to each other instantaneously without any exchange of energy, force, or classical information.

4. **Cosmological Implications – A Universe from "Nothing":**

Cosmologists propose that our universe originated from a quantum vacuum—essentially "nothing"—and that the total sum of matter and energy remains zero. This concept challenges conventional notions of material existence and reinforces the idea that the underlying fabric of reality is non-material in nature.

5. **Consciousness as the Foundation of Existence:**

The logical conclusion drawn from these observations is that **consciousness** is not merely an emergent property of the universe but rather its fundamental essence. The entire structure of reality can be regarded as a manifestation of consciousness, with energy and matter being secondary phenomena arising from it.

Two Realms of Reality

1. **The First Realm – The Observable World:**

This is the domain of everyday experience, governed by classical physics and causality. However, like ocean waves dependent on the ocean or the flow of a river dependent on its source, its reality is not independent but contingent.

2. **The Second Realm – The Foundational Consciousness:**

This realm is the fundamental reality, beyond energy, matter, and classical causation. It is not separate from the first realm but rather its very basis—**pure consciousness**.

Thus, we arrive at the understanding that both realms are ultimately unified in their foundation as a singular reality—**Consciousness itself**.

In Appendices 2 and 3, we will explore the philosophical and mystical perspectives on this foundational principle, further integrating these insights into our research.

Appendix Two

Appendix Two: Understanding the Universe (From a Philosophic Perspective)

In this appendix, we will first examine the prehistoric origins of human intellectual efforts to understand the surrounding environment and the natural forces that influenced life. We will then trace the historical periods that led to demythologization. Following this, we will provide a general definition of philosophy, introduce several key philosophical viewpoints, and finally explore the possibilities of understanding the universe.

The Background of Human Intellectual Efforts

Findings from paleontology suggest that as early as 100,000 years ago, humans buried their dead with personal belongings, indicating not only an awareness of death but also early supernatural beliefs, including the concept of spirits. Cave paintings dating back over 70,000 years further demonstrate humanity's capacity for abstraction and a deep-rooted spiritual need to comprehend existence beyond material reality. At this stage, humans did not perceive themselves as separate from nature but rather saw all of existence as alive (animism). The distinctions between matter and mind, animate and inanimate, were nonexistent in their worldview.

Early humans encountered formidable natural forces—earthquakes, volcanoes, floods, thunderstorms, and wildfires—whose immense power inspired both awe and fear. While these forces could bring destruction, events like timely rainfall in a barren desert could mean survival. To make sense of these unpredictable occurrences, humans attributed them to spirits, ancestral forces, or great cosmic powers (Mana). They worshiped these forces, honored sacred objects, and gradually developed myths and deities. In parallel, shamans emerged as mediators between the spiritual and material realms, practicing rituals, healing arts, and magical ceremonies to influence these supernatural forces.

Some scholars argue that the origins of prehistoric religion stemmed from humanity's attempt to comprehend and exert control over nature's immense power, as well as from the confrontation with the profound mystery of death. However, about 10,000 years ago, a transformative shift occurred: humans began practicing agriculture and domesticating animals, leading to the rise of settlements and the emergence of early civilizations.

Archaeological studies indicate that by 5,000 years ago, irrigation systems had been developed in Mesopotamia (modern-day Iraq) and along the Nile River in Egypt, fostering new modes of production and social organization. The discovery of metallurgy, pottery, and other technologies facilitated the growth of villages and cities, and eventually, the establishment of the first kingdoms. As agriculture became more sophisticated, a specialized class of individuals—priests, scribes, healers, and astronomers—emerged. This class, often inheriting their status through lineage, assumed roles similar to their shamanic predecessors but also engaged in scientific and technological pursuits such as medicine, mathematics, botany, astronomy, and engineering.

These early scholars, by studying celestial movements, were able to measure time and predict seasonal changes, particularly the flooding of rivers crucial for agricultural societies. However, alongside their contributions to knowledge, they also perpetuated an aura of mysticism and superstition, consolidating their influence over kings and religious structures. The deification of rulers and the formulation of elaborate mythologies likely took place during this period, giving rise to gods associated with creation, destruction, the sun, and other natural phenomena.

The Shift Towards Rational Inquiry

Despite the dominance of mythology, the gradual advancement of knowledge, art, and literature in various civilizations prompted some thinkers to question the divine nature of the gods. They began to recognize human-like flaws in these deities—selfishness, envy, and unpredictability—leading them to suspect that the gods were merely projections of human psychology. - (see Note 171) This realization spurred an intellectual movement toward demythologization, wherein individuals sought to explain the world through reason rather than myth. The pursuit of knowledge free from religious and supernatural explanations became what the early Greek thinkers humbly called "philosophy"—the love of wisdom.

This movement, emerging around 3,000 years ago, flourished in ancient India, Iran, China, and Greece. Although each civilization developed its own unique approach, they all shared a commitment to understanding reality through inquiry rather than superstition.

Ancient Philosophical Traditions

India

Indian philosophy, deeply rooted in spiritual inquiry, found one of its earliest expressions in the *Upanishads*, which are among the oldest and most profound texts of philosophical thought. The *Upanishads* sought to transcend myths and gods, articulating the concept of a singular, all-encompassing reality known as *Brahman*, often equated with *consciousness* in modern philosophical discourse. In addition, Buddhism emerged as a significant school of thought, emphasizing inner awakening and liberation over theological speculation. While Buddhism did not deny divinity outright, it considered such discussions irrelevant, prioritizing individual enlightenment and ethical living instead.

Note 171- *Among those who addressed this theme, the Greek philosopher Xenophanes—who was a student of Socrates like Plato—argued that humans create gods in their own image, modeling them after their own characteristics. Similarly, the Roman philosopher Lucretius suggested that it was fear that first brought gods into the human imagination.*

China

Chinese philosophy gained prominence with the teachings of Confucius, whose central tenet was that "human nature is fundamentally good." This principle not only defined Chinese civilization but also provided a moral framework for governance and society. Around the same time, Lao Tzu introduced *Taoism*, encapsulated in his seminal work, the *Tao Te Ching*. He described the *Tao* ("Way") as the fundamental principle underlying all existence—a concept that resonates with contemporary notions of consciousness. The *Tao Te Ching* remains one of the most influential texts of wisdom and mysticism.

Iran

Zoroastrianism, which emerged approximately three to four thousand years ago, represented a significant departure from earlier belief systems. Founded by Zoroaster, this religion rejected superstition in favor of monotheism, emphasizing individual choice and moral responsibility. Unlike many other religious traditions, Zoroastrianism granted humans the freedom to determine their own destinies rather than adhering to predestined fate. It upheld the equality of men and women, as evidenced in the *Gathas* (sacred hymns), where no gender-specific pronouns exist. The wisdom, rationality, and emphasis on ethical living within Zoroastrian teachings positioned it as both a religion and a philosophy.

Zoroastrianism's impact extended beyond Iran, influencing Greek and Roman philosophy as well as later religious movements. One of its most notable adherents, Cyrus the Great, demonstrated these teachings in governance. According to historical accounts by Herodotus and Xenophon, as well as the *Cyrus Cylinder* (considered the first charter of human rights), Cyrus abolished slavery, promoted religious tolerance, and upheld the dignity of all people under his rule. His policies inspired later political thinkers, including those involved in drafting the U.S. Constitution.

Greece

Although philosophical inquiry existed in earlier civilizations, Greece is often credited as the birthplace of systematic philosophy. Greek philosophers did not call themselves "wise" but rather "lovers of wisdom," emphasizing intellectual humility. They sought to understand reality through reason and empirical observation rather than mythological narratives. - (see Note 172)

Greek philosophy is most famously associated with Socrates, Plato, and Aristotle, but it was shaped by many thinkers before them. This intellectual tradition laid the foundation for a continuous philosophical discourse that remains vibrant to this day.

Conclusion

Since our aim is not to provide an exhaustive study of philosophy, we encourage readers to explore the vast array of philosophical works available. In the next section, we will provide a general definition of philosophy and briefly examine key philosophical schools relevant to the second insight of the NUSAWVO school.

Note 172- *This is a fundamental and critical point for any truly philosophical inquiry: philosophy should not begin with religious assumptions or a predefined belief in God and then attempt to construct an understanding of existence around that premise. Such an approach reduces philosophy to mere repetition, cloaked in obscure and convoluted language, without genuine truth-seeking. Conversely, if philosophy begins its search for truth with an open and unprejudiced mind—free from predetermined theological assumptions—then one must be prepared for unexpected discoveries. If, through this process, the inquiry ultimately leads to a belief in God, such a conclusion will be all the more profound and groundbreaking.*

A General Definition of Philosophy and an Overview of Several Philosophical Schools

Philosophy, derived from the Greek words *Philo* (love) and *Sophia* (wisdom), literally means "the love of wisdom." It encompasses a vast body of knowledge that investigates the most fundamental and intricate questions about existence. Philosophy represents the intellectual pursuit of understanding oneself, the nature of being, and the individual's relationship with the world. It is the endeavor of a thoughtful person seeking autonomy over their own destiny.

At its core, philosophy attempts to answer profound and existential questions such as:

- Who am I?
- Where did I come from?
- Why am I here?
- Where am I going?
- What is the origin of the universe?
- Is there life after death?
- Is all of existence merely an illusion?
- What is the foundation of good and evil?
- Does God exist?
- Does life have inherent meaning?
- Is happiness an actual or illusory concept?

It is no coincidence that Plato, one of the greatest philosophers of all time, called philosophy a "dear pleasure" and traced its origins to wonder—wonder at existence itself and the fact that "being" *is*. Historically, philosophy once encompassed all sciences and arts, but over time, as specialized fields emerged, it retained only the most foundational inquiries. Today, philosophy's primary branches include:

- **Metaphysics** – The study of existence, reality, and the nature of being.
- **Epistemology** – The study of knowledge, its limits, and its justification.

- **Logic** – The study of reasoning, argumentation, and the principles of valid inference.
- **Ethics (Moral Philosophy)** – The study of morality, virtue, and moral responsibility.
- **Political Philosophy** – The study of justice, governance, and the organization of society.
- **Aesthetics** – The study of beauty, art, and sensory perception.
- **Value Theory** – The study of what is valuable, including moral, aesthetic, and personal values.

Two Fundamental Divisions in Philosophy

Philosophical inquiry today is broadly categorized into two essential branches:

1. **Metaphysics**, particularly its fascinating subfield, **ontology**, which explores the nature, origin, and purpose of existence.
2. **Epistemology**, which critically examines the mind as an instrument of knowledge and questions whether human cognition can truly grasp *reality* and *truth*. It investigates whether our perceptions are reliable reflections of an objective world or merely self-generated illusions.

Throughout history, different philosophical schools have offered varying responses to these inquiries, each containing fragments of truth. However, despite their differences, all philosophies can be classified into two fundamental approaches: **dogmatism** and **skepticism**. Since understanding these concepts is crucial for grasping the insights of the **NUSAWVO school**, we will briefly examine them below.

Dogmatism: A Restricted Approach to Inquiry

Dogmatism refers to any intellectual stance that restricts investigation and forces inquiry into predetermined channels—either due to ideological commitment, lack of motivation, limited knowledge, or environmental influences (such as religious fundamentalism or rigid ideological doctrines)- (see Note 173).

Note 173 - Ideology and Its Distinction from Philosophy
Ideology is a structured system of beliefs and ideals that guides the thoughts, actions, and aspirations of individuals or groups. It functions as a framework that directs behavior and intention, often presenting a set of "dos and don'ts." In this sense, ideology is an "actuating" mode of thought, focused on shaping the world through action. While ideologies often draw upon philosophical perspectives, they differ fundamentally from philosophy itself. The key distinctions between philosophy and ideology are:

1. ***Purpose**: The primary goal of philosophy is to understand existence and truth, while ideology seeks to bring about change in the world through practical action.*
2. ***Approach**: Philosophy is inherently skeptical and exploratory. It favors open-ended questioning and the search for wisdom, not rigid certainty. Ideology, on the other hand, is pragmatic, goal-oriented, and often less inclined toward the critical inquiry that defines philosophy. Due to its practical nature, ideology can sometimes become prone to dogmatism.*

*It is important to note that the NUSAWVO school, while embodying certain ideals and goals, is relatively distant from strict ideological "dos and don'ts" or coercion. For instance, in the **seventh insight** (seeking the truth), the school emphasizes a skeptical approach, encouraging individuals to personally search for truth, critically questioning all insights of the philosophy. In the **eighth insight** (adherence to truth, correctness, and persistence in truth), tolerance and ethical integrity are core principles, explicitly rejecting Machiavellianism and the notion that "the ends justify the means." Opposing viewpoints are not dismissed but valued. Moreover, in the **eleventh insight** (the power of love and compassion), the school introduces its method of action, strictly forbidding any form of violence or coercion. While NUSAWVO retains its philosophical core, it does contain a degree of ideological structure, as it envisions ideals and goals for human progress. For example, the **ninth insight** promotes the desire for human happiness, the **tenth insight** presents idealism as voluntary action, and the **sixth insight** introduces the concept of an untainted state as the foundation for moral values. In this sense, while NUSAWVO emphasizes philosophical inquiry, it also embodies a vision for progress and human flourishing.*

In dogmatic thinking, unproven or even inherently unprovable beliefs are accepted as absolute truths from the outset. Any evidence contradicting these assumptions is dismissed or met with hostility.

From both **ontological** and **epistemological** perspectives, any philosophical school—whether materialist, idealist, or otherwise—risks falling into dogmatism when it treats its assumptions as unquestionable and immune to challenge. However, dogmatism is not confined to philosophy; it can be found in science, art, politics, and various aspects of human culture.

Characteristics of Dogmatism

- Resistance to changing one's beliefs, even in the face of contradictory evidence. - (see Note 174)

Note 174- The Nature of Taboos and Their Distinction from Dogma *throughout history, human cultures have developed taboos, but these should not be confused with dogma.*

For example, the taboo against incest (prohibition of adultery with a forbidden person) is not the result of philosophical or ideological imposition but rather the outcome of hundreds of thousands—if not millions—of years of natural selection and social evolution. Remarkably, this taboo has emerged independently across diverse human societies, even in those that remained isolated for tens of thousands of years, such as the indigenous peoples of Australia and the Americas. The same principle is upheld with equal rigor in the interwoven civilizations of the Old World, spanning Europe, Africa, and Asia. Moreover, equivalent behaviors have been observed among certain animals, particularly great apes, suggesting a deep-rooted biological basis.

The core taboos of human societies are relatively few prohibitions against incest, murder, theft, and deceit being among the most universal. Unlike dogmatic beliefs, which are often imposed by religious, ideological, or philosophical systems, these taboos have emerged as fundamental mechanisms for maintaining social cohesion and survival. While dogma is often rigid and arbitrary, taboos reflect deeply ingrained behavioral patterns that have stood the test of time.

- Viewing the world in binary terms: those who agree with us vs. those who oppose us.
- Intolerance toward differing perspectives, leading to hostility and irrational reactions.
- Blind devotion to a belief system, sometimes to the point of self-sacrifice or violence.

Historical examples of dogmatism include:

- The belief in a flat Earth or the geocentric model of the universe, which were enforced dogmatically in the Middle Ages.
- The persecution of those who challenged these views, with dissenters being labeled heretics and executed.
- The rigid ideologies of **Nazism, fascism**, and **extreme Marxism**, which led to immense suffering in the 20th century.
- **Religious fundamentalism**, which propagate dogma through fear, repression, and promises of salvation.

Skepticism: A Commitment to Open Inquiry

In contrast to dogmatism, **skepticism** is an intellectual approach characterized by a willingness to acknowledge ignorance and maintain doubt regarding unproven or unprovable claims. Skeptics remain open to further questioning and refinement of knowledge, whether in science or philosophy. No answer is ever final, and continuous inquiry is encouraged.

One of the most famous skeptics in philosophy was **René Descartes**. Seeking to escape dogmatic assumptions and establish a firm foundation for knowledge, he famously declared:

"I doubt everything in the world and exclude no concept from this doubt. However, while I may doubt everything, I cannot doubt my own doubting, because at the very least, my existence as a thinking being is undeniable."

This reasoning led to his foundational statement: **"I think, therefore I exist"** (*Cogito, ergo sum*).

Varieties of Skepticism
Skepticism varies in scope, from **radical skepticism**, which doubts the possibility of certain or justified knowledge, to **moderate skepticism**, which acknowledges that while knowledge may be possible, it should always be critically examined.

One specific form of skepticism is **agnosticism**, which argues that we lack sufficient evidence to determine whether God exists. This principle can also extend to other major philosophical inquiries. In a broader sense, scientific inquiry often maintains a form of **epistemic humility**, acknowledging uncertainties in theories such as the Big Bang, quantum mechanics, or dark matter, while remaining open to new evidence and refinement.

The Relevance to the NUSAWVO School

As we delve into the themes explored in the second insight of the NUSAWVO school—"Understanding the world of being"—we will examine various philosophical perspectives and schools of thought. These will be analyzed through the lens of their ontological, epistemological, and methodological assumptions, ensuring that our inquiry remains free from dogmatic constraints and open to the boundless possibilities of truth.

Dualism (Dual Perception)
Dualism is a philosophical perspective that posits the mind and body as two distinct and independent entities that, despite their separateness, interact in a manner that remains philosophically unresolved.

Historically, Plato introduced this view, but it was René Descartes who formalized and revived it. According to Descartes, reality is composed of two fundamental substances: the "ordinary essence" (res extensa), characterized by spatial extension (length, width, and height), and the "mental essence" (res cogitans), defined by thought. Descartes argued that animals lack a mind or soul, and that only humans possess an immaterial soul, which resides in the body but remains independent of it. Upon death, the body perishes, but the soul persists, continuing its existence beyond the physical realm.

Despite its historical significance, dualism faces critical philosophical and scientific challenges, particularly in explaining the precise mechanism of mind-body interaction. One of the major challenges is the "mind-body problem"—the difficulty of explaining how two fundamentally different substances, mind and matter, interact. Many philosophers reject dualism, yet the public—whether consciously or unconsciously—often holds a dualistic perspective.

Criticism of Dualism from the Perspective of the NUSAWVO School

The NUSAWVO school advocates for an integrated and holistic philosophical framework, whereas dualism inherently introduces a problematic divide between mind and matter. The gap between these two realms has remained unresolved in dualistic thought.

At first glance, NUSAWVO might appear to propose a dualistic structure, as it recognizes two realms of existence. However, the crucial distinction lies in NUSAWVO's assertion that the first realm—the "objective everyday world," also termed the "perceptible world"—is ultimately illusory and inauthentic. Its existence is entirely dependent on the second realm: Consciousness. Thus, upon deeper examination, it becomes evident that the NUSAWVO school is not dualistic but rather aligns with monism—a worldview in which all reality is fundamentally one.

Monism

Before Plato, monism dominated both common belief and philosophical thought, with reality perceived as a single, unified essence. However, with the contributions of Plato, Aristotle, and later Descartes, a dualistic framework emerged, dividing reality into separate categories of mind and body. This separation, however, has consistently faced difficulties in explaining how these two entities interact.

Baruch Spinoza, a Dutch philosopher, became a leading proponent of monism. He argued that all existence is one substance—"God"—with various phenomena merely being different expressions of that singular essence. By adopting this perspective, Spinoza circumvented the problem of mind-body interaction and declared that mind and body are fundamentally one and the same. However, his deterministic framework raised concerns about human free will, as it implied that all events, including human choices, were predetermined by the nature of this singular essence.

Monism later evolved into two primary schools of thought: Materialism and Idealism.

1. **Materialism** asserts that matter (or, in modern terms, "energy") is the fundamental substance of reality. Consciousness and intelligence, according to materialists, emerge as byproducts of material interactions, making them secondary rather than independent entities. This school of thought maintains that the external world exists objectively, independent of human consciousness.
2. **Idealism**, on the other hand, posits that the fundamental nature of reality is consciousness rather than matter. While idealism does not deny the existence of matter, it regards matter as a derivative of consciousness, rather than the other way around.

Criticism of Monism from the Perspective of the NUSAWVO School

The NUSAWVO school embraces monism and rejects the mind-body or mind-matter duality. Among the two main monistic perspectives, NUSAWVO aligns more closely—though not entirely—with idealism, as it emphasizes consciousness as the fundamental reality while distinguishing itself from traditional idealist frameworks. NUSAWVO asserts that consciousness is not a product of matter, nor is it emergent, but rather an intrinsic property of existence itself, permeating all of reality, including quantum particles and even the simplest forms of life.

The School of Materialism

The school of materialism holds that the universe is fundamentally physical, composed of matter and energy governed by physical laws. According to this philosophy, awareness, intelligence, and the mind emerge as products of complex material interactions rather than existing independently. Materialism denies the primacy of consciousness, asserting that the external world exists objectively, independent of human perception or cognition.

In this view, it is not human intellect that creates the external world but rather billions of years of evolutionary processes that have shaped matter, eventually giving rise to life, animals, and, ultimately, human beings and their capacity for reason. Many materialists reject belief in God, asserting that the universe was not created with inherent purpose but simply exists as an eternal or self-arising reality. However, materialism does not necessarily entail atheism—some who believe in God also embrace materialism, considering the universe's origins divine while maintaining that its fundamental nature remains physical.

Criticism of Materialism from the NUSAWVO Perspective

From the NUSAWVO perspective, materialism is flawed because it denies the fundamental primacy of consciousness. NUSAWVO holds that:

1. **Consciousness is the foundation of existence.** Matter or energy is not independent; rather, its essence is grounded in a truth called consciousness. This consciousness is not an emergent property but an intrinsic aspect of existence itself.
2. Consciousness permeates all existence. Even at the quantum level, wave-particles exhibit forms of

responsiveness that suggest awareness is interwoven into the fabric of reality—not as 'intelligence' in the human sense, but as an inherent property of existence itself.

3. **Human self-awareness emerged through evolution.** NUSAWVO acknowledges that human intelligence developed over billions of years of evolution. However, it maintains that even the simplest life forms and fundamental particles possess some degree of awareness, indicating that consciousness is not exclusive to advanced life but is present at all levels of existence.
4. **The world exists independently, but its manifestation is relative.** While the world exists beyond human perception, its appearance depends on the sensory and cognitive faculties of each being. If humanity were to vanish, the fundamental reality would persist, but the world *as we define and perceive it*—through our sensory and intellectual framework—would lose its meaning.
5. **The NUSAWVO school believes in God but refrains from human-imposed definitions.** Unlike some branches of materialism that reject divinity, NUSAWVO affirms the concept of God while maintaining **absolute humility** in how it approaches the divine. It does not attribute human-like qualities, actions, or roles to God, including the act of creation. Thus, while NUSAWVO disagrees with materialism's denial of consciousness as primary, it shares a skepticism toward definitive claims about creation and teleological purpose. This is not due to disbelief in God, but rather NUSAWVO's commitment to **a non-anthropomorphic and non-attributive** understanding of the divine.

Conclusion

NUSAWVO transcends both dualism and materialism, advocating for a **monistic worldview** where consciousness is the fundamental reality. While it acknowledges the evolutionary emergence of human intelligence, it asserts that **awareness is intrinsic to all existence**. By neither anthropomorphizing God nor imposing human-centric narratives onto reality, NUSAWVO presents a distinct and nuanced alternative—a worldview that bridges elements of idealism with a deep commitment to transcendent awareness.

Idealism: A Philosophical Perspective

The term "idea" originates from the Greek word meaning form, appearance, pattern, or whole. Plato introduced this term into philosophy to denote a "general" or "universal" essence behind particular instances. He proposed that for every kind of entity in the world, there exists a rational and universal archetype. For example, all elephants are merely imperfect reflections or shadows of an ideal "elephant" that exists in a higher intellectual realm. In this view, the reality of individual elephants is ephemeral and derivative, whereas the ideal elephant possesses an immutable, rational existence beyond sensory perception.

Plato did not deny the existence of the sensory world but considered it changeable, perishable, and secondary to the unchanging realm of ideas. He argued that true knowledge arises not from sensory experience but from apprehending universal truths, which exist independently of our perceptions. In this sense, Plato is often regarded as the first idealist, though earlier philosophers also questioned the reliability of sensory experience.

This philosophical tradition continued after Plato, gaining momentum during the Renaissance and Enlightenment. Thinkers such as Leibniz, Berkeley, Schelling, Kant, Hegel, Fichte, and Schopenhauer significantly contributed to idealism, each approaching the concept from distinct perspectives. The evolution of idealist thought finds resonance in modern physics, where reality is increasingly understood in terms of fields, energy, and perception. These philosophers collectively emphasized reason, thought, and consciousness as the ultimate foundations of reality.

The Core of Idealism

Idealists argue that sensory qualities—such as color, sound, heat, taste, smell, and texture—do not exist inherently in objects but are constructed through the interaction between external stimuli, sense organs, the brain, and even cultural influences. Without the faculty of vision, the world would lack light and color; without hearing, there would be no sound. Similarly, different species experience reality differently. A bat, relying on echolocation, perceives its surroundings in a manner unimaginable to humans. An ant, experiencing extreme surface tension, might perceive a drop of water as an insurmountable sphere rather than a fluid.

One of the most notable idealists, George Berkeley, denied that matter has an independent existence. He asserted that only minds and ideas are real, suggesting that we exist as parts of a universal mind, which he equated with God. However, Berkeley's views have often been misunderstood. Walter Terence Stace clarifies Berkeley's stance in *The Philosophy of Hegel*:

"The multiplicity and movement of the world of sensations are not ultimately real. But this does not mean they do not exist at all. The claim is not that a train traveling from London to Bristol is actually motionless, or that having twenty pounds in one's pocket is equivalent to having one. Rather, Berkeley asserts that matter exists only in our minds and in the mind of God. The existence of the material world is not denied; rather, its reality is conceptualized differently. If one insists that a tangible, physical world exists, Berkeley does not reject it outright—he only reinterprets its nature. Dr. Samuel Johnson, in a famous anecdote, attempted to refute Berkeley's idealism by kicking a rock and proclaiming, "I refute it thus." However, this misrepresented Berkeley's point, which was not the denial of physical experience but the assertion that such experiences are dependent on perception."

Leibniz and the Monadology

Another influential idealist, Gottfried Wilhelm Leibniz, proposed that reality consists of indivisible, non-spatial entities called "monads." He argued that everything material is ultimately reducible to mental phenomena. According to Leibniz:

- Any composite object can be divided into its constituent parts.
- Each part can be further divided indefinitely.
- Since anything occupying space is divisible, fundamental reality must consist of entities that do not occupy space.
- These indivisible, non-material entities—monads—are the true building blocks of existence.

Modern physics, particularly quantum mechanics, lends intriguing support to Leibniz's argument, as matter is now understood in terms of wave-particle duality and probabilistic fields rather than solid, indivisible objects.

Contemporary Understanding of Idealism

Today, idealism posits that the external world is ultimately of a mental or conscious nature. The fundamental distinction within idealism lies in:

• **Subjective Idealism**: The belief that the external world depends entirely on human consciousness. In this view, reality, as we perceive it, ceases to exist without a perceiving mind.

• **Objective Idealism**: The belief that consciousness is the foundational reality, but the external world exists as an expression of a deeper awareness independent of individual minds.

Some variations, such as **panpsychism** and **pan-consciousness**, suggest that all matter possesses some degree of consciousness or mental properties.

NUSAWVO's Critique of Idealism

The NUSAWVO school recognizes the profound insights of idealism, particularly its assertion that the world of sensory perception is not ultimate reality but rather a dependent phenomenon. However, NUSAWVO critiques subjective idealism, arguing that while the world outside the human mind is independent of individual perception, its appearance and nature are deeply intertwined with human cognition.

Physics today suggests that the fundamental fabric of reality consists of wave-particles, dynamic fields of energy, and information. The intelligence inherent in these structures has evolved over billions of years, leading to complex forms of life and, ultimately, self-awareness. Every living being, from bacteria to humans, perceives reality differently based on its biological and cognitive faculties.

The world as perceived by a bacterium differs radically from that of a butterfly, a dog, or a human. The evolutionary process has shaped an immense variety of sensory and cognitive systems, each constructing reality in its own way. Thus, while the external world exists independently of individual perception, its form and meaning are always mediated by the observer's sensory and cognitive framework.

A key question remains: What is the world in the absence of perception? Could it be dynamic energy, vibrating wave-particles, or something more fundamental—an "Energy-Consciousness" that underlies and gives rise to all existence?

Conclusion

Even objective idealism acknowledges that perception and reality are interconnected. NUSAWVO extends this perspective, suggesting that the external world is neither fully independent nor entirely dependent on perception. Instead, it posits a deeper ontological structure—consciousness as the fundamental essence of existence. This viewpoint bridges the gap between idealism and contemporary physics, offering a nuanced understanding of reality that transcends traditional philosophical dichotomies.

Animism

Animism is one of the oldest known religious and spiritual frameworks, dating back to prehistoric times. While often categorized as a religion, animism differs fundamentally from the major world religions due to its decentralized and localized nature. It is sometimes mistakenly conflated with philosophical or mystical schools such as **pantheism** and **panentheism** (both of which will be discussed later). However, animism is distinct in its understanding of spiritual presence, which makes it necessary to explore its core ideas here.

The term **animism** originates from the Latin word *anima*, meaning "soul," "spirit," or "life." At its core, animism is the belief that all aspects of nature—humans, animals, plants, rivers, mountains, celestial bodies, natural phenomena such as thunder and wind, and even human-made objects, shadows, and spoken words—possess a **spirit** or **life force**. In this worldview, there is no strict distinction between **physical reality** and the **spiritual essence** of objects and elements of nature.

Animism is widely observed in indigenous societies across Asia, Africa, Northern Europe, the Americas, and Oceania. Some animist traditions include a belief in a **higher power or divine will**, acknowledging a **creator** or universal presence, though animism in its purest form does not necessarily posit a single unifying deity. This distinction sets it apart from **pantheism** and **panentheism**, which conceptualize reality as an aspect or emanation of a singular divine principle.

A crucial aspect of animist societies is the presence of **spiritual intermediaries**, often referred to as **magicians, healers, or shamans**. Anthropologists use the term **shaman** (originally specific to Siberian native traditions but now applied broadly) to describe individuals who act as mediators between the **spirit world** and **everyday existence**. These figures, often viewed with both reverence and fear, are believed to **balance the visible and invisible realms**. They engage in spiritual healing, guidance, and ritual practices. In some societies, particularly among the indigenous peoples of the **south of Iran**, they are known as **"Baba Zar" or "Mama Zar"**.

From a mystical perspective, shamans are sometimes considered **mystics** due to their claimed ability to access **higher dimensions of reality**. Some assert that they can transcend ordinary perception and **enter other realms**. The American anthropologist **Carlos Castaneda**, through his field studies among indigenous communities in Mexico, documented aspects of shamanic culture. His works, particularly those featuring the teachings of **Don Juan Matthews**, provide insight into the complex cosmology of shamanic traditions.

Criticism of Animism from the Perspective of the NUSAWVO School

The **NUSAWVO school** does not endorse the belief in **separate souls or individual spirits** inhabiting various objects and beings. Instead, it posits a **single reality—consciousness**—which manifests itself in seemingly distinct, independent forms, whether **living or non-living**. From this standpoint, **all existence is grounded in a unified conscious reality rather than dispersed souls within objects**.

NUSAWVO does believe in God but does not attribute characteristics such as creatorhood or personal agency to it. Therefore, while it aligns with some schools of animism in acknowledging divinity, it fundamentally differs by rejecting any specific attributions or personifications of God.

Pantheism

Pantheism is a philosophical doctrine that equates the entire universe with a single spiritual essence—God (from "pan" meaning "all" and "theos" meaning "God"). It asserts that the universe and God are identical, rejecting any notion of a personal, anthropomorphic deity.

Elements of pantheistic thought can be traced back to ancient philosophical traditions, particularly in Neoplatonism, where thinkers like Plotinus emphasized the unity of all existence. However, it was Spinoza who systematized pantheism into a comprehensive metaphysical framework, equating God with the totality of existence in his *Ethics* through the concept of *Deus sive Natura* ("God or Nature").

Pantheism has been widely critiqued by theologians for dissolving the distinction between God and the cosmos, challenging traditional views of divine transcendence. Many religious scholars view it as incompatible with theistic traditions, while philosophers debate its implications for free will and moral agency—if all things are God, can individuals truly act independently?

In Islamic philosophy, certain mystical thinkers, including Ibn Arabi, proposed ideas resembling pantheism, though within an Islamic metaphysical framework. His doctrine of *wahdat al-wujud* ("unity of existence") later influenced thinkers such as Mulla Sadra, who expanded on the dynamic nature of being through his theory of substantial motion. However, these perspectives are often classified as monistic rather than strictly pantheistic.

Pantheism encompasses a range of interpretations, including **panpsychism**, which attributes consciousness or awareness to the universe as a whole. The distinction between these concepts is important:

- **Pantheism** = "All is God."
- **Panpsychism** = "All has mind/consciousness."

Hindu philosophy, particularly the **Vedas, Upanishads, and Bhagavad Gita**, contains strong elements of pantheistic thought, linking it to Eastern spiritual traditions.

Difference Between Pantheism and Animism

While pantheism sees everything as one unified divine reality, animism envisions the world as inhabited by multiple, independent spiritual entities. Animism does not necessarily posit an overarching divine principle but instead recognizes the presence of spirits within nature and objects.

Criticism of Pantheism from the NUSAWVO Perspective

While NUSAWVO acknowledges the concept of God, it refrains from assigning any attributes, descriptions, or definitions to it—including existence. Unlike pantheism, which equates God with the universe, NUSAWVO makes no such identification. Furthermore, while NUSAWVO defines the world as **consciousness**, it does not equate this with "mind" in the way panpsychism does. This distinction sets NUSAWVO apart from both pantheism and panpsychism, as it neither attributes divinity to the world nor assigns characteristics to God in any form.

Panentheism

Panentheism is a philosophical belief that God is both immanent within the world and transcendent beyond it. In this view, God permeates all existence but is not identical to the universe, maintaining independence from it. Panentheism differs from pantheism by asserting that while divinity is present throughout reality, it also extends beyond the physical realm, thus retaining a higher, ineffable aspect that transcends the material world.

The roots of panentheistic thought can be traced back to various religious and philosophical traditions. In Hindu philosophy, the notion of Brahman encapsulates a divine essence that is both within the universe and beyond it, manifesting in all forms while remaining ultimate and transcendent. Similarly, aspects of panentheism are woven into Jewish Kabbalistic thought, where the concept of divine emanation (Sefirot) suggests that all existence unfolds from a higher, unified source yet retains a transcendent nature. In Islamic mysticism, thinkers like Ibn Arabi explore the intricate relationship between divine presence and creation, emphasizing that while God is intimately involved with the world, God's essence remains distinct and beyond full human comprehension.

Some Native American spiritual traditions also resonate with panentheistic themes, where the sacred is viewed as immanent within nature, but always connected to a transcendent, unseen force that sustains the cosmos. This worldview fosters a deep reverence for the earth while acknowledging that the divine, though present in all things, is not confined to the material world.

Difference Between Pantheism and Panentheism

- **Pantheism** states that God and the universe are one.
- **Panentheism** asserts that God is beyond the universe but also present within it.

Criticism of Panentheism from the NUSAWVO Perspective

While NUSAWVO recognizes the philosophical depth of panentheism, it ultimately diverges in its approach. Since NUSAWVO does not assign any attributes or characteristics to God, it does not accept the notions of divine immanence or transcendence. Instead of framing existence as infused with divinity, NUSAWVO emphasizes **consciousness as the nature of the world**, without attributing it to a divine essence. This distinction sets NUSAWVO apart from both pantheism and panentheism.

School of Pragmatism

The school of pragmatism emphasizes the **"authenticity of action."**

This philosophical movement primarily addresses epistemological concerns, particularly those related to the search for "truth." According to pragmatism, ideas and beliefs hold value only when they prove useful and applicable in practical life. In this sense, truth is determined by its beneficial consequences, rather than by its correspondence to an objective reality. Ultimately, everything is validated through its practical results.

The foremost figures in pragmatism include American philosophers William James, John Dewey, and Charles Sanders Peirce. Pragmatism positions itself against idealism and purely intellectual inquiries, which often engage in abstract concepts that are detached from immediate practical utility. In pragmatist philosophy, truth is defined by what is beneficial from a human perspective. If an idea or belief leads to a favorable and effective outcome, it is regarded as valid—at least until circumstances necessitate its revision. Consequently, pragmatism denies the existence of an absolute reality independent of human perception.

A common philosophical belief posits that truth exists independently of human recognition. Pragmatism, however, asserts that truth is not independent but rather contingent upon its usefulness to human beings. The validity of an idea is determined solely by its practical effectiveness, not by its alignment with an external, immutable reality. In this view, concepts and beliefs are neither inherently true nor false; they acquire their truth-value only when subjected to real-world testing and practical application. An idea may be useful for a time, making it temporarily "true," but if it later fails to yield satisfactory results, it is deemed false and abandoned. Truth, therefore, is not static but evolves alongside human knowledge and experience.

For instance, Newtonian mechanics was considered the definitive truth in physics for centuries. However, as scientific progress raised new questions that Newtonian physics could not answer, Einstein's theory of relativity emerged, offering solutions where Newtonian mechanics fell short. While Newtonian physics remains applicable in many contexts, relativity has supplanted it in fields such as high-speed dynamics.

English philosopher John Stuart Mill even suggested that mathematical truths are not absolute; rather, they are confirmed by accumulated human experience. - (see Note 175) Similarly, William James emphasized that truth must have practical consequences, asserting that even merchants should be able to grasp it. Instead of evaluating concepts through abstract reasoning, pragmatism urges us to test them experientially and assess their tangible effects.

Nevertheless, pragmatism has faced criticism. Will Durant argued that pragmatism allows individuals to embrace even the most misguided ideas as "truth" if they bring comfort and happiness. He contended that fleeting personal benefit cannot validate an idea; only an "eternal and collective benefit" can do so—yet such an absolute measure of truth remains elusive. Friedrich Nietzsche, in a similar vein, suggested that many so-called "truths" may in fact be merely the most useful errors we have encountered. From this perspective, truth could be understood as a shifting social construct rather than an objective reality. Durant further likened truth to a stained-glass dome, where different observers perceive different hues and patterns. Thus, truth may be nothing more than the common denominator of human errors.

Note 175*- It may seem surprising, but even mathematics is ultimately a function of experience! For example, the fact that the sum of a triangle's interior angles equals 180 degrees is not an absolute, independent truth—it arises because the ratio of our size to the Earth's size makes the planet appear flat. Now, if there were beings living on another sphere—perhaps far larger than us yet residing on a much smaller sphere—then the interior angles of their triangles would exceed 180 degrees. Even on this planet, large-scale mapping surveys must account for the Earth's curvature in their triangulations.*

Summary of the School of Pragmatism

Pragmatism offers valuable insights by asserting that any "hypothetical truth" or philosophy must be tested in practical life and evaluated based on its real-world consequences. In this view, any ideology or philosophy claiming absolute truth—especially if it derives solely from abstract ideas without consideration of human nature, historical context, and scientific progress—will ultimately be put to the test. Over time, its flaws will become evident, and if it proves detrimental to society, it will be discarded as a failed intellectual experiment. A clear example of this principle is the ideological gap between the ideals of communism and its real-world outcomes. In the 19th and 20th centuries, communism promised liberation for the working class and the overthrow of capitalism.

It sought to abolish private ownership of production, dismantle the wage system, and establish a proletarian dictatorship. Initially, many found these ideas appealing and viewed them as a pathway to human freedom. However, reality sharply diverged from these aspirations. Two major strains of communism emerged. In Western Europe, moderate currents evolved into social democracy, working within democratic frameworks to promote social justice. These movements abandoned radical measures such as expropriation and one-party rule, instead advocating for wealth redistribution, labor protections, and social welfare. Conversely, radical communist movements in other parts of the world remained committed to Marxist-Leninist principles, leading to totalitarian regimes.

These regimes suppressed dissent, eliminated political opposition, and ultimately succumbed to corruption, economic inefficiency, and public discontent. - (see Note 176)

Note 176- *The English philosopher* ***Karl Popper*** *once remarked about Marxism:* ***"Marxism is merely one event—one of the many mistakes we have made in our endless struggle to build a better and freer world."***

Over time, communist states either collapsed, reformed, or abandoned their original ideologies altogether. - (see Note 177)

The failure of radical communism illustrates pragmatism's key claim: theories and ideologies must be evaluated by their practical consequences. When an ideology disregards human nature and empirical realities, it will ultimately fail, regardless of how appealing its abstract ideals may seem.

Criticism of Pragmatism from the Perspective of NUSAWVO

While NUSAWVO aligns with idealism—specifically "Pan-consciousness"—it also integrates aspects of a pragmatic approach. How so?

NUSAWVO acknowledges the existence of God but refrains from ascribing any specific attributes to God. Instead, it posits that consciousness is the fundamental essence of the universe. This belief in God, when understood in such a manner, may yield practical outcomes:

1. **Monistic View of Consciousness**: By adopting a monistic view of consciousness, NUSAWVO eliminates the mind-matter duality, fostering a unified understanding of both natural and supernatural phenomena. This approach helps dispel superstitions, although skeptics may dismiss it as merely another form of superstition.

Note 177- *It is estimated that in the second half of the 20th century, due to the implementation of communist policies in China, around **40 million** people perished from starvation. However, it is important to note that these policies have long since been reformed, and China's current economic structure more closely resembles a capitalist system.*

2. **Human Freedom and Volition**: Unlike pantheism, NUSAWVO affirms human freedom and volition, avoiding the deterministic implications often associated with the "unity of existence." Some mystics may critique this stance as naive or superficial.
3. **Avoidance of Ideological Distortion**: By refusing to assign attributes to God, NUSAWVO prevents ideological distortions and monopolization of religious concepts. However, some atheists may argue that entirely dispensing with the concept of God would be a more effective solution.
4. **Balanced Perspective on Theism**: This approach mitigates the influence of extreme anti-theistic ideologies, offering a balanced perspective in an era of widespread rejection of God. Yet, militant atheists may perceive this as a veiled form of religious apologetics.
5. **Intimate Relationship with the Divine**: By distancing itself from conventional theological constructs, NUSAWVO facilitates a more profound, intimate, and authentic relationship with the divine. However, traditional religious adherents may find this perspective unconventional or even heretical.
6. **Ethical Imperative**: By emphasizing consciousness as the singular truth of the universe, NUSAWVO fosters deep moral awareness. It reframes the "Golden Rule"—"Do unto others as you would have them do unto you"—in a more profound manner: since all beings are manifestations of the same fundamental consciousness, ethical treatment of others becomes an intrinsic moral imperative. - (see Note 178)

Note 178- If, as ***John Stuart Mill*** *suggested,* ***"even mathematics requires human experiential confirmation, and that experience itself is derived from humanity's inherited nature,"*** *then there can be no doubt that moral conscience, too, is rooted in human nature rather than in a philosopher's efforts to artificially shape it. In this light, theoretical inquiries are only valid* ***if*** *they correspond to real and noble elements already present within human nature—not the other way around.*

Thus, while NUSAWVO is not explicitly pragmatist, it embodies a form of pragmatic idealism—one that does not emerge solely from logical deductions, philosophical discourse, or scientific inquiry, but rather from direct, personal experience (or at least has been perceived as such, if not illusionary) with meaning. The theoretical framework of NUSAWVO is shaped by these experiences, blending intuition with intellectual rigor to construct a cohesive worldview.

Ultimately, all theoretical and existential inquiries within NUSAWVO converge on three fundamental principles: - (see Note 179)

1. **Meaning Transcends Interpretation**: Meaning transcends interpretation and philosophizing. Our knowledge of meaning is an internal, inexpressible perception rooted in personal relationships. Any attempt to formulate meaning weakens this relationship.
2. **Pure Consciousness as the Origin of Being**: The ultimate and true origin of being is pure consciousness.
3. **Intrinsic Human Nobility**: Human nature is noble and good; when all its aspects operate in balance and harmony, they awaken a person's happiness and self-awareness.

This idealistic yet pragmatic approach serves as the foundation for all insights within the NUSAWVO school.

Note 179- *There is no doubt that none of these insights emerged from nothing; rather, they have taken shape within the shared cultural consciousness of humanity, deeply intertwined with an inherited faith in devotion. It is in this context that the* ***NUSAWVO school*** *owes its perspectives to the pure and sincere presence of humble, well-intentioned individuals—those with moderate, non-dogmatic religious beliefs, free from prejudice.*

At the same time, certain aspects of this philosophy uphold moral ideals—such as an unwavering commitment to truth, integrity, and ethical conduct in all endeavors, particularly in politics, where it rejects Machiavellianism - (see Note 180)—regardless of external conditions. In this sense, it echoes Kant's assertion that certain truths exist independently of human perception and do not require validation through their utility. However, pragmatism ultimately holds the final word, for even these eternal truths emerge from the essence of true human nature, where spirituality is deeply embedded. Thus, they are neither abstract nor detached from human existence.

Hypothesis of the "Unknowable"

Before concluding our discussion on philosophy—specifically the subjects of ontology and epistemology—we must address a critical issue: the limitations and capabilities of the human mind in understanding existence. Many argue that the human mind will never grasp the true nature of reality and that existence itself is fundamentally beyond our comprehension. In other words, reality is "unknowable." However, in the *Philosophy of Hegel*- a book written by Walter Terrence Stace, not Hegel himself – under sections 63 to 69, Stace challenges this view. With great clarity and conviction, he argues against the hypothesis of the "unknowable," demonstrating that the human mind is indeed capable of recognizing and understanding the truth of existence.

Note 180- *Machiavellianism refers to a set of political principles and methods proposed by the Italian philosopher* ***Niccolò Machiavelli*** *to guide rulers in their pursuit of power. He viewed political action primarily as a means to acquire and maintain power, justifying the use of any means necessary to achieve this goal—* ***"the end justifies the means."*** *According to Machiavelli, a ruler who wishes to succeed must not fear the use of evil or violence; for the sake of maintaining power, the state is permitted to commit acts such as murder, betrayal, assassination, and deception—even when these actions contradict morality, honor, and justice.*

Stace makes a crucial distinction between what is unknown, what is unrecognizable, and what is unknowable. It is undeniable that there are countless things in the universe that remain unknown to us and may always remain so. Many of these are also unrecognizable meaning that, due to external limitations such as distance or lack of resources, we may never become aware of them. For example, a piece of gold on a distant planet orbiting an invisible star would be unrecognizable simply because we lack the means to detect it. However, this does not mean it is unknowable in principle. If those external limitations were removed, we could recognize it just as easily as a gold coin in our pocket.

In contrast, the notion of the unknowable suggests that certain aspects of existence are inherently beyond the grasp of the human mind—not due to practical limitations, but due to the fundamental nature of our cognition. This view, often associated with philosophers like Kant and Spencer, suggests that there are truths permanently and absolutely inaccessible to human understanding. However, Stace argues that this idea is contradictory.

He asserts that any act of recognition depends on concepts. To know something means to apply an appropriate concept to it. But if something were truly unknowable, we could not even conceive of it, let alone assert its existence. The mere act of positing that something is unknowable presupposes some degree of recognition—otherwise, we could not speak of it at all. This creates an inherent paradox: if something were truly beyond all recognition, we would not even be able to classify it as unknowable.

Furthermore, some argue that while we may know something exists, we may never know its essence. However, Stace challenges this distinction. In philosophy, existence and essence are not separate categories; existence is an inseparable aspect of a thing's essence. If we are aware of something's existence, even in a limited way, we necessarily grasp some aspect of its essence. Therefore, the claim that something is absolutely unknowable becomes untenable.

Stace also provides a Hegelian argument regarding the nature of limits: a boundary can only be recognized if we are also aware of what lies beyond it. No one can perceive the edge of a line without recognizing the space beyond it. Likewise, if our knowledge had a strict limit, we could only become aware of that limit by knowing something beyond it—otherwise, we would not recognize it as a limit in the first place. Kant's theory of the unknowable overlooks this fundamental point, assuming we can know a limit while remaining entirely ignorant of what lies beyond it.

Some may argue that acknowledging the unknowable does not contradict the possibility of its existence. However, Stace refutes this by emphasizing that the very category of existence applies universally to all things that exist, including the supposedly unrecognizable. If something exists, then in principle, it is subject to human recognition, because existence and recognition are interwoven. To claim otherwise—that something can exist but remain forever beyond recognition—is an incoherent contradiction.

Finally, Stace challenges the notion that the human mind is incapable of understanding infinite or absolute truths. He argues that such a belief is a mere superstition. The very concept of existence is tied to our capacity for recognition. If something exists, it must, in some way, be knowable. This assertion directly opposes the idea that any fundamental truth of existence can be wholly and permanently beyond human comprehension.

In conclusion, the hypothesis of the "unknowable" is not only philosophically problematic but also inconsistent with the nature of existence and the human mind's capacity for understanding. Whether something is infinite or absolute, if it exists, it must, at least in principle, be knowable.

Now, we come to the conclusion of our discussion on philosophy. Like many, you might find yourself surprised by some of the opinions presented here, particularly those of the idealists. But do not be alarmed—this reaction is common. As Einstein once remarked: "Isn't philosophy like something written entirely in honey? It's astonishing when you first think about it, but if you try to look at it again, it's entirely gone, leaving only a soft paste behind."

Similarly, the English philosopher Francis Bradley observed: "The philosophy of metaphysics seeks altered evidence for what we instinctively believe. Yet the search for such evidence is, in itself, a kind of instinct." Nietzsche, the German philosopher, added: "Philosophers pretend that their opinions are the result of a natural transformation of arguments presented with cold, indifferent neutrality and sanctity... But in truth, they are the product of judgments, images, or suggestions that existed in them before, often born from their heart's desires."

Will Durant, the American philosopher and historian, said: "For every truth, there are multiple fictions, but a lie can only be proven by an expert." And Michio Kaku, the American physicist, humorously remarked: "Physicists tell a fictional story where a university president, after reviewing the budget allocations for physics, mathematics, and philosophy departments, says: 'Why do physicists always need such expensive equipment, while the mathematics department requires nothing but paper, pencils, and a trash can? And the philosophy department? It doesn't even need a trash can!'"

Whether it's honey or hemlock, distorted or valid proofs, hidden desires of the heart or mental possibilities, truth or lies, serious or humorous... the judgment lies with you.

In any case, our exploration of philosophy ends here. While we acknowledge that many philosophical schools were not mentioned whether related to the second insight of the NUSAWVO school or not, we recognize that this book cannot encompass all views. We leave the pursuit of deeper truths to those who seek them earnestly.

Appendix Three

Appendix Three: Understanding the Universe (From a Mystical Perspective)

In the previous discussion, we identified two general approaches to understanding the world and its truth: logical-rational cognition and intuitive cognition. We also noted that three primary sources contribute to this pursuit. The first two—scientific and philosophical perspectives—fall under logical-rational cognition. The third source, the mystical perspective, is unique in that while its theoretical aspects belong to logical-rational knowledge, its practical application—through meditation, mystical customs, and rituals—falls within the realm of intuitive cognition. In this section, we will focus on the mystical viewpoint within the framework of logical-rational cognition.

The Mystical Concept of Truth

Before proceeding, we must first define "truth" in its mystical-philosophical sense. Within this framework, truth refers to the unique essence underlying all things—the deeper nature concealed behind the tangible appearances of daily phenomena. It is not accessible through ordinary sensory perception, yet it remains within the realm of human cognition.

For example, from a modern scientific perspective, one could argue that the fundamental essence of everything in our everyday world (the first realm) is energy—or perhaps even consciousness-energy—and that the various forms and manifestations we observe are merely expressions of this essence. Delving deeper, it could ultimately be said that the singular essence, or truth, of all existence is consciousness itself.

The distinction between truth and reality is also clear. By definition:

- **Truth** refers to the fundamental essence of a thing.
- **Reality** pertains to objective events or things that "happen."

Etymologically, "truth" conveys correctness, while "reality" originates from the notion of occurrence. This distinction is essential for understanding the relationship between existence, reality, and truth.

As discussed in the second insight of the scientific and philosophical appendices (Nos. 1 and 2), the NUSAWVO school allegorically divides the world of existence into two realms:

1. **The First Realm** – Our everyday world, whose reality is undisputed. However, based on our discussions, we argue that this world—whether understood as energy or consciousness-energy—lacks true, independent existence; it is merely a manifestation.
2. **The Second Realm (Basis Realm)** – This is consciousness itself, the essence and ultimate truth of the universe in its wholeness. While both realms can be studied through rational and logical means (scientific, philosophical, and mystical inquiry), a deep understanding of the second realm requires not only theoretical knowledge but also practical intuition and experiential mysticism.

What is Practical Mysticism?

Practical mysticism is a non-linear, non-argumentative form of cognition, often described as clairvoyance or intuition. It is a mode of perception that operates independently of sensory input and rational processes, offering a direct experience of the essence and truth of being.

In this approach, the seeker retreats from the distractions of daily life through meditation, disciplined exercises, and physical and mental austerities. Over time, this practice quiets inner dialogue and suspends sensory activity, creating the conditions for a different mode of perception to emerge.

With sufficient dedication, the seeker may begin to experience direct insights—perceptions that are often understood as an unmediated comprehension of the universe's truth. In essence, this process removes all barriers between the experiencer and the experienced, culminating in an experience of oneness—an inseparable unity where no separation remains.

At the culmination of this journey, the mystic describes a state of oceanic transcendence, where the distinction between self and other dissolves, individuality becomes unintelligible, and the perception of separate entities disappears entirely. Ultimately, only unity and wholeness remain.

The Mystic Across Cultures

A seeker of the second realm is generally referred to as a mystic, though different cultures and traditions have assigned them various names, such as:

- Magician
- Shaman
- Guru
- Saint
- Healer
- Sage
- Seeker
- Walker
- Monk
- Sufi
- Baba (Mama) Zar
- Aarif
- Seer

Mysticism Through Time and Its Relationship with Human Nature

From a historical perspective, mysticism is as ancient as human existence itself. The roots of mystical experiences likely stretch back hundreds of thousands of years, possibly marking the very beginning of belief systems such as animism. It is reasonable to assume that early humans—like their modern counterparts—encountered extraordinary experiences in dreams, - (see Note 181) where they found themselves flying, engaging in battles, witnessing their own deaths or those of loved ones, or even meeting the departed. Upon waking, they were left astonished by the nature of these experiences, struggling to comprehend their reality. These encounters may have been among the earliest catalysts for the belief in a world beyond the conventional one—perhaps a world more authentic than the fleeting realm of daily existence. This perspective may have even led to the notion that what is commonly perceived as reality is itself a kind of dream.

The discovery of hallucinogenic plants likely further expanded these explorations, enabling early humans to consciously induce altered states of perception. However, as human understanding evolved, it became clear that similar experiences could be achieved without reliance on external substances. Through disciplined practices—endurance of hardships, meditative techniques, and mental and physical austerities—they found ways to access these states naturally.

Note 181- *Dreaming is not exclusive to modern humans or even the earliest humans—animals dream as well. Scientists have discovered that nearly all animals sleep, and many—including mammals, birds, and reptiles—experience dreams. Like humans, they enter the REM (Rapid Eye Movement) stage of sleep, during which their eyes move rapidly. In this stage, the brain (whether human or animal) emits distinctive waves that can be recorded using an electroencephalogram. These waves are markedly different from those of other sleep stages, revealing the unique nature of dream activity.*

These practices were not seen as altering consciousness in the traditional sense; rather, they were viewed as a return to a state of awareness that had been obscured by the distractions of the material world. In this context, what might seem like an "altered" state is in fact a rediscovery of the original, unaltered state of consciousness that existed before the layers of conditioning and distraction took hold. These experiences, initially perceived as controlled dreams or altered states, eventually became gateways to deeper knowledge. The insights gained from such encounters—when combined with empirical knowledge—formed the wisdom of ancient sages. These custodians of humanity's earliest understanding of existence were, in essence, guiding others back to the original state of awareness—one that transcends the distractions of daily life and reveals the deeper truths underlying all of existence. As civilizations developed—particularly over the last five thousand years—mystical traditions gradually intertwined with the emerging philosophical and religious currents. In China and India, mystical philosophies such as Daoism, Hinduism, and Buddhism took shape. In Persia (Iran), the rise of Mithraism, with its Seal Rituals, followed by the more refined and mystical Zoroastrian traditions, and later the deeply mystical teachings of Mani, provided structured disciplines and rituals for the pursuit of truth.

In ancient Greece, the blending of philosophy and mysticism produced profound results. Even before Socrates, mystical philosophers secluded themselves, practicing long periods of meditative retreat under the guidance of chosen mentors. Some engaged in incubation rituals, - (see Note 182) withdrawing into caves to induce mystical states. Pythagoras and Parmenides—whom some regard as the true founders of Greek philosophy—were among those who sought fundamental knowledge through unaltered states of awareness.

Note 182- *Peter Kingsley, the English philosopher, has traced the presence of healers even among ancient Greek philosophers. In his book* Dark Corners of Wisdom, *he states that some mystics in ancient Greece were known as* Iatromantis *—"healers." Their knowledge differed from conventional wisdom, as*

Greek philosophy, shaped by these mystical undercurrents, gave rise to intellectual movements that continue to influence human thought. With the emergence of the Milesian and Eleatic schools, Socrates, Plato, and subsequent generations of thinkers, philosophy became an enduring pursuit of truth. Among the various philosophical traditions, idealism can be seen as an intricate attempt to define and structure "truth"—the inherently elusive essence of existence. Yet, despite all its intellectual rigor, idealism may ultimately be another interpretation of the simple yet profound realization of the ancient sages: that the everyday world is but a mirage, an illusion, and that another, deeper truth exists beyond it—one that can not only be believed in but directly perceived and experienced through the fullness of one's being.

The Varieties of Mystical Thought

Traditionally, mystics have been classified into different categories, particularly regarding their perspectives on divinity. Some mystics do not adhere to conventional notions of God, or their understanding diverges significantly from theological definitions. Others, however, maintain a deeply theistic outlook.

Non-theistic mystics seek a direct, experiential understanding of the ultimate truth of the universe—an undifferentiated unity that transcends all distinctions. They refer to this truth by various names, such as "consciousness," "power," "intention," "volition," or "nirvana."

(continued from Note 182)- *it was acquired through seclusion in a practice called* Incubation. *This state of retreat led them to a heightened awareness—an ecstatic condition. It took place in special caves under the guidance of an experienced spiritual master.*

The Iatromantis *were said to connect with an indivisible unity—a state beyond the five senses, beyond time and space. Yet, they believed that what lies beyond time and space is not separate from anything; rather, in its very essence, it is separate from separation itself.*

Theistic mystics, on the other hand, regard the world as an illusion but identify the ultimate truth with God. Their pursuit is not merely intellectual but deeply relational; they seek not only to know but to unite with the Divine. For them, the path of mysticism is a journey toward surrender—where the self dissolves in an intimate, transcendent union with the Eternal.

Despite these differences, both traditions share a fundamental quest: the search for something beyond the ephemeral concerns of daily life. Whether approached through reason or devotion, through metaphysical contemplation or direct experience, all mystical paths converge upon the realization that there is a truth greater than the fleeting nature of worldly affairs. In time, seekers from different traditions may come to recognize that they have been striving toward the same essence, albeit from different angles dictated by culture, temperament, and belief. - (see Note 183)

Yet, beyond these philosophical distinctions, one truth remains evident: the human need to seek meaning beyond the ordinary is innate and universal. It is this longing that compels individuals to look beyond their limitations, to heal their sense of existential solitude, and to enrich their lives with a deeper understanding. This need is not incidental but rather an essential aspect of human nature—one that cannot be ignored or dismissed. - (see Note 184)

Note 183- *Mystics who believe in God regard awareness as a sign of the divine. Thus, according to their claim, they focus directly on the source and origin of awareness—God—or, as they refer to it, "love" or "the truth of truth."*

Note 184 *-As emphasized in the insights of this school, human nature consists of four fundamental aspects. The first three—the physical, emotional, and mental—govern daily life. However, the fourth aspect, the* spiritual, *functions like an umbilical cord, connecting the deepest layers of our existence directly to the ultimate truth of the universe and the meaning concealed within it.*

From the perspective of this school, the harmonious and balanced integration of these four aspects ultimately leads to what is called full-scale human nature *or* sublime human nature*—a state that encompasses all the essential elements of true happiness.*

The Role of Mysticism Within and Beyond Religion

Throughout history, tensions have sometimes arisen between institutionalized religion and mysticism. Within major religious traditions—particularly those of the past few millennia—mystics have often been viewed with suspicion, and divergences in spiritual interpretation have led to conflict. However, it is important to recognize that mysticism predates organized religions, including the Abrahamic faiths. Moreover, many prophets themselves engaged in mystical practices, using introspection, solitude, and altered states of consciousness to receive their revelations.

History has shown that whenever religious institutions have imposed excessive rigidity on practical aspects of faith, the human spirit—ever longing for freedom—has gravitated toward mysticism, sometimes even to extreme forms of detachment. This movement, though unpredictable, is a recurring pattern in human spiritual evolution.

The NUSAWVO Perspective on Mysticism

The NUSAWVO school approaches mysticism through its second insight, advocating for a balanced integration of rational and experiential methods. In the pursuit of truth—particularly when exploring the "second realm"—NUSAWVO emphasizes combining logical inquiry (scientific, philosophical, and mystical studies) with disciplined mystical practices. These practices include well-established meditative traditions such as yoga and Tai Chi, which serve as pathways to direct experience.

However, the NUSAWVO school asserts that the second realm is not separate from our being; it is, in fact, our true self. Therefore, exhaustive philosophical and mystical efforts are not always necessary to experience it. With proper mental preparation, even reading the introduction of an enlightening book, listening to mystical or spiritual music—such as *"Loulian"* by Shahram Nazeri—or engaging with other forms of inspiration can spark intuition and insight. In an instant, the seeker may uncover the secret of the ultimate self, experience oneness with existence, and realize that their true nature is part of the essence of consciousness—beyond time, space, and limitations. As it is said: "The one who has not realized their true self—so-called pure consciousness—is like a fish that has not found water.

NUSAWVO firmly rejects the use of opiates and hallucinogens, seeing them as misguided shortcuts that lead only to illusion, suffering, and self-destruction. True mystical realization, in this school's view, does not arise from chemical manipulation but from the refinement of consciousness through disciplined effort. - (see Note 185) Furthermore, NUSAWVO asserts that seekers must never surrender their volition or intellect to external authorities—whether sages, elders, or spiritual leaders. The path to truth must remain an autonomous and deeply personal journey.

Ultimately, NUSAWVO upholds belief in meaning—what may be called God—but refrains from defining or conceptualizing it. At the same time, it honors the sincere efforts of mystics across cultures and humbly listens to the wisdom of those who have walked the path of direct experience.

Conclusion

This appendix has briefly examined mysticism from a theoretical and rational perspective. However, it must be acknowledged that true mystical research cannot rely on intellectual analysis alone—it requires direct experience. The well-known methods of meditation and inner discipline remain open to any seeker who wishes to embark on this journey.

Note 185- *Gandhi once said that pure goals can only be achieved through pure means and that the relationship between ends and means is like that between a tree and its seed. According to the NUSAWVO school, mysticism that emerges from the heart of opiates is not only unworthy but also a betrayal of its very purpose.*

Since discussions about mysticism and philosophy can never fully capture the essence of what they describe, it is fitting to conclude with the words of those who have walked the mystical path. Thus, as a farewell, we turn to the sayings and experiences of mystics and Sufis throughout history, in honor of the noble and timeless human quest for truth.

On Duality and Perception (*The Upanishads*)
The Upanishads contain many teachings on duality, perception, and the nature of existence. The below quotes reflect themes found in the *Mandukya Upanishad*, which discusses the nature of reality and consciousness. Here's a relevant excerpt from the *Mandukya Upanishad* (translated by Eknath Easwaran, 1987):

Mandukya Upanishad (1.2):
"In the state of deep sleep, when all the external senses are at rest, the individual is in the state of pure consciousness. This is the same as the state of the eternal, indivisible reality."

The *Maitri Upanishad* also explores duality and unity:
"Where there is duality, as it were, there one sees another, but where everything is as the Self, nothing remains to be seen." (*Maitri Upanishad 6.34*)

On Transcendence and Letting Go (*Chuang Tzu*)
Chuang Tzu's Zhuangzi is a Daoist text that emphasizes spontaneity, non-attachment, and transcendence. Here's an excerpt from the *Zhuangzi* that relates to the theme of transcending the body and perception:

Zhuangzi (Chapter 2, "The Great Happiness"):
"When one lets go of the body, and the mind no longer clings to its perceptions, one reaches the state where the mind and the body are no longer distinct. The self that once existed in the body is now without form."

This conveys the idea of merging with the larger reality, a theme central to *Chuang Tzu's* teachings.

Lao Tzu's Sayings (*Dao De Jing*)
The *Dao De Jing* (Tao Te Ching) is the foundational text of Daoism, attributed to Lao Tzu. The quotes below come from the core teachings of Lao Tzu on the Dao (the Way) and the unity of opposites.

Dao De Jing, Chapter 1:
"The Dao that can be told is not the eternal Dao. The name that can be named is not the eternal name. The nameless is the beginning of heaven and earth."

Dao De Jing, Chapter 2:
"When people see things as beautiful, other things become ugly. When people see things as good, other things become bad. Being and non-being produce each other."

Dao De Jing, Chapter 11:
"Thirty spokes converge at the wheel's hub, but it is the empty space that makes the wheel useful. Clay is shaped into a vessel, but it is the hollow within that makes the jar. Doors and windows are cut into walls, but it is the empty space that makes the room a dwelling."

Dao De Jing, Chapter 25:
"There is something undifferentiated and yet complete, which existed before heaven and earth. How still it is! How profound it is! It is the mother of the world."

Attribution:
Mandukya Upanishad translation by Eknath Easwaran, 1987.
Zhuangzi translation by Martin Palmer, 1996.
Dao De Jing (Tao Te Ching) translation by James Legge, 1891.

The Toltec Shamans and Their Worldview: A Reflection on Castaneda's Work

In recent years, our understanding of shamanic beliefs—particularly those of a group known as the *Toltecs*—has significantly expanded, largely due to the research of the American anthropologist *Carlos Castaneda*. Through Castaneda's published accounts, we gain insight into the thoughts and practices of a Yaqui shaman named *Don Juan Matus*, who was Castaneda's teacher for thirteen years.

According to Don Juan and the lineage of seekers to which he belonged, reality is divided into two fundamental realms. The first is called *Tonal*—which, in the framework of the NUSAWVO school, corresponds to the first realm. The second is *Nagual*, equivalent to the second realm in NUSAWVO.

A Summary of Toltec Shamanic Teachings (Based on the Works of Carlos Castaneda)

In the Toltec tradition, there is an understanding of two realms that constitute the full spectrum of human perception: **Nagual** and **Tonal**.

1. **Nagual**: This concept refers to a boundless, ineffable state of being. It represents an infinite and eternal stillness, a realm that encompasses all consciousness, sensations, and existence. Beings in this realm are like vessels adrift in a vast sea, connected by the "thread of life." However, these beings often become distracted by the external world, a world known as **Tonal**, and lose awareness of their original essence.
2. **Tonal**: In contrast, the Tonal is the realm of form, logic, and order. It encompasses everything that we can categorize or rationalize, from birth to death. It is shaped by societal structures, knowledge, and tangible experiences. Once the life force departs from a being, the elements of

consciousness return to Nagual, where they dissolve back into the undifferentiated source.

The Path of the Seeker

For the seeker, engaging with the unknown (Nagual) is akin to a kind of "death"—an expansion of awareness beyond the confines of the Tonal. However, this process is not about annihilation but transformation, where the consciousness remains intact and continues to evolve.

This journey requires years of preparation. A seeker must learn to refine their rational faculties, their Tonal, so that they are not overwhelmed by the boundless unknown. The aim is not to eliminate the logical mind but to transcend its limitations, allowing the seeker to experience a broader and deeper reality.

Will (Volition) as the Gateway

The key to navigating the space between Tonal and Nagual is **volition**—the power of will. By refining one's internal state and silencing inner dialogue, the seeker creates the conditions to encounter Nagual. The practice is not about forcefully rejecting logic but about purifying the mental clutter and cultivating the capacity to move beyond it.

The Perception Bubble

Shamans believe that each person lives within a "bubble of perception," a structure shaped by individual experiences and societal conditioning. When we are born, this bubble is open, allowing for a broader range of experiences. However, over time, this bubble gradually closes, and our perception becomes limited and self-referential, trapping us within a narrow worldview.

A master shaman helps to break open this bubble, allowing the seeker to experience the world from a new perspective. This process requires confronting paradoxes and impossibilities, as well as allowing one's awareness to break free from its fixed boundaries. By doing so, the seeker gains access to a new way of perceiving reality, one that transcends the limitations of logic.

The Secret of Luminous Beings

In Toltec teachings, there is a central mystery known as the **secret of luminous beings**: All living entities are essentially sentient awareness, existing as consciousness itself. While rational thought is an important tool, it is only one center of perception. True mastery comes when one recognizes that logic is not the sole means of understanding reality. Instead, one must embrace all forms of perception, integrating both the logical and the non-logical, to reach a fuller understanding of existence.

Castaneda's Personal Journey

Carlos Castaneda, in his work, recounts his personal experience of breaking through his perception bubble, describing a moment when he felt his sense of self dissolve into countless fragments. During this experience, he came to realize that his former identity was merely a construct—an assemblage of independent perceptions that were bound together by the force of life. Upon reassembling these fragments, he experienced a profound sense of unity, no longer observing the world from the outside, but becoming a part of it.

Closing Thoughts

The Toltec understanding of reality shares significant parallels with the foundational insights of the NUSAWVO school. Just as the Toltec magicians emphasize the importance of refining perception, purifying logic, and preparing for the encounter with the deeper, ineffable realm of reality, NUSAWVO also advocates for these practices. In both traditions, the structured world (Tonal) is not to be abandoned but perfected, and the formless world (Nagual) is not to be conquered but entered with reverence and mindfulness. Volition—the conscious force of will—serves as the bridge between these realms, enabling transformation and growth.

For the seeker, the path ahead is one of preparation. It involves stripping away illusions, silencing unnecessary mental chatter, and cultivating the discipline required to cross the threshold into deeper awareness. Only by achieving this state of readiness can one witness the opening of the perception bubble and step into the fullness of existence.

Alan Watts' Perspective on Mysticism

Alan Watts, a prominent 20th-century philosopher, explored the nature of reality, consciousness, and the self with profound clarity. He frequently emphasized that the individual ego is a construct of the mind, an illusion that separates us from the true flow of existence. Watts suggested that our awareness is not isolated but a part of the universe's continuous exploration of itself. This view resonates with the NUSAWVO school's understanding of consciousness, which holds that existence is inseparable from awareness, and all that exists is an expression of this awareness.

Watts was also critical of rigid, dualistic thinking. He encouraged a view of life as a dynamic, ever-evolving process rather than a fixed set of truths. This aligns with the NUSAWVO worldview, which advocates for a direct experience of being that transcends intellectual comprehension and embraces the continuous unfolding of reality. His view that "the menu is not the meal" underscores the distinction between conceptual understanding and actual experience, a sentiment shared by NUSAWVO in its emphasis on experiential awareness over theoretical knowledge.

Furthermore, Watts' reflections on spontaneity and the illusion of control provide valuable insights into the nature of existence. He suggested that permanence, in any form, stifles life's vitality. This aligns with the NUSAWVO rejection of static structures and the belief that meaning emerges through consciousness, constantly evolving. Watts' call to "trust the universe" and flow with life rather than resist it mirrors the NUSAWVO emphasis on surrendering the ego and recognizing the interconnectedness of all things.

In essence, the philosophy of Alan Watts offers valuable insights that complement the NUSAWVO school's perspective, reinforcing the idea that true understanding comes not from clinging to concepts but from an experiential realization of consciousness as the ultimate reality.

Key Concepts from Alan Watts

- **Interconnectedness of Humans and the World**:

Watts highlighted the deep interconnectedness between humans and their environment. He pointed out that, like a physicist's measurement of a particle, our observation of the world inevitably alters it, reflecting the inseparable nature of perception and the universe.

- **Perception and Reality**:

Watts explained that our sensory experience is shaped by how our brains interpret external stimuli. Just as sound is only perceived when vibrations interact with our ears, the world exists only through our perception. This aligns with NUSAWVO's view of consciousness as the filter through which reality is experienced.

- **Harmony and the Nature of Existence**:

He suggested that existence is a relationship rather than a series of isolated events, emphasizing the interdependence between the observer and the observed. Similarly, NUSAWVO views reality as a co-creation between the self and the universe.

- **Expanding Awareness**:

Watts noted that traditional views of awareness often overlook the more expansive, interconnected awareness that encompasses all experiences, even those outside focused attention. This expanded awareness mirrors the NUSAWVO understanding of consciousness as not limited to the ego but all-encompassing.

- **Spiritual Realization**:

Watts argued that spiritual enlightenment is not about acquiring something new but recognizing what is already within us. This resonates with NUSAWVO's belief that the realization of our true nature requires no attainment, only recognition.

- **The Play of Life**:

He compared life to a game, where we often forget that we are both the player and the observer. In this way, NUSAWVO's perspective also emphasizes the unity of self and the world, where life's roles and perceptions are both illusions and realities that we transcend.

- **Oneness of Existence**:

Watts believed there is no fundamental separation between the self and the universe; rather, everything is connected as parts of a greater whole. This is reflected in the NUSAWVO worldview, which sees the individual as an expression of the universe itself.

The following passages are translated excerpts from the words and narrations of the renowned Sufi Mansour Hallaj, who was martyred approximately 1,100 years ago.

Once, Shibli asked Hallaj, "O Sheikh, how is the path to God Almighty?"
Hallaj replied, "It is but two steps, and you have arrived: strike the world upon the faces of its lovers and surrender the hereafter to its seekers."

Ibrahim bin Muhammad Nahrwani recounts: "I asked Hallaj about monotheism, and he said, 'If a servant proclaims the oneness of his Lord, he affirms his own existence, and whoever affirms himself has fallen into hidden shirk (associating partners with God). If he expresses monotheism through my tongue, then it is a favor upon me. But what have I to do with monotheism, my brother?'"

Abdullah bin Taher Azodi narrates:
"I was engaged in a debate with a Jew in the market of Baghdad when Hossein bin Mansour Hallaj passed by. He cast an intense gaze upon me and said, 'Your dog is barking,' and quickly left. Once I resolved the argument, I sought out Hallaj and entered his presence. He turned away from me. I apologized; he accepted, and then he said:
'My son, all religions originate from God, the Exalted. If one condemns another's faith, he has not truly chosen his own. Judaism, Christianity, Islam, and other religions are but different names; their essence is one.
I have examined all creeds deeply and found them to be rooted in the same truth, diverging only in branches. Do not seek a particular religion for mankind, for it will only hinder him from reaching the immutable principle. What one must seek is the fundamental truth in which all heights of thought are revealed.'"

During his execution, when they severed his leg, he said: "I once traveled the earth with this foot; now I need but one more step to pass beyond the universe. If you can, cut that step as well."

Hallaj also said:
"Knowledge is beyond all heights, beyond the ultimate, beyond effort, beyond secrets, beyond reports, beyond perception. All these things—whatever I have named—once did not exist; they were

brought into being later. And that which comes into existence cannot be found except within a place. But that which has eternally existed before directions, causes, and means—how can directions contain Him, and how can limits bind Him?

O seeker! In the realm of existence, there is nothing smaller than a particle, and yet you do not even grasp it. How, then, can one who does not know a particle comprehend the essence of Truth, which is even subtler? A true mystic is the one who sees, and knowledge is that which endures."

In his final prayer, moments before execution, Hallaj uttered:
"Lord! You manifest from all directions, yet You are bound by none.
Lord! By Your right upon my being and by my being upon Your truth,
By Your eternal reality that has encompassed my transient existence,
Grant me the grace to thank You for the blessings You have bestowed.
For You have unveiled to me the beauty of Your secrets, hidden from others—
Allow me to express gratitude for this gift.

These people, your servants, have gathered to take my life, believing they do so in service of Your religion and in pursuit of Your nearness.
Forgive them. For if You had revealed to them, what You have shown me,
Or concealed from me what You have hidden from them,
This fate would not have befallen me.
I praise You for all that You have done, and for all that You will do."

Even as they severed his hands and feet, Hallaj continued speaking. A man approached him and asked:
"What is Sufism?"
Hallaj replied, "The simplest form of it is what you see now."

"I see no difference between the blessings and the trials given by God," he added.
"And what is the highest form of Sufism?" the man asked.
Hallaj answered, "You will witness it tomorrow."

The next day, his ashes were scattered into the Tigris.

Jalal al-Din Rumi, one of history's greatest mystics, later reflected on Hallaj's statement, 'I am the Truth', *in his book* Fihi Ma Fihi*:*
"When people hear 'I am the Truth,' they think it is arrogance. But in reality, it is the greatest humility.
To say 'I am God's servant' affirms two existences: oneself and God.
But the one who says 'I am the Truth' has annihilated himself—he has given himself to the wind.
To say 'I am the Truth' means 'I am nothing; all is Him. There is no existence but God.'
This is true humility. But people do not understand.
A man who serves God still sees himself, his actions, and his Lord—he is not yet drowned.
The one who is truly drowned has no movement or action left in him—his movements are the movements of the ocean itself."

Resources

Philosophy & Psychology

- **Will Durant** – *The Pleasures of Philosophy*
  Publisher: Simon and Schuster, 1953.
- **Jonathan Dancy** – *Berkeley: An Introduction*
  Publisher: Basil Blackwell, 1987.
- **Norman L. Munn** – *Psychology: The Fundamentals of Human Adjustment*
  Publisher: Houghton Mifflin, 1971.
- **C. B. Macpherson** – *The Real World of Democracy*
  Publisher: Oxford University Press, 1965.
- **Walter Terence Stace** – *The Philosophy of Hegel*
  Publisher: Dover Publications, 1955.

Science & Physics

- **Albert Jacquard & Axel Kahn** – *L'avenir n'est pas écrit* (*The Future Is Not Written*)
  Publisher: Stock, 2001.
- **Hubert Reeves** – *Atomes de silence* (*Atoms of Silence*)
  Publisher: Éditions du Seuil, 1994.
- **Hubert Reeves** – *L'Origine de l'univers* (*The Origin of the Universe*)
  Publisher: Fayard, 2004.
- **Michio Kaku** – *Parallel Worlds: A Journey Through Creation, Higher Dimensions, and the Future of the Cosmos*
  Publisher: Doubleday, 2004.
- **Fred Alan Wolf & Bob Toben** – *Space-Time and Beyond: Toward an Explanation of the Unexplainable*
  Publisher: Bantam Books, 1982.
- **Lyall Watson** – *Supernature: A Natural History of the Supernatural*
  Publisher: Anchor Books, 1973.

- **A. I. Oparin** – *The Origin of Life*
  Publisher: Dover Publications, 1953.
- **Morin, Edgar.** *La Méthode de la méthode : le manuscrit perdu.* Actes Sud, 2024.

Spirituality & Mysticism

- **Alan Watts** – *The Wisdom of Insecurity: A Message for an Age of Anxiety*
  Publisher: Vintage Books, 1951.
- **Fritjof Capra** – *The Tao of Physics: An Exploration of the Parallels Between Modern Physics and Eastern Mysticism*
  Publisher: Shambhala Publications, 1975.
- **Thich Nhat Hanh** – *The Miracle of Mindfulness: An Introduction to the Practice of Meditation*
  Publisher: Beacon Press, 1975.
- **Rumi** – *Fihi Ma Fihi* (*It Is What It Is*)
  Publisher: Various editions available; originally compiled in the 13th century.
- **Michael Talbot** – *The Holographic Universe, Publisher: HarperCollins,1991*

Social Sciences & Ethics

- **Desmond Morris** – *The Human Zoo*
  Publisher: Jonathan Cape, 1969.
- **Howard Selsam** – *Socialism and Ethics*
  Publisher: International Publishers, 1943.
- **René Sédillot** – *Histoire des socialismes* (*The History of Socialisms*)
  Publisher: Librairie Fayard, 1978.
- **Konrad Lorenz** – *On Aggression*
  Publisher: Harcourt Brace & World, 1966.

Esotericism & Indigenous Knowledge

- **Carlos Castaneda** – *Tales of Power*
  Publisher: Simon and Schuster, 1974.
- **Carlos Castaneda** – *The Teachings of Don Juan: A Yaqui Way of Knowledge*
  Publisher: University of California Press, 1968.
- **Lothar R. Lütkehaus** – *Carlos Castaneda und die Lehren des Don Juan* (*Carlos Castaneda and the Teachings of Don Juan*)
  Publisher: Fischer Taschenbuch Verlag, 1982.
- **Graciela Corvalán** – *Der Weg der Tolteken* (*The Way of the Toltecs*)
  Publisher: Fischer Taschenbuch Verlag, 1987.

Other Works

- **Bill Bryson** – *A Short History of Nearly Everything*
  Publisher: Broadway Books, 2003.
- **Steven Lowe** – *The Thousand-Year Questions*
  Publisher: [Publisher information not available].

www.ingramcontent.com/pod-product-compliance
Lightning Source LLC
LaVergne TN
LVHW090256210926
848480LV00001B/35